W9-BRI-167

WEST GREENLAND CURRENT

EAST GREENLAND CURRENT

NORWEGIAN CURRENT

LABRADOR C.

IRMINGER CURRENT

NORTH ATLANTIC DRIFT

PORTUGAL CURRENT

GULF STREAM

NORTH ATLANTIC OCEAN

FLORIDA C.

CANARY CURRENT

ANTILLES CURRENT

NORTH EQUATORIAL CURRENT

GUINEA CURRENT

SOUTH EQUATORIAL CURRENT

S.W. MONSOON

N. EQUATORIAL C.

EQUATORIAL COUNTER C.

SOMALI C.

SOUTH EQUATORIAL CURRENT

INDIAN OCEAN

BENGUELA C.

AGULHAS C.

CURRENT

BRAZIL C.

SOUTH ATLANTIC OCEAN

HUMBOLDT CURRENT

FALKLAND C.

WEST WIND DRIFT

ANTARCTIC CIRCUMPOLAR CURRENT

EAST WIND DRIFT

PRINCIPLES OF OCEANOGRAPHY

SECOND EDITION

PRINCIPLES OF OCEANOGRAPHY

SECOND EDITION

Richard A. Davis, Jr.

University of South Florida

ADDISON-WESLEY PUBLISHING COMPANY

Reading, Massachusetts
Menlo Park, California
London
Amsterdam
Don Mills, Ontario
Sydney

Third printing, July 1978

Copyright © 1977, 1972 by Addison-Wesley Publishing Company, Inc. Philippines copyright 1977, 1972 by Addison-Wesley Publishing Company, Inc.

ISBN 0-201-01464-5
FGHIJKLMN-HA-898765432

To Mary Ann,
for her encouragement and understanding

PREFACE TO THE SECOND EDITION

Since the publication of the first edition of this text, there have been great strides made in our knowledge of the oceans. These include the addition of much data from all the various marine disciplines, but the highlight of the past few years is the wealth of data on ocean-basin history provided by the Deep Sea Drilling Project (DSDP). As a result, a comprehensive and reasonable explanation of ocean/land-mass relationships and dynamics has been formulated.

The new plate tectonics/continental drift data, new instrumentation for collection of various marine data, and especially the excellent constructive comments by numerous professors who have used *Principles of Oceanography* have indicated that a second edition is both warranted and necessary. The overall philosophy and organization are the same as in the original edition.

The reader will notice that some chapters are nearly unchanged, whereas others are quite different in content. The topics of continental drift and plate tectonics are expanded; the physical oceanography section is a bit more comprehensive; chemistry has been abbreviated; and the biological section is more community and ecologically oriented.

Virtually all these modifications are, at least in part, the result of the many comments provided by individuals who have used the first edition in their introductory oceanography courses. The author extends his sincere appreciation to all these people for their suggestions. Particular acknowledgement is due M. P. Wiess (Northern Illinois University) and Paul Kirst (Miami-Dade Junior College) for their extensive and unsolicited assistance. Much benefit was also gained from suggestions by Edward C. Roy (Trinity University) on the biology section, Francis Birch, Jr. (University of New Hampshire) on continental drift and geophysics, and A. L. Reesman (Vanderbilt University) on marine chemistry. The staff of Addison-Wesley Publishing Company were also of considerable assistance to the author in preparing this revised edition. Special appreciation is extended to Wanda B. Evans for her clerical assistance in the revision.

Tampa, Florida   R.A.D.
January 1977

PREFACE TO THE FIRST EDITION

Oceanography is commonly mistermed the "science of the seas." It is really much more than that because oceanography is not really a single science but a conglomeration of all sciences as applied to the marine world. Unfortunately, this particular quest for knowledge about our planet has been neglected until rather recent times. If we look at a world map, however, it is apparent that the vast majority of our earth is covered by salt water.

A common excuse for our relative lack of emphasis on the oceans is that man does not inhabit this part of the earth, and we need to concentrate on areas that more directly influence our species. There are, however, at least two strong arguments for accelerating our investigations of the sea, one of which is fast becoming essential. First, because of the ever expanding world population, we are going to become more and more dependent on the oceans for the natural resources that they can provide. These include food, minerals, and other raw materials for industry, water supply, and perhaps space for habitation. Transportion and defense are other important uses of the ocean. A second argument is that rather than concentrate so much of our efforts on outer space, we might better benefit from an expansion of our knowledge of inner space on the planet Earth. The latter argument may be interpreted as a typical "sour grapes" attitude taken by many ocean scientists, but it is worthy of consideration.

Man's recent strides in the ocean sciences should not be overlooked, because there are many. Federal, state, and local governments as well as industry and academic institutions are continually increasing their studies of the oceans; however, many people do not think the rate of increase is sufficient. During a single recent year (1969), many significant events occurred: the Deep Sea Drilling Project, Sealab, and Tektite projects are all worthy of mention, although it is difficult to evaluate their merit at this time. There is also a prominent trend in colleges and universities toward establishing or expanding oceanography programs. Nearly every academic institution on or near the coast has, or is planning, an oceanography program. In fact many of the so-called "landlocked" universities also have programs of their own or in cooperation with coastal universities. At least two midwest institutions, the Universities of Michigan and Wisconsin, offer the Ph.D. degree in oceanography.

Perhaps more significant than the many professional programs in the marine sciences is the tremendous growth and development of oceanography as a part of the general undergraduate curriculum in institutions concerned with general education. This is particularly true in the various general science requirements and teacher preparation programs. Each year dozens of schools add a general oceanography course to their offerings largely because of the interest shown by students and the public in general. It is this type of course for which this text is designed.

Although my background is primarily in geology, I have been exposed to marine science and was in fact hired at Western Michigan University to teach a general course in oceanography. At that time I had a pretty good idea of what I thought should be included in such a course. My search for a textbook that included this material was fruitless, and so I began compiling an extensive set of notes around my initial outline. This book is based for the most part on these notes and the general oceanography course as it is taught at Western Michigan University.

The book is designed for use in a one-semester course in general oceanography and is written to acquaint students with the principles of physical, chemical, biological, and geological oceanography. It is purposely broad in scope and shallow in depth for a variety of reasons. Most students take a general oceanography course as part of their general science requirement, as a supplementary course for a specific science major, or as part of a teacher preparation program. These students usually do not take any other courses in oceanography. This book should provide such people with an appreciation for the complexities of the marine sciences as well as a broad background that can be related to other science courses.

I have briefly mentioned what the book *is*; it is therefore proper to say what it is *not*. The text is not designed for a first course in a professional oceanography curriculum or as reference material in oceanography. The material presented only scratches the surface and is intended to stimulate one's interest in the field rather than render one an authority on the oceans.

The "broad brush" treatment of subject matter and the general brevity of the book are designed to serve another purpose. The great volumes of data that are currently being added to our knowledge of the oceans cause many of today's ideas to be altered when tomorrow arrives. In order to avoid this as much as possible, most of the subject matter discussed deals with principles and rather well-established information. This approach allows the instructor to add supplementary information without negating a great deal of the material in the book.

   Most people who will be teaching from this book will not be professional oceanographers. In most cases the task will fall on a person trained in a particular scientific discipline, probably geology, biology, or geography. The brief and general nature of the book makes it possible for the instructor to supplement the text material with in-depth coverage of areas that he wishes to stress and still cover all the broad aspects of oceanography in the book. This flexibility is necessary for the wide range of general oceanography courses that may exist or be developed in the near future.

   Like most first-level books, it contains very little information that is original with the author. Most of the data were taken from the many excellent specialty books and articles in the oceanographic sciences. The publishers and authors of these materials are gratefully acknowledged for their cooperation in making illustrations available.

   The Department of Geology at Western Michigan University provided various services in the preparation of the original manuscript. R. N. Passero, W. T. Straw, A. L. Reesman, and R. W. Pippen read various chapters. Their comments and assistance are gratefully acknowledged. Special thanks are due to Mrs. Janet Niewoonder and Miss Christine Robinette, who typed most of the original manuscript, and to Mrs. Odessa Straw and Mrs. Sandy Yellich, who typed the final copy.

Kalamazoo, Michigan   R.A.D.
December 1971

CONTENTS

INTRODUCTION: BRIEF HISTORY OF OCEANOGRAPHY

OCEANOGRAPHY PRIOR TO THE *CHALLENGER* EXPEDITION

Ships have served man as means of transportation for all of recorded history. Throughout the centuries that preceded the Christian Era, the Mediterranean Sea and vicinity was the primary area of travel (Fig. I.1). As time passed the ancients explored some of the African and European coastal areas. Early Phoenicians sailed as far as the Sargasso Sea in the central Atlantic, and around the African continent.

The first records of any observations in the sea that might be considered scientific are from the fourth century B.C. Pytheas determined latitude and longitude while journeying to the British Isles and also related tides to the moon. According to legend a contemporary of his, Alexander the Great, descended into the sea under a large bell and observed marine life. Two centuries later, the geographer Strabo observed and collected data on tides in the Mediterranean. He is also credited with a depth sounding to nearly 2000 meters, although his method is a mystery. At approximately the same time in history, Eratosthenes postulated a spherical earth with a circumference of nearly 40,000 kilometers. Others also advocated that the earth was a sphere.

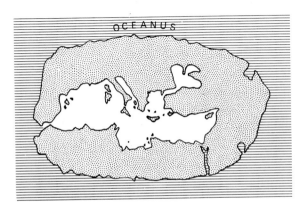

FIG. I.1 *Map of the world as pictured by Hecaecus, 500 B.C. (After P. Groen, 1967, The Waters of the Sea, London: Van Nostrand, p. 2.)*

1

The many long centuries of the Dark Ages did nothing to continue the progress made by these early scholars. In fact most people rejected the idea of a spherical earth. The Vikings, however, who were considered barbarians by the rest of Europe, made several discoveries including Iceland, Greenland, and North America during the tenth and eleventh centuries.

Virtually nothing of major oceanographic importance took place until the establishment of a school for navigation early in the fifteenth century by Prince Henry the Navigator of Portugal. His navigation center at Sagres was responsible for most of the great discovery voyages of the late fifteenth and early sixteenth centuries, including those of Diaz (1487), Columbus (1492), Vasco da Gama (1499), Balboa (1513), and Magellan, who completed the first circumnavigation of the globe in 1522. These and other significant voyages that followed were undertaken strictly for the sake of exploration. No scientific observations were purposefully made or recorded. However, a contemporary of the above mariners, Ponce de Leon, did make an important oceanographic discovery. While sailing south in the vicinity of what is now Cape Canaveral off the northeastern coast of Florida, he noted that his ship was not making progress while under sail. This was the first known observation of the northerly flowing current known as the Gulf Stream.

During the next 200 years there was no significant expansion of man's scientific interest in the sea. New lands were discovered and new routes to old ones were charted. The sea was used simply for transportation and as a source of income for whalers and fishermen. Finally, in 1769, Benjamin Franklin compiled and published a chart of the Gulf Stream (Fig. I.2). This is generally considered the first oceanographic publication. Captain James Cook, who explored the Pacific between 1768 and 1779, was the first to take a naturalist on a voyage. The scientist found, among other things, that rather deep waters were very cold compared to those at the surface.

The nineteenth century marked the formal beginning of man's scientific curiosity about the sea. William Eaton collected biological specimens and took soundings from a ship in 1804. In the same year the first test of light penetration was made by a curious sailor from the deck of the U.S. frigate President. He attached a white china dinner plate to a line and noted its disappearance from sight at a depth of 45.5 meters. It was only three years later, in 1807, that the United States established the first federal scientific organization: The Coast Survey, known for a long time as the Coast and Geodetic Survey. This was recently incorporated into ESSA (Environmental Sciences Service Administration), which has subsequently become part of NOAA (National Oceanographic and Atmospheric Administration).

There are several people who might qualify as the founder of oceanography. Edward Forbes, M. F. Maury, Charles Darwin, and others made significant early contributions to the sciences of the sea. Edward Forbes

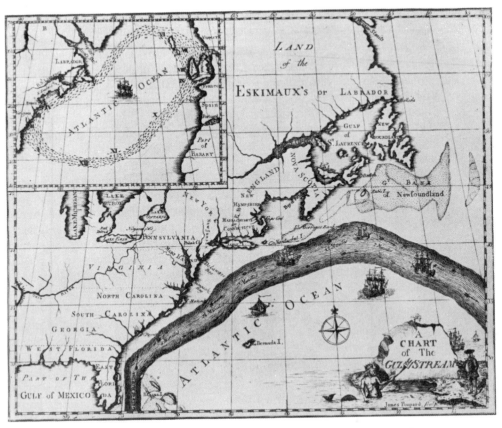

FIG. I.2 *Chart of the Gulf Stream published in 1769 by Benjamin Franklin. This is generally considered the first oceanographic publication. The chart is now in the collection of the American Philosophical Society.*

(1815–1854) is perhaps most widely known for his contributions to paleontology, but he also qualifies as a pioneer in oceanography. He exhibited brilliance as a child and is given credit for a manuscript on British natural history written at age twelve. Although he performed poorly as a medical student, Forbes excelled in fields of natural history. He was a professor of botany, curator of paleontology, and finally was appointed to the Chair of Natural History at the University of Edinburgh. In 1834 he worked on systematic dredging of the Irish Sea and also the Aegean Sea. Forbes was also responsible for the discovery of zonation of life in the sea; however, his investigations were limited to shallow ocean margins. He believed that lack of light and great pressures prohibited organisms from living at great depths, and he fostered the belief that below 600 meters there is a general absence of life.

Charles Darwin made significant contributions to our knowledge of the seas, although he is better known for his theories of evolution. He served as naturalist on the *H.M.S. Beagle* from 1831–1836, and it was on this voyage that he collected data for his famous theories. He also observed hundreds of coral reefs in the Pacific and made rather careful observations of their form. The result was his theory of coral reef development, and atolls in particular. Although sternly tested over the years, his theory is still accepted. It is discussed in detail in Chapter 22.

Matthew Fontaine Maury (1806–1873), an American naval officer, is considered by most historians as the first full-time oceanographer (Fig. I.3). He spent most of his productive life collecting data from the seas, whereas Forbes and others prior to Maury's time were engaged in many other types of investigations. Maury became a full-time oceanographer by default. He was a shipboard naval officer who did a great deal of work with navigation routes, but a permanent injury, incurred while surveying southern harbors, lamed Lt.

FIG. **I.3** *Lt. Matthew Fontaine Maury of the U.S. Navy is considered by most to be the first full-time oceanographer.*

Maury and rendered him unfit for sea duty. Because of his great interest in navigation and oceanographic data collection, he was assigned to the Navy's Depot of Charts and Instruments in 1841 (this agency has since evolved into the U.S. Navy Hydrographic Office). During the several years Maury was at the Depot, he accumulated nearly a half-million shipboard observations of wind, currents, and other data. These were compiled and made available as aids to navigation. Maury was also instrumental in organizing the first international convention for marine navigation and meteorology at Brussels in 1853. Perhaps the most famous contribution bearing the name of M. F. Maury is his book, *Physical Geography of the Sea*, published in 1855. This was the first textbook in oceanography.

During the 1840s and 1850s, there were other isolated events worthy of mention. In 1840 James C. Ross sounded to a depth of 7560 meters in the South Atlantic. A naval lieutenant later exceeded this when he payed out 10,500 meters of line without reaching the bottom. Both soundings are now considered inaccurate, because of the techniques used; however, they were the first indications of the great depth of the ocean floor. As late as the end of the Civil War, man had little knowledge of ocean-floor topography. An interest in the topography was generated by the need for a transatlantic cable.

OCEANOGRAPHY COMES OF AGE

Charles Wyville Thompson was born in Scotland in 1830. After graduation from the University of Edinburgh he became lecturer and eventually professor of botany at the University of Aberdeen. In 1868 he became one of Forbes' successors to the Chair of Natural History at his alma mater, Edinburgh. Perhaps it was Forbes' interest in marine biology that led Thompson and an associate, W. B. Carpenter, to spend two months on a British ship attempting to collect marine organisms and chemical data from the sea. Bad weather prohibited much success; however, contrary to Forbes' belief, they found a diversity of life even at depths of 1700 meters. The British Admiralty was impressed and for the following summers (1869 and 1870) provided Thompson and his colleague with a more substantial vessel. These were the first cruises organized for the sole purpose of scientific marine investigations.

With the encouragement of the Royal Society, the British Admiralty provided a ship and crew to make an around-the-world expedition for the purpose of examining biological and physical conditions of the sea. The H.M.S. *Challenger* was chosen to be the ship (Fig. I.4), with Charles W. Thompson as the director of a scientific staff which also included three naturalists, a chemist, and a secretary.

The *Challenger* left port in December of 1872 and spent more than three years at sea, returning in May of 1876 (Fig. I.5). During this time the ship traveled nearly 69,000 nautical miles (125,800 kilometers) and established

FIG. **I.4** *The* H.M.S. Challenger, *which undertook the first and most famous voyage for scientific purposes.*

362 observation stations. Collections were made of organisms, bottom samples, and water samples. Data were recorded for depths, temperature, currents, and weather at each station.

The tremendous amount of data collected by the *Challenger* Expedition took 19 years to synthesize and has resulted in the 50-volume *Challenger Report.* A total of 4717 new species was collected and described, and the collections are still studied by marine biologists all over the world. The chemical analysis of 77 water samples has yet to be equaled in magnitude and detail by a similar report.

This voyage and the information gathered by its scientific crew is without doubt the most famous of its kind. Not only was a tremendous supply of knowledge made available, but the voyage was a great stimulus to other nations interested in similar endeavors. In the minds of most oceanographers, the sciences of the sea "came of age" with the culmination of the *Challenger* Expedition.

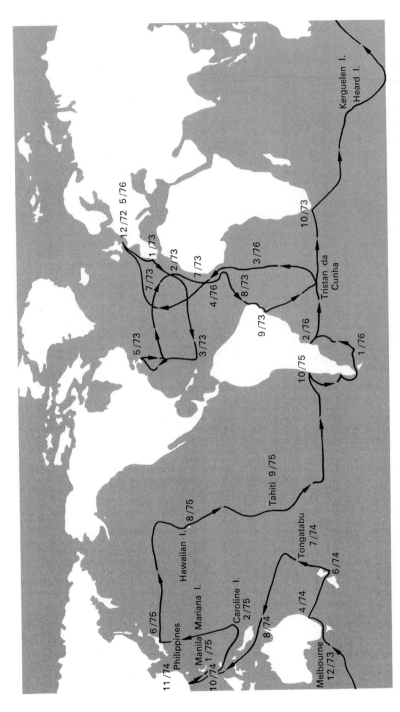

FIG. I.5 *Route traveled by the H.M.S. Challenger during its historic voyage. (After W. J. Cromie, 1962, Exploring the Secrets of the Sea, Englewood Cliffs: Prentice-Hall, p. 38.)*

The same year that the *Challenger* set sail on its voyage (1872), the first marine science laboratory was established: the Zoological Station of Naples. A year later Louis Agassiz set up a summer laboratory on Penikese Island off Cape Cod, Massachusetts. It was abandoned shortly thereafter but the present Marine Biological Laboratory of Woods Hole, which was founded in 1888, is considered a direct descendant. In 1879 the Marine Biological Association of the United Kingdom was started at Plymouth. It is now one of the most famous and productive marine stations in the world.

The U.S. Bureau of Commercial Fisheries commissioned the vessel *Albatross* in 1882. This ship, 72 meters long, was the first to be designed and constructed specifically as a research vessel.

The next few decades saw the establishment of many oceanographic laboratories in Europe and North America. Some of the more notable are: Oceanographic Museum of Monaco, established in 1910 by the Prince of Monaco; Hopkins Marine Station (Pacific Grove, California) in 1892 by Stanford University; Scripps Institution of Oceanography (La Jolla, California) in 1905; and the Lamont Geological Observatory (Palisades, New York) after World War II.

Techniques for data collection were also being improved and a wide variety of devices was being developed for sampling. One such device is the closing net, credited to the Italian scientist Palumbo. The net is designed to be lowered in a closed position, opened for sampling, and then closed again. It was first used successfully in 1884 in the Pacific.

Nets that allow collection of the microscopic organisms we call plankton were devised at about the same time. These nets enabled Victor Hensen to make quantitative studies of these tiny organisms on a German expedition in 1889.

The year 1905 produced two of the most noteworthy devices used in oceanography. Frederick Nansen designed an apparatus for taking water samples at a given depth. Now known as the Nansen bottle, this device is used by virtually every oceanographic vessel in existence. Equally important was V. W. Ekman's current meter, which allowed investigators to determine the direction and speed of a current at depth. These devices are considered in detail in later chapters.

During World War I, oceanography was not utilized for the war effort, and progress in basic science of the seas was curtailed due to lack of funds, manpower, and ships, and because of hazardous wartime seas. After the war, advances were made in a variety of phases of oceanography. One particular effort in undersea exploration stands out among all the others. William Beebe and Otis Barton developed the now famous bathysphere (Fig. I.6) in 1934. In it they descended to a depth of 930 meters and pioneered the way for future deep-sea voyages.

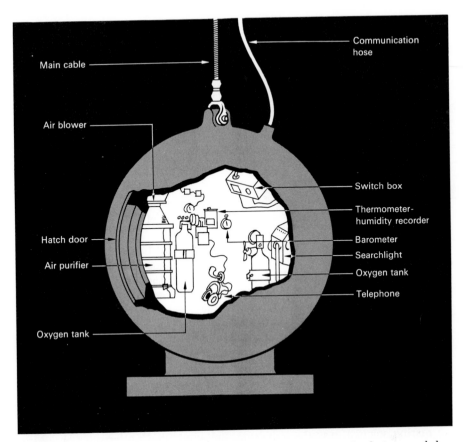

FIG. I.6 *Bathysphere used by William Beebe and Otis Barton in their record descent.*

WORLD WAR II TO THE PRESENT

Despite war's deleterious effects, technological advances made in wartime can be useful for many years afterward. During World War II, the U.S. Navy played a major role, particularly in the Pacific. As a result, there were significant advances that have since been applied in academic research.

For example, the echo-sounding type of depth recorder, first developed and put into use during the 1920s, was improved and extensively used during World War II. This was one of the first electronic devices in oceanographic studies. It measures depth by timing a sonic impulse from the source to the bottom and back again. Continuous records trace out the ocean bottom profile as the ship passes over. This and improved similar devices have provided us with a tremendous volume of depth data and have led to rather

FIG. I.7 *Jacques-Yves Cousteau, coinventor of the aqualung and one of the foremost investigators of the natural history of the oceans. (Courtesy of the Cousteau Society.)*

detailed bathymetric maps of the ocean basins. Another applicaton of sonic techniques, in which explosives are used to produce seismic vibration, was used in the 1940s to study the crust of the earth under the sea.

During this time, two other important events contributed to the development of oceanography. In 1942, H. U. Sverdrup, M. W. Johnson, and R. H. Fleming collaborated on the first comprehensive oceanographic text, *The Oceans.* The book, which exceeds 1000 pages in length, is devoted to all phases of oceanography and is still the most comprehensive treatment of the subject under one cover. Were it not for rapid strides in the past two decades which have rendered the book out of date, it would still be widely used for university teaching.

Another brilliant development, and one of the most significant advances in marine science, was the aqualung. Self-contained underwater breathing apparatus (SCUBA) was first invented in 1865 by Denayrouze. It was reinvented and perfected in 1943 in the Mediterranean by Jacques-Yves Cousteau (Fig. I.7) and Emile Gagnan. Until that time, people were restricted in underwater mobility by air hoses or other apparatus that limited or hindered movement. Experience has since demonstrated that no method of sampling from the sea can compare with actual observation by the person collecting the samples. We now have a tool that opens up the inner shelf, reefs, and other shallow environments to direct observation. An increasing number of oceanographers regard diving as a basic skill needed for research.

The study of the oceans received a significant boost in 1950 with the establishment of the National Science Foundation (NSF) by Congress to provide support for basic research in science. This agency has financed a large amount of oceanographic research.

The 1960s saw the development of deep-water submersibles. The *Trieste I*, developed by the Piccards, was the predecessor of the great variety of submersibles in existence today. It was rebuilt as *Trieste II* (Fig. I.8), and two

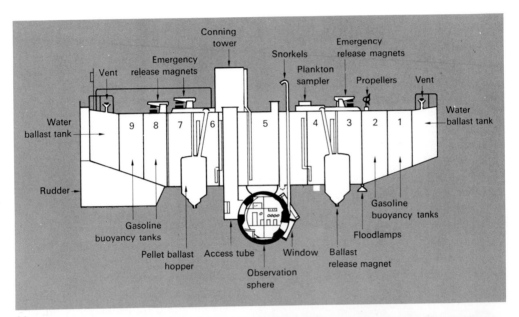

(a)

(b)

FIG. I.8 (a) Cross section of the Trieste I, which was the first of the modern deep-sea vehicles. It was modified and rebuilt as Trieste II (b), which made the world-record descent. (Photo courtesy of the U.S. Naval Electronics Laboratory.)

(a)

FIG. **I.9** (a) The DR/V Star II and (b) the DR/V Shelf Diver, two of the many recent types of deep-sea research vehicles. (Photos courtesy of (a) L. H. Somers and (b) the U.S. Naval Electronics Laboratory.)

(b)

men made a record descent in the Marianas Trench to a depth of 11,000 meters in 1960. This feat proved that such descents were possible and also demonstrated their value to science. Now there are dozens of different kinds of submersibles (Fig. I.9) designed for a particular depth or to do a certain type of data collecting. Such vehicles may prove to be as valuable for deep-water exploration as SCUBA is for shallow water.

During the 1960s, two areas of ocean exploration were undertaken which have provided a wealth of data from heretofore inaccessible parts of the ocean basins. Although geophysical techniques in general, and seismology in particular, had been applied to the investigation of the sea prior to this period, there was great expansion and refinement of these tools. By modifying techniques used on land for oil exploration, it has become possible to determine the geometry and composition of sediment and rock layers on the continental margins and ocean basins as well. Directly related to this was the development of drilling and coring capabilities at oceanic water depths. In the 1960s, the *Glomar Challenger* and its sister ships were developed, allowing the crust under the ocean to be sampled directly. With the possible exception of deep trenches, all parts of the ocean floor now can be drilled and cored for study.

The combination of these significant technological advances has not only provided the scientist with considerable data on the nature and history of the ocean basins, it also has provided extreme benefit for resource exploration of the ocean, particularly in the petroleum industry.

SELECTED REFERENCES

Burton, M., et al., 1962, *Seas, Maps, and Men: An Atlas-History of Man's Exploration of the Oceans*, Garden City, N.Y.: Doubleday. Excellent and well-illustrated historical treatment of ocean exploration. However, not much of the book is devoted to scientific investigation of the sea.

Coker, R. E., 1947, *This Great and Wide Sea*, Chapel Hill: University of North Carolina Press, Chapters 2–4. Fairly good section on the history and development of oceanography, with a strong slant toward marine biology.

Committee on Oceanography, 1972, *Oceanography, 1960 to 1970*. No. 11, "A History of Oceanography: A Brief Account of the Development of Oceanography in the United States," Washington, D.C.: National Academy of Sciences—National Research Council Committee on Oceanography. A series of papers on problems and new techniques in the measurement of various physical and chemical properties of seawater. The most comprehensive treatment of the subject up to the date of publication.

Cowen, R. C., 1960, *Frontiers of the Sea: The Story of Oceanographic Exploration*, Garden City, N.Y.: Doubleday, Chapter 1. A general book written for the natural science "bug" rather than university students. One good chapter on the historical development of sea sciences.

Deacon, G. E. R., 1962, *Oceans—An Atlas-History of Man's Exploration of the Deep*, London: Paul Hamlyn, pp. 17–73. Nicely illustrated and thorough chapter on man's exploration of the sea, but without much emphasis on scientific data collecting or sampling.

Deacon, Margaret, 1971, *Scientists and the Sea, 1650–1900: A Study of Marine Science*, New York: Academic Press. The most comprehensive history of oceanography for the period included.

Gordon, D. L. (ed.), 1970, *Man and the Sea; Classical Accounts of Marine Exploration*, Garden City, N.Y.: Natural History Press. A very comprehensive collection of articles on ocean sciences and exploration ranging from the Bible to the Deep Sea Drilling Project.

Guherlet, M. L., 1964, *Explorers of the Sea*, New York: Ronald Press. Excellent book on oceanographic history, with special emphasis on the famous and important scientific voyages. Lack of illustrations detracts somewhat from the book's merit.

Schlee, Susan, 1973, *The Edge of an Unfamiliar World: A History of Oceanography*, New York: E. P. Dutton. Very readable and interesting treatment of the history of oceanography up to and including the Deep Sea Drilling Project.

OCEAN BASINS

PART I

MAJOR WATER BODIES OF THE WORLD $\quad$ 1

The marine waters of the world may be divided into three, four, or five oceans, depending on how the term "ocean" is defined. Geography textbooks usually name five oceans: Atlantic, Pacific, Indian, Arctic, and Antarctic (or Southern). There are many reasons why oceanographers do not agree with this division and usually consider only three or perhaps four oceans.

In order to establish which bodies of marine water are oceans, we must first consider the distinguishing characteristics of oceans. Size might be the first thing that comes to mind; an ocean must be quite large. Obviously all five water bodies mentioned above are large and occupy a significant percentage of the earth's surface. In addition, an ocean must have features that set it apart from neighboring oceans, because all of these water bodies are interconnected. Such features could include currents, water masses, submarine ridges, and at least some definable land boundaries. One look at a world map (inside cover) will confirm this last point: the Atlantic, Pacific, Indian, and Arctic Oceans are clearly defined by surrounding land masses, with the Indian Ocean showing the poorest definition. The Antarctic or Southern Ocean, however, is not surrounded by land masses, but itself surrounds a land mass (Antarctica). As a result, no land boundary between the Antarctic Ocean and the southern part of the Atlantic, Pacific, and Indian Oceans is evident. The northern boundary of the Antarctic, therefore, could be determined only by current or water characteristics. Current patterns, which might be thought to define such a boundary, cannot be distinctly separated in the Antarctic. Water masses may be difficult to distinguish and in this respect, too, the Antarctic does not at all lend itself to being considered a separate ocean. In this book, the water surrounding the Antarctic continent will be considered as the southern extent of the Atlantic, Pacific, and Indian Oceans.

The Arctic or North Polar Sea is considered an ocean by many but here will be placed in the category of a **mediterranean*** or marginal sea, as it is considerably smaller than the three true oceans.

------
* Words in boldface type are defined in the Glossary (Appendix A).

DISTRIBUTION OF LAND AND WATER

Of the total surface area of the earth, water occupies more than twice as much as land; more precisely, the earth's surface is 71 percent water and 29 percent land. Marine waters cover 361 million square kilometers (Table 1.1). The distribution of this water shows a lack of uniformity. A quick glance at a globe or world map also shows that there is considerably more land in the Northern Hemisphere than in the Southern. The water-to-land ratio in the Northern Hemisphere is 1.5:1.0, whereas in the Southern Hemisphere it is about 4:1. Only between the parallels of 45°N and 70°N does land area exceed water area.

Volumetrically, 97 percent or 1.4 billion cubic kilometers of all water on earth is marine; that is, it is in oceans and seas. This leaves only 3 percent of all water, whether liquid, solid, or gas, in other environments. Lakes, rivers, and ground water account for more than 2 percent; snow and ice (glaciers) are less than 1 percent; and atmospheric water is 0.00057 percent of the total.

In comparing size and shape of the major water bodies, we notice that each has its own features. Although the Atlantic and Indian Oceans are not much different in size, the shapes contrast. Even within the Atlantic, the configuration of the northern part is considerably different from that of the southern part. The North Atlantic is irregular in outline, with many large, enclosed water bodies adjacent, while the South Atlantic is bordered by rather smooth coastlines.

TABLE 1.1 *Various earth dimensions.*

Dimension	Magnitude
Mass	6×10^{27} kg
Volume	1.1×10^{12} km^3
Circumference	40×10^3 km
Equatorial radius	6378 km
Polar radius	6356 km
Surface area	510×10^6 km^2
Land surface area	150×10^6 km^2
Ocean surface area	360×10^6 km^2
Volume of marine waters	1.4×10^9 km^3

TABLE 1.2 *Statistics on oceans and other selected water bodies.*

Water body	Area (10^6 km^2)	Volume (10^6 km^3)	Mean depth (m)
Atlantic	82.4	323.6	3926
Pacific	165.3	707.6	4282
Indian	73.4	291.0	3963
Mediterranean Sea	2.9	4.2	1429
Gulf of California	0.16	0.13	813

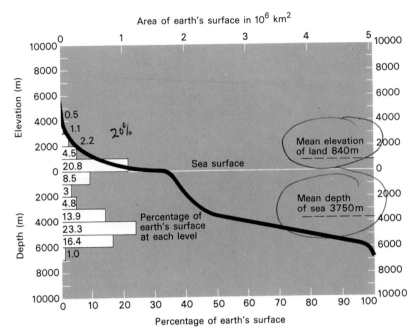

FIG. 1.1 *Hypsographic curve showing the percentage of the earth's surface at a given level and the cumulative percentage of the earth's surface (heavy line). (After H. U. Sverdrup, M. W. Johnson, and R. H. Fleming, 1942, The Oceans, Englewood Cliffs: Prentice-Hall, p. 19. Copyright renewed 1970.)*

In area, the Atlantic covers more than 82 million square kilometers (Table 1.2). The length is nearly 12,000 kilometers, and although the ocean is sinuous in shape, the width remains fairly constant at about 3300 kilometers.

The average depth of the Atlantic is nearly 4000 meters; in this respect, it is similar to the Pacific and Indian Oceans. Its greatest depth exceeds 9000 meters in the Puerto Rico Trench.

The Indian Ocean is shaped like a triangle and covers about the same area as the Atlantic (73 million square kilometers). Very little information has been available on any aspect of the Indian Ocean until the 1960s. In 1965 an International Geophysical Year project collected tremendous quantities of data on all aspects of this water body. One of the notable products of the venture was a revised physiographic map of the Indian Ocean; it was published in *National Geographic*, October 1967.

Approximately circular in shape, the Pacific covers an area more than equal to the combined areas of the Indian and Atlantic Oceans. All three oceans have mean depths within a range of about 300 meters (Table 1.2). Note that the mean depth of each ocean approximates that of the marine environment as a whole (3800 m). There is considerable difference between the mean elevation of the land masses and the mean depth of the oceans (Fig. 1.1).

In Table 1.2, several land-encompassed basins are not included as part of the three oceans. Some of them are large, containing at least 3 million square kilometers: the Arctic Sea is 13.6 million square kilometers; the deep basin on the Pacific side of Asia contains 8 million; the Caribbean and Gulf of Mexico together contain 4 million; and the Mediterranean about 2.9 million square kilometers.

GENERAL STRUCTURE OF THE EARTH

A cross section of the earth shows three distinct zones—the **crust, mantle,** and **core** (Fig. 1.2). The thickness and general composition of these zones are known from the way earthquake waves are transmitted. Scientists studying these waves can determine the density of each layer and thus make an intelligent estimate of the composition.

The crust, which ranges from about 3 to 50 kilometers in thickness and comprises 0.4 percent of the earth's volume, is composed of material with lower density than the mantle or core. Unlike the mantle and core, the crust is differentiated into two separate general rock types with different densities. **Simatic rock,** so named because of its high silicon and magnesium content, is also referred to as oceanic or basaltic rock because its general composition is similar to that of **basalt.** The other material in the crust is called **sialic** rock, which is high in silicon and aluminum. Sialic rock is sometimes termed continental or **granitic;** it is in the form of discontinuous masses (continents)

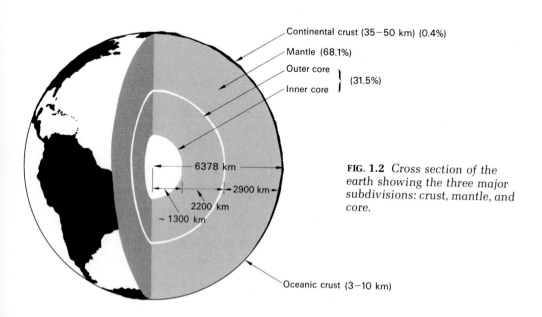

Continental crust (35–50 km) (0.4%)

Mantle (68.1%)

Outer core
Inner core } (31.5%)

6378 km

2900 km

2200 km

~1300 km

Oceanic crust (3–10 km)

FIG. 1.2 *Cross section of the earth showing the three major subdivisions: crust, mantle, and core.*

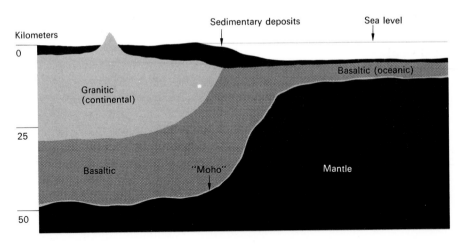

FIG. **1.3** *Relationships within the earth's crust.*

resting on higher-density, continuous, oceanic crust (Fig. 1.3). Continental masses have a density of about 2.7, whereas the density of the oceanic crust is 2.9.

The crust shows great range in thickness, with the areas under the continents being the thickest (up to 50 kilometers). The oceans overlie only 3 to 10 kilometers of crustal material. Although basaltic material is common on continents, coming from sources in the lower crust under the continental blocks, continental-type rocks in the ocean basins are extremely rare.

MOHOLE DRILLING PROJECT

In the early 1960s, a deep-drilling project was conceived and planned with the purpose of drilling down through the earth's crust into the upper mantle. Not much was then known about the mantle except that it has a density of about 3.3 and is probably composed of **peridotite**, or **eclogite**. It is separated from the crust by an abrupt change in density known as the Mohorovicic discontinuity, or Moho. It was hoped that "Project Mohole" would enable us to find out the characteristics of the lower crust, the Moho, and the upper mantle. The hole was to be drilled over thin oceanic crust from a specially designed drilling barge, *Cuss I*. The barge was designed so that it could maintain a fixed position by manipulation of four large outboard motors, one on each side of the barge. There were, however, a variety of problems involved in drilling with over two miles of water between the drilling rig and the hole. A test hole was drilled near Guadalupe Island off the west coast of Baja California during 1961. The hole penetrated the crust at 3780 meters below sea level and retrieved almost 200 meters of crust material. The drilling of a real "Mohole" was never attempted, however, as funds for support were not appropriated by Congress.

ORIGIN AND AGE OF OCEAN BASINS

The age of the earth is thought to be in the vicinity of 5 billion years. Rocks from continents have been dated at 3.9 billion years. Does this imply then that ocean basins are of similar age? Many scientists do not concur with this reasoning. However, there are those who support the idea of the "permanancy of the ocean basins." In other words, they believe that the ocean basins have existed essentially in their present configuration since the crust was formed.

Contrary to this concept are theories of an expanding earth and continental drift. According to both these ideas, the present ocean basins were formed some time after the crust. The theory of continental drift, which is considered in detail in a later chapter, is currently being subjected to considerable scrutiny.

Until initiation of the Deep Sea Drilling Project in 1968 (see Chapter 25), our investigations of the oceans had provided little evidence to support any hypotheses about their age and history. Theories were based on gross crustal relationships between continental and oceanic portions of the crust and on seismic data. Oceanic sediments are relatively undisturbed, so that the old sediments are not brought near the surface of the crust as they are on continents. Previous sampling techniques penetrated only the upper few tens of meters of sediment.

Estimations of ocean-basin ages have been made on the basis of the total thickness of sediment and rates of deposition. The range is considerable, due to wide disagreement on the average rate of sediment accumulation and the total thickness of sediment present. These estimates range from more than a billion years to about 200 million years. Data from the Deep Sea Drilling Project indicate that the age of the ocean basins as we know them conform closely to the latter figure.

PHYSIOGRAPHY OF THE OCEANS   *2*

Ocean floor surfaces, like their continental counterparts, can best be described in terms of a few recurrent features, or in terms of provinces. These physiographic provinces are recognized by their relief, depth below sea level, general slope of the major surface, and composition. In general, their arrangement within an individual ocean is more predictable than are analogous provinces on continental surfaces. As a result it is possible to generalize to a greater extent about the ocean basins than about continents.

TECHNIQUES

The configuration of ocean-bottom surfaces can be illustrated in various ways. By far the most common are the **profile section** and the **bathymetric map**, which is analogous to the topographic map except that contoured depth values increase *below* sea level. **Physiographic charts** are also used, particularly when there is at least a fair amount of relief with irregular topography. Excellent examples are those of the North and South Atlantic and the Indian Ocean prepared by Bruce Heezen and Marie Tharp (Fig. 2.1).

Regardless of the method used to illustrate bottom configuration, a tremendous amount of data is necessary to provide a reasonably accurate representation. Since shortly before World War II, great strides have been made in obtaining accurate depth data. Most oceanographic research vessels are now equipped with a Precision Depth Recorder (PDR) unit, which is a type of echo-sounding apparatus that provides depth soundings accurate to one meter in 1500. Naturally such data greatly improve the detail and accuracy of profiles and bathymetric maps. Prior to echo-sounding devices, use of wire soundings was standard procedure, but they were slow, tedious, costly, and inaccurate. Because of these drawbacks, soundings were made at rather wide intervals, whereas modern PDR units yield continuous records (Fig. 2.2). The difference between a profile constructed from wire soundings and one from electronic records (PDR) is shown in Fig. 2.3.

A word of caution is necessary about interpreting bottom profiles. Although any given profile may show considerable **relief**, there is usually great

FIG. 2.1 Physiographic diagram of a portion of the northwest Atlantic Ocean showing various major provinces. (After W. E. Yasso, 1965, Oceanography, New York: Holt, Rinehart and Winston; original physiographic diagram by B. C. Heezen and M. Tharp, Geological Society of America.)

(a)

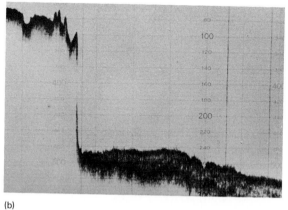

(b)

FIG. 2.2 (a) PDR unit with chart showing trace of bottom profile. (b) PDR record showing steep cliff (scale in feet). (Photos courtesy of (a) L. H. Somers and (b) J. L. Hough.)

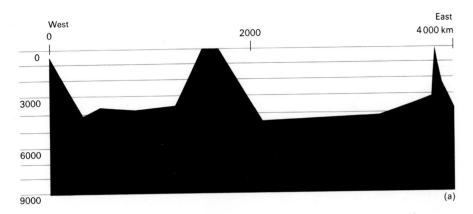

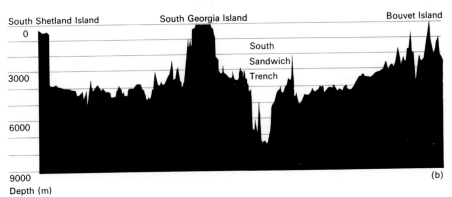

FIG. 2.3 *Two versions of bottom topography across a traverse in the South Atlantic Ocean. (a) Profile based on 13 wire soundings shows little detail compared to (b) the profile based on 1300 sonic soundings. (After Sverdrup, et al., 1942, p. 18. Copyright renewed 1970.)*

vertical exaggeration due to the limiting size of illustrations. For instance, in parts (a) and (b) of Fig. 2.3, the sharp-crested topographic highs shown on the profile are not really that sharp on the ocean floor. By examining and comparing the vertical and horizontal scales, it is evident that the **vertical exaggeration** is nearly 200:1. This means that the vertical scale is 200 times that of the horizontal; that is, for each kilometer on the vertical scale the equivalent distance represents 200 kilometers on the horizontal scale. The best way to visualize such a vertical exaggeration is to compare an exaggerated profile with one which is true to scale, or has no exaggeration (Fig. 2.4).

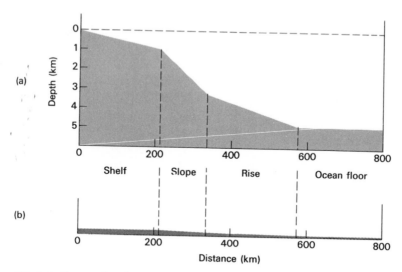

FIG. 2.4 *Comparison between (a) profile across the continental margin which shows vertical exaggeration and (b) similar profile with no exaggeration.*

Most bottom profiles show vertical exaggerations ranging from tens to a few hundred to one. Such distortions should be kept in mind at all times to avoid misconceptions of the true character of ocean-bottom topography.

CONTINENTAL MARGIN PROVINCES

The continental margin contains those physiographic provinces that are part of the oceanic realm but that are closely related to the continental masses. They form the boundary zone between the continental and oceanic crusts. Continental margin provinces from the shoreline seaward are the continental shelf, continental slope, and continental rise. The shelf and slope are structurally, stratigraphically, and petrologically like the continent, because they are actually extensions of the continent. These two provinces are collectively called the **continental terrace**. The continental rise is composed largely of sediments derived from the continents. The margins occupy a significant portion of the ocean floor, as shown in Fig. 2.5.

Continental shelf

The continental shelf is the most thoroughly studied of the oceanic provinces because of its proximity to land and its shallow depths. It abounds with life and is economically important especially because of fisheries and petroleum.

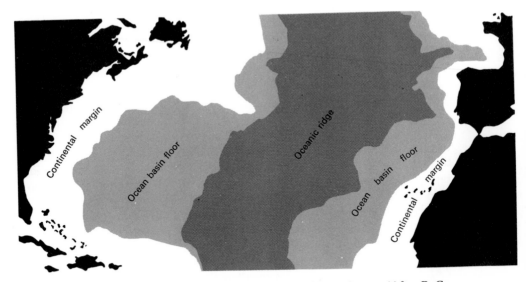

FIG. 2.5 *Major morphologic provinces of the North Atlantic Ocean. (After B. C. Heezen, et al., 1959,* The Floors of the Oceans—I. The North Atlantic, *Geological Society of America, p. 16.)*

Current exploration is expected to result in occupation of the shelf by human beings for long periods of time.

As the name implies, the continental shelf is adjacent to a continental land mass and is somewhat shelflike in configuration, i.e., rather flat. The shelf is fairly easy to define and recognize from a bathymetric chart. Years ago the shelf was defined as the platform area extending from the shore of a continent to a depth of 200 meters. Such a generalization is not precise, as the maximum depth for the shelf may range from less than 35 meters to more than 240 meters. Many factors control the depth of water over the shelf. Some areas have been glaciated, so that the bottom material has been redistributed, and others have been subjected to **tectonic** activity. Such shelves may have an average maximum depth of more than 200 meters. Toward the other extreme are shelf areas where coral reefs grow or areas adjacent to the mouths of large rivers. Depths in these locations may be less than 35 meters. The average depth of the continental shelf all over the world is 128 meters.

The preferred criterion for defining the seaward edge of the continental shelf is a marked change in slope. There is a distinct gradient change from the shelf to the relatively steep continental slope. Although there is some range, the gradient on the shelf is about 1:500 or about 2 meters/kilometer; expressed in degrees this is 00°07′. Such a gradient is essentially horizontal when drawn without exaggeration. The inner shelf is generally slightly steeper than the outer shelf, probably because the inner shelf is influenced to a greater extent by modern sedimentation.

Width of the continental shelf has considerable range, from being virtually absent to a maximum of nearly 1500 kilometers. Shelves off glaciated areas or adjacent to large rivers are wide, whereas a shelf bordering a mountainous area is commonly narrow. Relief on the shelf is generally low, with local ranges less than 20 meters. Positive features which contribute to this relief are the hummocks of glaciated shelves, submerged relict reefs as in the Gulf of Mexico, and features which are extensions of terrestrial topography. **Submarine canyons**, shallow basins, and longitudinal troughs provide topographic lows on the shelf area.

The southern California–northern Mexico shelf is unique in that it is quite narrow and geologically complex. It is usually referred to as a **continental borderland** and is geologically similar to a basin and range type structure. Many small, shallow basins are present, because of numerous faults (Fig. 2.6). Several islands have also resulted there from the complex structural relationships. As in other areas, this borderland is a seaward continuation of similar geologic relationships on the adjacent land mass.

The Grand Banks of Newfoundland is an example of a broad shelf (200–500 kilometers) that has been glaciated. Topography is irregular with

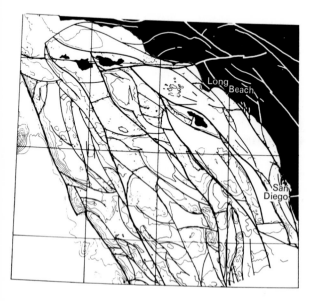

FIG. 2.6 *Complex faulting in the continental borderland area off southern California. The heavy lines denote faults, whereas the other lines represent bathymetry. (After K. O. Emery, 1960, The Sea Off Southern California, New York: Wiley, p. 79.)*

local high relief and the edge of the shelf reaches a maximum depth of 200 meters in the northern area.

Many shelf areas in high latitudes contain a series of longitudinal channels which are up to a hundred meters or so deep, usually extending below the maximum depth of the seaward shelf margin. These features are common in Greenland and the Scandinavian countries, where they consistently parallel the shoreline. H. W. Menard of the Scripps Institution of Oceanography has suggested that their presence in high latitudes and their shape indicate a probable glacial origin caused by ice lobes moving along the low areas.

Another type of shelf valley is the drowned valley, which is commonly normal to the shore and is a seaward extension of rivers. Many of these are kept open by tidal action. On the New England shelf is the Laurentian Channel, a continuation of the St. Lawrence River; it is up to 380 meters deep. Similar features are present off the Gulf of Maine and the Hudson River.

Although not part of the true continental shelf, marginal plateaus are best considered at this point. They are broad, rather flat features that are deeper than the true shelf and separated from it by a minor slope. The Blake Plateau off the southeastern United States (Fig. 2.1) is the best known of these features, although similar features are present off the eastern coast of southern South America and off New Zealand.

Continental slope

Seaward from the continental shelf and inherently related to it lies the relatively steep continental slope. It is easily recognized from traces of bottom profiles by the distinct gradient change at the seaward margin of the continental shelf (Fig. 2.7) and a similar, but less pronounced and locally subtle gradient change where the slope merges with the continental rise. The continental slope, like the shelf, has a considerable range in gradient from about 1:2 to 1:40. The latter figure is used by B. C. Heezen as the lowest gradient of the continental slope for the purpose of defining this province in areas where it is difficult to distinguish the slope from the continental rise. The average continental slope is inclined only 4°17' from the horizontal. Such a slope is actually quite gentle—thus the previous caution about viewing greatly exaggerated profile illustrations.

Continental slopes extend to an average depth of 3660 meters, but in areas where the slope continues into deep sea trenches the depth may reach 8200 meters. Relief on the slope may reach more than a thousand meters, particularly in the vicinity of **submarine canyons**. The continental slope is cut by these features in all parts of the world. At many localities they extend

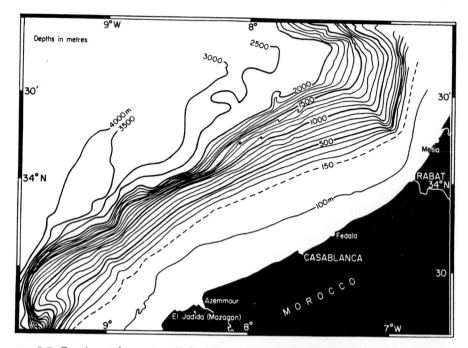

FIG. 2.7 *Continental margin off the coast of Morocco, showing the continental slope gradient in comparison to that of the shelf and rise. (From A. Guilcher, 1963, "Continental Shelf and Slope," The Sea, M. N. Hill (ed.), New York: Wiley, vol. 3, p. 288.)*

into the continental shelf and across at least part of the continental rise. Submarine canyons will be discussed in detail in a later chapter.

The gradient of the continental slope is lowest off the mouths of large rivers, where it may be as low as 1°. Faulted coasts have slopes which average nearly 6° and attain a maximum of more than 45°. One such slope is off the coast of Ceylon. Equally steep slopes which are also fault scarps are present off the west coast of Baja California and in the Bartlett Trough area of Santiago. In the borderland area of southern California the continental slope may be a **dip slope**, with its scarp paralleling the dip of Tertiary strata. The average gradient of the continental slope is different in each ocean: in the Pacific it is 5°20′, in the Atlantic it is 3°34′, and in the Indian it is slightly less at 2°55′.

Origin of shelf and slope

Most theories on the origin of the continental margin apply equally to the shelf and slope because it is impossible to separate one from the other genetically. An explanation of the origin of one of the provinces also applies to the other.

An often cited hypothesis about the origin of the shelf and slope is that proposed in 1919 by Douglas W. Johnson, a famous coastal geomorphologist. According to his theory, the inner shelf is a relatively flat, abraded platform of bedrock, a wave-cut terrace; and the outer shelf is a constructed feature derived from rivers and the abraded products of the inner shelf (Fig. 2.8), a wave-built terrace. The depth of the wave-cut surface would correspond to wave base (the effective depth of water motion in waves). Years ago scientists believed that wave base extended to depths of a hundred meters or more; then for a time wave base was thought to be at a maximum depth of only several meters. Recent data have confirmed earlier ideas and it is now recognized that considerable water motion under large waves exists at about 200 meters. These facts, along with data derived from sub-bottom profilers

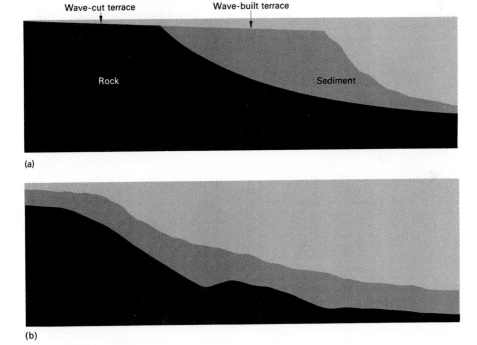

FIG. 2.8 *Wave-built and wave-cut terrace concepts of continental margin origin. Sketch (a) can be compared with an actual profile (b) from Nova Scotia. (After K. O. Emery, 1968, "Shallow Structure of the Continental Shelves and Slopes,"* Southeastern Geology **9**, 182.)

(devices that detect layering in ocean sediments), indicate that the shelf and slope at some localities are wave formed.

A second explanation of the origin of continental shelves is that the outer part of the present shelf and the slope are the result of planation of marine islands, with the slope being the seaward side of the island. Between the island and the continent a basin existed which was filled in by land-derived sediment (Fig. 2.9). The entire surface was rather uniform and controlled by wave base.

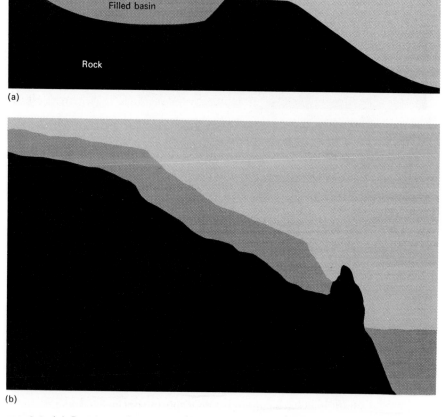

FIG. 2.9 *(a) Continental margin caused by sediment filling in behind an island or reef. (b) An actual profile off southwestern Florida shows sediment that accumulated behind a Cretaceous reef. (After Emery, 1968, p. 179.)*

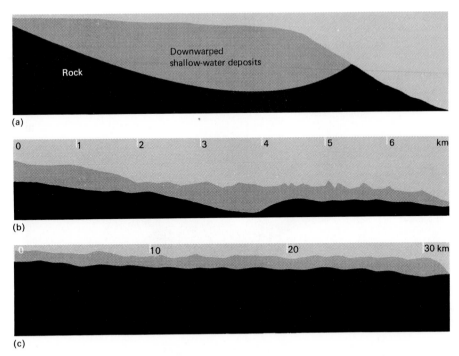

FIG. 2.10 (a) Schematic diagram of downwarping of shallow-water deposits as a cause for the origin of the continental margin. Profiles from (b) southern California and (c) northwestern Africa coasts show somewhat similar geometry. (After Emery, 1968, p. 186.)

The tremendous amount of sediment that is now being contributed to the continental shelf and slope by the Mississippi and other large rivers has led to the speculation that the continental margin may be the result of the coalescence of many large river deltas. Geological and topographic data do not support this thesis; however, a related theory seems to have some merit. Surface and core samples indicate that the shelf and slope wedge is composed primarily of sediments typical of deposition in shallow water. Structurally, there is evidence of downwarping (Fig. 2.10), as on the northern coast of the Gulf of Mexico and the eastern coast of North America. Both of these areas are adjacent to broad Tertiary coastal plains, with the strata dipping gently toward the sea. The continental margin is really just a seaward extension of this coastal plain. Relict longitudinal submarine ridges on the shelf are analogous to the gently sloped **cuestas** of the coastal plain. Other areas of the world probably have similar characteristics, but they are not as well known as the Gulf and Atlantic areas.

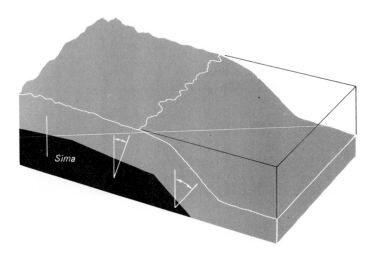

FIG. **2.11** *Flexure at the continental slope at a
place where there is little sedimentation. (After
R. S. Dietz, 1964, "Origin of Continental Slopes,"
Am. Sci. **52**, 53.)*

In some other regions of the world the margin seems to be related to subsidence; however, there is no wedge of sediment thick enough to have caused the downwarping (Fig. 2.11). Such a margin lacks a shelf or has a narrow one. The southern coast of France adjacent to the Alps has this type of flexured margin; it is probably related to the Tertiary mountain building during which the Alps were folded. This orogenic event provided the movement which seems to have continued through Pleistocene times in the area around Nice. A somewhat similar margin is present off western Africa; there a sediment wedge is being built over the flexure.

Faulting has been important in the formation of continental margins in many areas. Perhaps the best known area is the southern California borderland. This margin is narrow and does not have the smooth profile characteristic of the Atlantic and Gulf Coast margins. The California area was faulted during the Tertiary, and this tectonic activity has continued through modern times (the earthquakes in the San Andreas fault zone are an example of modern activity). There are numerous high-angle block faults on land and the adjacent submarine borderland (Fig. 2.12) that present an irregular profile. Sediment has filled and is filling in the small basins between uplifted fault blocks. To many geologists this type of margin represents an immature coast, whereas the Atlantic and Gulf are mature margins.

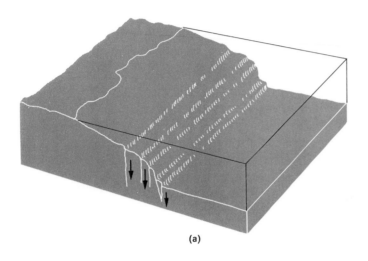

(a)

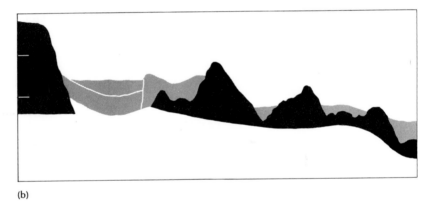

(b)

FIG. 2.12 (a) Fault-block origin of continental margin caused by high-angle normal faults. (b) Profile from Baja California, Mexico, showing structure from the California borderland area. (After Emery, 1968, p. 176.)

The above are the commonly cited theories for the origin of the continental shelf and slope couplet. Phenomena such as glaciers, tidal currents, turbidity currents, and slumping also affect the shelf and slope, but they are of relatively minor genetic significance. Quite obviously no single theory can be applied to all margins. Several factors are apparently present in shelf and slope evolution, such as downwarping, thick sediment accumulation, and faulting. They may be present singly or in combination.

Continental rise

The remaining portion of the continental margin is the most seaward province, the continental rise. It is the most difficult of the three margin provinces to recognize and define, and is also apparently absent in several areas. Its apparent absence, however, may be the result of its subtle definition. The gradient of the continental rise ranges from 1:50 to 1:800 and averages 1:150. Its shoreward boundary is easily recognized in most areas; however the gradient decreases away from this margin and its boundary with the abyssal plain is commonly obscured by a gentle gradient. A general rule of thumb is that the gradient of the rise is more than 1:1000, whereas that of the abyssal plain is less than 1:1000.

Relief on the rise is low, generally less than 20 meters. Exceptions are the seaward extensions of submarine canyons and scattered seamounts. The depth of the rise may be as little as 1370 meters and as much as 3960 meters, which is near the average depth of the ocean floor. The continental rise is fairly well developed in the Atlantic and Indian Oceans but is narrow or lacking in the Pacific. It may be as much as hundreds of kilometers wide in the Atlantic. The famous East Pacific Rise should not be confused with the continental rise. It is actually part of the oceanic ridge system, which is discussed in a later part of this chapter.

The origin of the continental rise is fairly simple and consistent throughout the oceans of the world. It consists of a wedge of sediment derived primarily from the continents and adjacent margin. Slumping, sliding and **turbidity currents** carry material down the slope and deposit it at the base. Many seismic profiles show the sediments of the rise overlapping the base of the slope (Fig. 2.13). The fairly wide range in the thickness of the rise sediments explains this similar range in the depth of the rise. Some areas have a thick prism of rise sediments which extends to only about a kilometer vertically from the shelf.

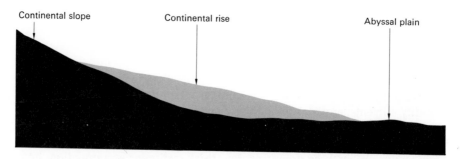

Continental slope   Continental rise   Abyssal plain

FIG. 2.13 *The relationship between the continental rise and the adjacent continental slope and abyssal plain.*

OCEAN BASIN PROVINCES

Abyssal floor

In the terminology of Bruce C. Heezen, the abyssal floor consists of the abyssal plain and the abyssal hills, two distinct topographic areas. The abyssal plain is that part of the ocean floor where the gradient is less than 1:1000. Abyssal hills are small, rather low hills which occur in clusters.

Features of the abyssal plain are those that early oceanographers once believed to be characteristic of the entire ocean floor: that is, a flat, virtually featureless plain with a fairly constant depth. Until shortly after World War II there was some debate over the existence of flat, featureless plains on the ocean floor. Then in 1947 and 1948 two expeditions documented such plains by echo soundings in the Atlantic and Indian Oceans. Actually the abyssal floor (both hills and plains) comprises only about one-third of the Atlantic and Indian Oceans but nearly three-quarters of the Pacific. The depth is fairly uniform throughout each individual abyssal plain but there may be a difference in depth of several hundred meters between one plain and another.

The origin of abyssal plains is fairly clear. Most evidence indicates that they formed by the spreading out of turbidity currents and the ponding of the sediment in basins on the ocean floor. Many places have been discovered where submarine canyons extend across the continental rise to the ocean floor, thereby providing a path for turbidity currents. **Pelagic sediments** and **oozes** supplement the above source by providing a thin and rather uniform layer of sediment on the ocean floor.

Abyssal plains are not restricted to the three oceans but are also present in many of the mediterraneans, such as the Gulf of Mexico, near Antarctica and the Arctic Basin. The latter has an especially uniform abyssal plain.

The abyssal hills province is a bedrock crust covered with a thin veneer of pelagic deep-sea sediments. Here there are many hills which range from a few to hundreds of meters in height and up to 8 or 10 kilometers in diameter. Such features also occur as isolated hills on abyssal plains. The abyssal hill province is at the seaward part of the abyssal floor. There is a positive correlation between abundance of abyssal hills and lack of terrigenous influx via turbidity currents. This might be the result of a sill or ridge preventing sediment movement by blocking turbidity currents, or a combination of a large area and small supply of sediment. The latter is the situation in the Pacific Ocean, where the abyssal hills comprise most of the abyssal floor. In the Atlantic and Indian Oceans the reverse is the case.

Oceanic rises

The oceanic rise province consists of a rather large area which is elevated above the abyssal floor and is distinctly separated from the continental margin and oceanic ridge. In area the rise covers several thousand square kilometers, and it is elevated 300 meters or more above the surrounding abyssal floor. There is a wide range of relief on the various oceanic rises of the world. Also, oceanic rises are aseismic in contrast to their neighbors, the oceanic ridges. The Bermuda Rise of the eastern North Atlantic Ocean is by far the best known of these features. Rises are thought to result from the gentle upwarping of the ocean floor (Fig. 2.1).

It should be emphasized that the term "rise" is applied to three different provinces of the oceans, and the reader should make a distinct effort to sort these out in his mind. The continental rise is part of the continental margin; the oceanic rise may have a similar topographic expression but is encompassed by the abyssal floor. The third, and as yet unmentioned rise, is the East Pacific Rise, which is really not a rise at all but part of the oceanic ridge system which will be discussed later in this chapter. (Unfortunately, the name became entrenched in the literature before it was known that this feature was part of the ridge system.)

Oceanic ridges

The oceanic ridge system is a more or less continuous and largely submarine mountain range which extends some 65,000 kilometers around the globe. It is a large area of high relief and seismic activity. The first ridge recognized and studied was the Mid-Atlantic Ridge in the North Atlantic. Since that time we have learned that the ridge in the middle of the Atlantic extends around the world, although not necessarily in a central position in every ocean. As a result of the original designation, the term "mid-oceanic ridge" is still used in many texts and scientific papers. It is misleading and should be avoided. As the map (Fig. 2.14) shows, the oceanic ridge "runs aground" in some parts of the world. In general, the province is quite wide, usually several hundred kilometers throughout its extent.

The oceanic ridge has a variety of characteristics unique in the oceans. The most obvious is the high relief and rugged topography, particularly near the median part of the ridge. Volcanism is widespread, and in fact the volcanically active Icelandic area lies directly on the crest of the ridge. Beneath the ridge, the crust appears to be abnormally thick for the ocean basin, achieving a maximum beneath the crest of the ridge. Actually the use of the term "crust" here is in the loose sense. Some interpretations would designate the ridges as neither true crust nor true mantle (see Chapter 24). The heat flow along the crest of the ridge is anomalously high and it decreases toward the flanks. Associated with the volcanism is the usual high seismic activity and also a high level of gravity and magnetic anomalies. There are many faults, with systems oriented both parallel to and normal to the ridge. The crest of the ridge is characterized by a narrow, high-relief **rift valley** that ranges from 13 to 48 kilometers in width. The Indian Ocean

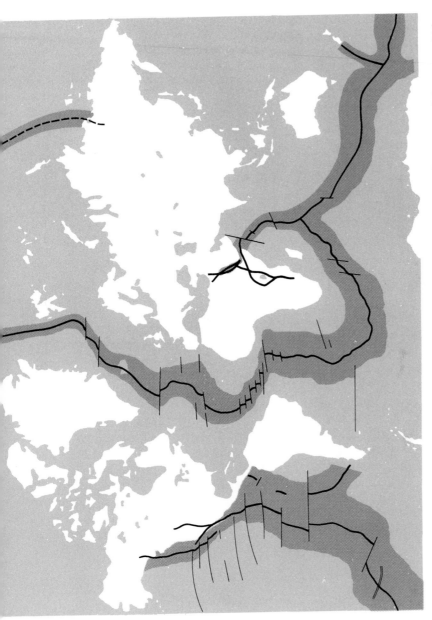

FIG. 2.14 *World distribution of the oceanic ridge, which is continuous for more than 60,000 kilometers. (After M. J. Keen, 1968, Introduction to Marine Geology, New York: Pergamon Press, p. 48.)*

portion of the ridge can be traced continuously to the rift valleys of Africa (Fig. 2.14). Profiles of both the African and oceanic rift zones are similar but subaerial erosion has smoothed the African profile, making it less rugged. In the oceanic ridge, there is evidence of subsidence and **isostatic** adjustment in the form of high-angle faults, tilted volcanic peaks, and some gentle downwarping.

It is convenient to subdivide the oceanic ridge into crest and flank provinces. The crest is an area of high relief which is composed of rift mountains and faulted plateaus; it ranges from 80 to 320 kilometers in width. The crest, in turn, can be subdivided into the rift valley, rift mountains, and high fractured plateau (Fig. 2.15). All three are areas of high and rugged relief with small, sediment-filled valleys.

The flank provinces are less rugged, have less relief, and show a greater number of filled valleys than are present near the crest. Heezen applies the terms lower, middle, and upper steps to the flank areas. These three subdivisions exhibit subtle boundaries and are difficult to recognize individually. There is a general increase in depth away from the crest area, and each step province has a fairly regular depth range.

Associated with the oceanic ridge province and undoubtedly genetically related to it are a large number of great faults. Most of these faults are oriented normal to subnormal with the ridge system. The best documented

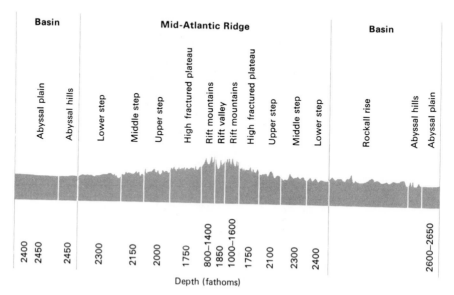

FIG. **2.15** *Subdivisions across the Mid-Atlantic Ridge. (After Heezen, et al., 1959, p. 90.)*

faults are in the Pacific, but as data are collected, similar faults are being found throughout the world. They range from several hundred to more than a thousand kilometers long, and the fracture zones may be several tens of kilometers wide. Investigation of these faults indicates that movement is primarily horizontal, with little vertical dislocation. The Mendicino Fault of the Pacific has apparently undergone about 1100 kilometers of movement. This is the maximum recorded movement along such fracture zones. The origin of these faults will be discussed in Chapter 25.

The origin of the oceanic ridge and associated features is still subject to speculation. The currently favored theory is related to sea-floor spreading and new concepts of global tectonics (see Chapters 3 and 25). It has been postulated that the ridge represents the upwelling portion of **convection cells**, which then spread laterally. Many ridge characteristics, such as high heat flow, vulcanism, mineralogy, and seismic activity, tend to support this idea. Fault zones normal to the ridge could then be the result of differential lateral movement within the crustal plates.

Trenches

The deepest parts of the ocean basins are elongate, narrow, and arcuate features called trenches. These extremely deep areas are present in all three oceans but are by far most common in the Pacific Ocean (Fig. 2.16). Trenches are more than 6100 meters deep and range widely in depth. The deepest known place in the world is 11,515 meters below sea level. This is known as the Challenger Deep, located in the Marianas Trench in the western Pacific. The depth was recorded in 1962 by the *H.M.S. Cook*. In time, deeper spots will probably be found as more exploration takes place.

Trenches are up to a few thousand kilometers long and less than 240 kilometers wide. They are invariably associated with **island arc** systems of active volcanos. These island arcs are located on the continent side of the trenches. The areal shape of trenches is arcuate with the convex side toward the ocean basin. In the Pacific these trenches and island arcs ring the ocean immediately shoreward of the **andesite line** (Fig. 2.16).

Trenches are asymmetrical in profile view, with the seaward side having a lower gradient than the side adjacent to the island arc. Other characteristics of trenches include associated negative gravity anomalies, intense seismic activity, and low heat flow. These features have been used to support various theories concerning the origin of trenches.

In 1932, Vening-Meinesz was the first person to partially descend in a trench area. He made gravity observations with a pendulum mounted in a submarine. From his observations he formulated a theory for the origin of trenches that involves a downwarping of the crust. The theory suggests that the downwarping could be due to underthrusting of the oceanic crust be-

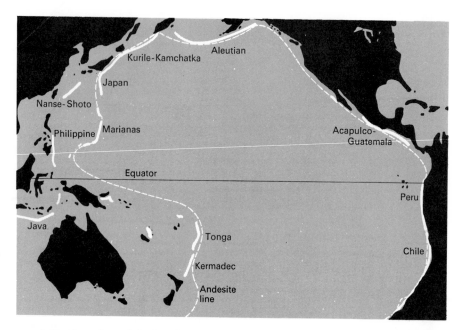

FIG. 2.16 *Trenches of the Pacific Ocean showing their position with respect to the andesite line. Note the general arcuate shape which is convex toward the ocean basin.*

neath the continental crust. Although Vening-Meinesz did not suggest it, this phenomenon could result from the descending part of **crustal plates**. Such an origin is compatible with the features and characteristics of trenches. Faulting has also been suggested as a possible origin. Geologists thought for a long time that trenches represented modern geosynclines. However, recent data on the size and shape of geosynclines and the type and amount of sediment in them discredit this hypothesis.

Seamounts and guyots

There are many small features of the ocean floor that exhibit high relief. Grouped in clusters or isolated, they are referred to as seamounts and guyots. Both seem to have a volcanic origin.

A **seamount** is at least one kilometer in height and is circular to elliptical in plan view. Although seamounts are found in all ocean basins, they are most abundant in the Pacific, particularly in the vicinity of fault zones (Fig. 2.17). Thousands have been recognized and more are continually being discovered.

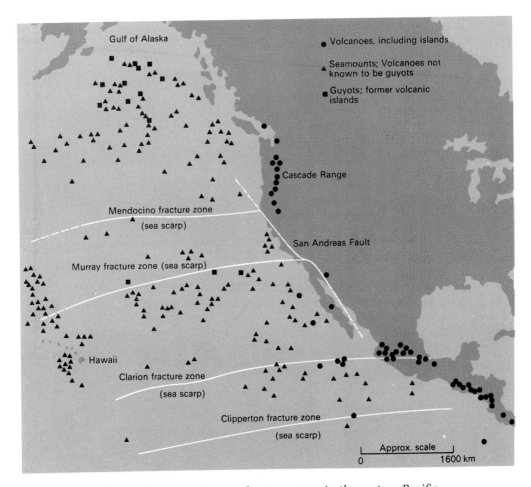

FIG. 2.17 *Distribution of seamounts near fracture zones in the eastern Pacific Ocean. (After H. W. Menard, 1955, "Deformation of the Northeastern Pacific Basin and the West Coast of North America," Bull. Geol. Soc. Am.* **66**, *1152.)*

Guyots or tablemounts are like seamounts but have a flat-topped profile, in contrast to the fairly peaked top of seamounts (Fig. 2.18). These features were first discovered by the late H. H. Hess of Princeton University while he was on duty in the Pacific with the Navy during World War II. The flat top of guyots is not easy to explain. There is a general depth at which this flat surface occurs: about one kilometer. The simplest explanation and that which was originally advocated is truncation by wave activity. This explanation was supported by remains of shallow-water organisms and features on

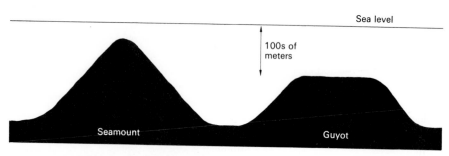

FIG. 2.18 *Profiles of a seamount and a guyot. The diameter at the base of each is a few kilometers.*

the flat surface of the guyot. However, the hypothesis suggests a sea-level drop far surpassing anything that could be attributed to glaciation and Pleistocene changes.

As more data were collected, the mystery of the deep, flat surfaces of guyots was solved. Profiles show what are apparently depression rings around the base, indicating subsidence. Some guyots have subsided to as much as twice the average depth of a kilometer. Also a few have their flat surfaces tilted with respect to sea level, indicating differential subsidence (Fig. 2.19).

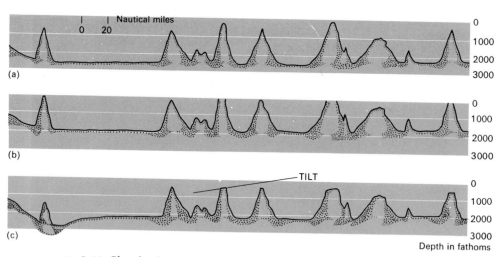

FIG. 2.19 *Sketch of seamounts and guyots showing possible explanation for the truncation on the latter: (a) formation of seamounts, (b) uplift along the East Pacific Rise and truncation, and (c) eventual subsidence and tilting. The deep at the left represents the Aleutian Trench. (After H. W. Menard, 1964,* Marine Geology of the Pacific, *New York: McGraw-Hill, p. 106.)*

SELECTED REFERENCES

Emery, K. O., 1960, *The Sea Off Southern California—A Modern Habitat of Petroleum*, New York: Wiley, pp. 1–61. A detailed description of the unique ocean floor features of the southern California borderland area.

Emery, K. O., and Elazar Uchupi, 1972, *Western North Atlantic Ocean: Topography, Rocks, Structure, Water, Life and Sediments*, American Association of Petroleum Geologists Memoir 17. Excellent treatment of the detailed physiography of the eastern continental margin of North America.

Guilcher, Andre, 1958, *Coastal and Submarine Morphology*, New York: Wiley. Fairly good treatment of subject but later information has made it out of date. Most examples are from the European area of the North Atlantic.

Heezen, B. C., *et al.*, 1959, *The Floors of the Oceans—I. The North Atlantic*, Geological Society of America Special Paper 65. Most comprehensive and best-illustrated coverage of the ocean floor to date. Only the North Atlantic is covered; however, many of the same features are present in all ocean basins.

Hill, M. N. (ed.), 1963, *The Sea—Ideas and Observations*, New York: Wiley, Vol. 3, Chapters 12–18. Detailed treatment of submarine morphology by experts in each of the various provinces.

Menard, H. W., 1964, *Marine Geology of the Pacific*, New York: McGraw-Hill. The most comprehensive treatment available on the Pacific Ocean floor, which has many features that set it apart from the other ocean basins.

Murray, John, and Johan Hjort, 1912, *The Depths of the Oceans*, London: Macmillan, Chapter 4. Out of date but of value and interest because it shows the care and detail with which many of the early workers proceeded. Authored by one of the scientists from the *Challenger* Expedition.

Poldervaart, Arie (ed.), 1955, *The Crust of the Earth*, Geological Society of America Special Paper 62, pp. 7–19. Brief and general article on ocean basin provinces. The book as a whole is a classic on the earth's crust and is pertinent to a good understanding of the oceans, ocean basins, and their origin.

Shepard, F. P., 1972, *Submarine Geology* (3rd edition), New York: Harper and Row, Chapter 9. Discussion of general characteristics of continental-margin topography plus a global tour of the continental margins.

3 DISTRIBUTION OF CONTINENTS AND OCEAN BASINS

One of the most fundamental problems in science is the distribution of land masses on the surface of the earth. More specifically, why are the continents arranged and shaped as they are on the globe?

During the latter half of the nineteenth century, geologists noted gross similarities in the structural and paleontological characteristics of the land masses in the Southern Hemisphere. Such comparisons led Edward Seuss to postulate that these masses were at one time joined together and subsequently separated. Seuss hypothetically assembled the southern continents, including India, into one megacontinent he called **Gondwanaland**. Named after an important geological province of India, it represented the distribution of land in the Southern Hemisphere prior to the present distribution. This began the long-debated and investigated theory of drifting continents. Only in the last decade or so have reasonable explanations for the distribution of continents and ocean basins been formulated.

WEGENER'S THEORY OF CONTINENTAL DRIFT

Seuss's Gondwanaland was based on little data, and his idea included no mechanism for breaking up or moving the land masses. In 1912 a German meteorologist, Alfred Wegener, proposed the first comprehensive theory of continental drift. Although Wegener conceived the idea in 1910, it was not made public until 1912 and was not published in its entirety until 1915 as *Die Entstehung der Kontinente und Ozeane*. The English translation, *The Origin of Continents and Oceans*, was first published in 1924 and is now also available in paperback form.

The shapes of eastern South America and western Africa were what originally attracted Wegener to the idea of drift. He found many geological features that could be explained by drift and argued that if continents could move vertically (mountain building, earthquakes, and so on), why not horizontally? Wegener hypothetically reassembled all the continents into one supercontinent he called **Pangaea**. According to his theory, this land mass split apart about 200 million years ago (early Mesozoic times). The earth's rotation provided forces necessary for the breakup and subsequent drift. Wegener's reconstruction and evolution of the present distribution of continents is shown in Fig. 3.1.

As Pangaea broke up, there were two major fractures where great separation took place: between the Americas and Europe-Africa and also between Australia-Antarctica and Africa-India. These fractures became quite large as drift proceeded and eventually they formed the Atlantic and Indian Oceans. The original ocean, that is, the predrift ocean, would be an ancestor to the present Pacific Ocean, which was actually getting smaller as drift proceeded.

The shallow Tethys Sea separated India from the rest of Asia until Cenozoic times. According to Wegener's theory, India was considerably foreshortened after drifting to the north across the present Indian Ocean (Fig. 3.1). When Asia proper stopped the subcontinent of India, the resulting compression formed the folded Himalaya Mountains. In a somewhat similar fashion the other major Tertiary mountain ranges of the world were formed. These include the Alps, Andes, Rockies, and others. Wegener believed that these complexly folded areas were the result of frictional drag as the continents moved over the basaltic crust. If one notes the location of these young folded mountain systems, it is apparent that each is located on or near what was a seaward-moving margin of the land mass (Fig. 3.1c).

As the continents moved they left detached pieces of sialic material in the form of islands. This part of Wegener's hypothesis was suggested by the high-latitude islands adjacent to North and South America. The Australia and New Zealand relationship is an apparent exception, but Wegener explained it by having Australia move to the southeast and then back up, leaving New Zealand behind.

The preceding paragraphs are a general summary of Wegener's hypothesis. For a decade or so after the theory was published there was considerable controversy over its merit. The mere thought of moving continents seemed unbelievable even to scientists. As more data were collected, geologists pointed to inaccurate reconstructions by Wegener and alternative explanations for his evidences of drift. The forces of rotation did not satisfy physicists as a mechanism, and most scientists discounted the theory as unfounded and undemonstrable.

The past 10 or 15 years, however, have seen a general reversal of opinion on continental drift. A poll of geologists taken now would probably show a substantial majority who favor the idea of drift. Similar trends are also evident among biologists, physicists, and others. Most of this change in opinion is the result of much recent data in support of drift, particularly from oceanic and paleomagnetic studies. This chapter will discuss the subject of continental drift using a variety of lines of evidence to support its existence.

48

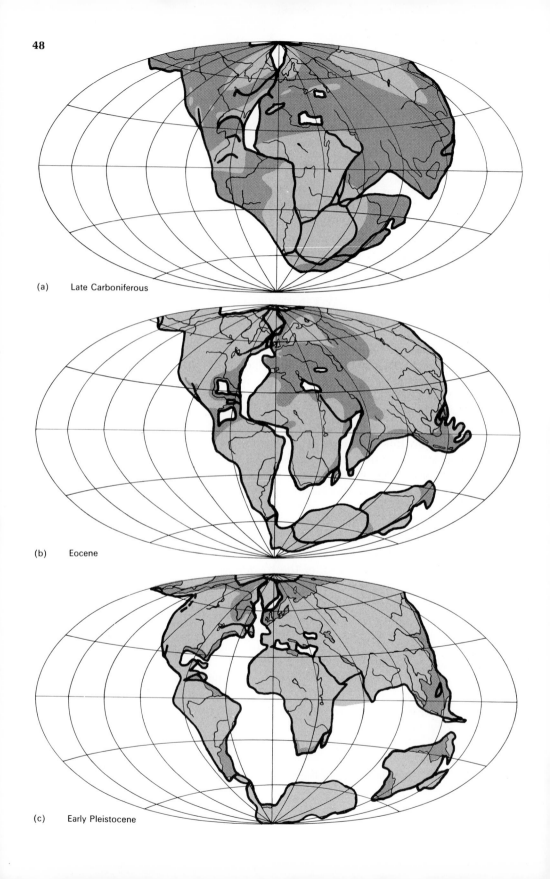

(a)   Late Carboniferous

(b)   Eocene

(c)   Early Pleistocene

◀ FIG. **3.1** *Reconstruction of the world's land masses as they appeared prior to and since the initiation of continental drift. Shallow seas are shown in the dark pattern. (After A. Wegener, 1966,* The Origin of the Continents and Oceans, *New York: Dover, p. 18. Translation of the 1915 original.)*

The final chapter in the book will essentially pick up where this chapter concludes. The background provided by the intervening chapters will enable the reader to fully grasp the current concepts in global tectonics as they relate to continental drift. Most of these concepts are relatively new and result from much recent data in the areas of marine geology and geophysics.

EVIDENCE FOR CONTINENTAL DRIFT
New techniques, expeditions to previously unexplored lands, and reevaluation of old data have led to considerable recent literature in favor of continental drift. Since the middle 1950s there have been several symposia on drift, and many of the papers and discussions have been published. The exploration of Antarctica has been particularly revealing by providing geological data which possibly link it to other land masses in the Southern Hemisphere.

If we are to take a purely objective view of the evidence, we should keep in mind that the evidence in favor of drift tells us that drift *could have* taken place, not that it *did*. As yet we have only circumstantial evidence. In order to establish drift as *fact*, a positive connection between land masses must be demonstrated.

Transoceanic structural connections
One of the most common types of evidence cited as support for continental drift is the presence of numerous structural connections, particularly across the Atlantic. J. Tuzo Wilson of the University of Toronto, a vigorous advocate of drift, has shown that the famous Great Glen Fault of Scotland has a North American counterpart in New England and the Maritime Provinces of Canada. According to Wilson, this North American system of faults, known as the Cabot Fault, would be an extension of the Great Glen if North America were moved back to its supposed predrift position. The geological ages of the systems correspond as well.

Similar structural connections are postulated for the Hebrides Mountains in Scotland and the Labrador fold belt trend, and for the Cape Mountains of South Africa and the Argentinian Sierras (Fig. 3.2). Also, recent work on faults in Brazil and western Africa has suggested a number of connections. Greenland and Ellesmere Island were believed by Wegener to have

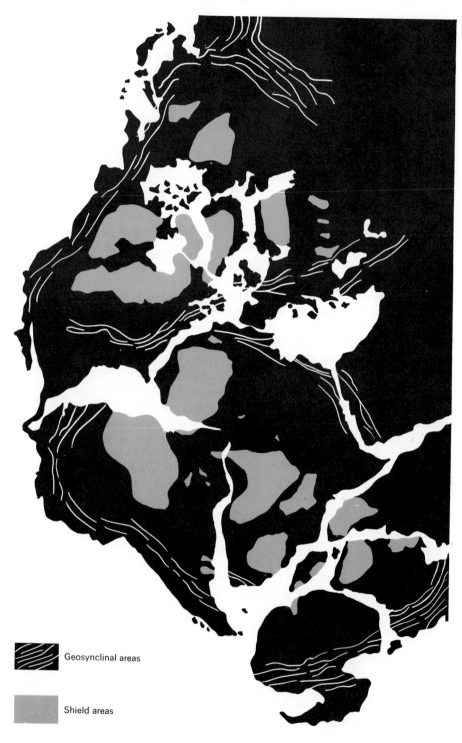

Geosynclinal areas

Shield areas

◀ FIG. **3.**2 *Geologic connections across various continental blocks when they are placed in their supposed predrift positions. (After P. M. Hurley, 1968,* Symposium on the Primitive Earth, *Miami University, Oxford, Ohio.)*

been separated by a fault, and recent evidence actually indicates the presence of such a fault. The Canadian government is in fact attempting to determine if these land masses are currently moving apart. Only about 30 miles separates the large islands, which makes this survey much simpler than transoceanic surveys.

In addition to transatlantic connections, there are similar features across the Indian Ocean. In particular, Malagasy (Madagascar) and India have a series of folded gneisses that match rather well. As more information is obtained about the structural provinces of Antarctica, a strong indication is growing that it also contains features which can be matched with other land masses. One of these is the structural trend from Chile across Cape Horn to northwestern Antarctica (Fig. 3.2).

It is obvious from the above examples that there are many structural features which can be correlated across oceans by geographic position in a predrift reconstruction and also by geologic age. Antidrifters could discount these examples by attributing them to mere chance. But the single most convincing fact relating to drift is the complete absence to date of any evidence of similar connections formed after the beginning of the Cretaceous period. This, of course, coincides with Wegener's early Mesozoic date for the split of Pangaea.

Late Paleozoic glaciation

Evidence of widespread glaciation on continents of the Southern Hemisphere during Permo-Carboniferous time has been recognized for some time. Although the glaciation took place about 250 million years ago, ample evidence firmly establishes not only its distribution but also the direction of ice movement from the center of accumulation. This evidence is in the form of moraines and glacial grooves and striations on bedrock surfaces.

The glaciers in question were present in Argentina, most of southern Africa, India, southern Australia and, according to recent evidence, also in Antarctica (Fig. 3.3). A first reaction might be that much of this glaciated area is within latitudes where one would least expect to find glaciers. However, paleoclimatologists (scientists who study ancient climates) have known for a long time that the earth's climatic belts have changed through time. Thus the

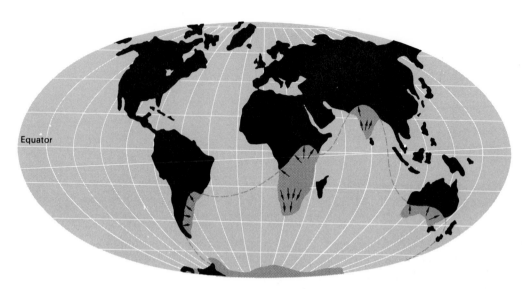

FIG. 3.3 Distribution of Permo-Carboniferous glaciation in the Southern Hemisphere, with arrows indicating direction of ice movement. (Modified from H. Takeuchi, et al., 1967, Debate About the Earth, San Francisco: Freeman, Cooper, p. 49.)

presence of the equator in Africa today does not exclude the possibility that the area has been glaciated in the past. Nevertheless, the wide scattering of these glaciated land masses does present a problem. They extend over 110° of latitude (Fig. 3.3), so that even if we placed the south pole at the center of their distribution, the glaciers would have had to extend into the low latitude region. The problem is complicated by the fact that we find no evidence of a similar glaciation in the Northern Hemisphere.

If, however, the land masses are moved to the predrift position and then the south pole is placed at the center of the glaciated area, the situation becomes much more compatible with a climatic distribution similar to our own. The predrift north pole would have been in the ocean and therefore there would have been no glaciation in the Northern Hemisphere. It is significant to note that while glaciers were covering these land masses, semitropical coal swamps covered much of the present-day Northern Hemisphere.

Paleontological evidence

One of the most debated issues in the controversy concerning continental drift is the similarity of fossil plant and animal assemblages in widely separated land masses. The argument is basically whether these similar fossils, which lived during similar times in the geologic past, are best explained by continental drift or by land bridges which provided passageways for migration.

A fossil can provide evidence of drift only if it is of the type that would be prevented from migrating, or at least be slowed considerably, by an ocean basin. The most likely possibilities are land plants and animals, or shallow-water organisms from the continental shelf. The most widely discussed such organisms belong to the *Glossopteris* flora. These plants are associated with coal measures and are present in all continents of the Southern Hemisphere. They are found in rocks from Early Carboniferous to Middle Cretaceous times and include 58 species. Their leaves look like ferns, although they are now considered seed plants. Because of the tremendous geographic spread in the distribution of the *Glossopteris* flora within a narrow range in geologic time, they have been used to suggest drift.

A certain kind of lemur is found today in India, southeast Asia, and Africa, but those who oppose drift have proposed a land bridge to account for its distribution. There are many other terrestrial organisms that are spread over vast areas of the world. This evidence for drift is circumstantial, admittedly; however, there is no evidence for the past existence of any land bridges in the Atlantic or Indian Oceans.

Shallow-water marine fossils have also provided some ammunition for the proponents of drift. Paleozoic corals occur in belts which are apparently displaced when correlated across the present land masses. It has long been known that certain marine fossils, particularly trilobites, from the Appalachian area of North America are more closely related to those of the British Isles than to those of the Rocky Mountains. A suggested reason for this similarity is that the Appalachian and British trilobites once occupied the same shallow sea. If we could prove migration along this seaway from the British Isles through the Appalachians or vice versa, then this would establish that these two areas were once continuous.

Paleomagnetism

One type of evidence which is fast becoming a forceful argument in favor of continental drift is found in the magnetic properties of rocks. Everyone is aware that the earth possesses a magnetic field; our compass needles align themselves with it. This magnetic field has apparently been present since the earth assumed its present state and shape. Certain minerals which crystallize from magma are magnetic. When they are formed their magnetic fields align with the magnetic field of the earth, just as iron filings align themselves in the field of a bar magnet. A similar effect is produced when sands containing the mineral magnetite are deposited. Both of these phenomena will preserve the orientation of the earth's magnetic field as it was at the time of their formation. Then by determining the age of the rocks and their remanent magnetism, we can obtain a plot of the earth's magnetic field and poles through geologic time. It is, of course, necessary to recognize metamorphism, which would recrystallize and thus reorient magnetic minerals, and structural deformation which would physically rearrange the grain orientation.

What then can paleomagnetism actually tell us that relates to continental drift? It has been known for a long time that the earth's magnetic field gradually changes orientation with respect to the earth's surface. As a result it would be expected that the Cambrian poles or Pennsylvanian poles would be considerably different from present-day magnetic poles. If, however, the continents have always occupied their present positions on the earth's surface, the magnetic pole for any given time in the geologic past should plot at the same location when plotted with data from any continent. Such is not the case. Magnetic poles prior to the Tertiary Period show a wide scatter when plotted from data obtained on different land masses; after the beginning of the Tertiary they show some agreement.

Most investigations show the ancient poles on North America and Europe shifting through the Pacific Ocean from Precambrian times to the present. There is a distinct separation between plots based on data from European rocks and those from North America (Fig. 3.4). The separation is nearly 30 degrees, but if these continents were moved to their supposed predrift position, there would be close coincidence of the plotted poles.

When pole positions from the geologic past are compared with paleoclimatological data, there is further evidence for continental drift. There should be agreement between ancient poles and the latitudes determined from such features as coal seams, coral reefs, evaporite basins, and so on. These are commonly used to determine ancient climatic belts. Such reconstructions show some agreement of climatic and magnetic data on North America during the Paleozoic Era. However, there are many disagreements in other places, particularly in Africa and Australia. Even in the case of North America, this evidence does not refute drift if we assume that the drift did not begin until Mesozoic times.

We have greatly simplified the above arguments by assuming that the magnetic and rotational poles were always approximately coincident. It is possible that there has been considerable shifting of one type pole with respect to the other.

The reader should make special note that while polar wandering and continental drift are commonly investigated and discussed together, one phenomenon may occur without the other. Most scientists agree that the poles have wandered; however, such movement is not dependent upon or necessarily associated with continental drift. It is the plots of pole positions with respect to the continents at a given period in geologic time which is critical to the argument.

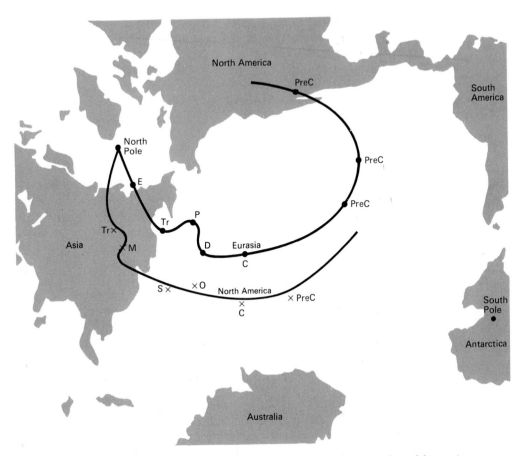

FIG. 3.4 *Magnetic pole positions from North America and Eurasia as plotted for various periods of geologic time. Key to abbreviations: PreC, Precambrian era; C, Cambrian period; O, Ordovician period; S, Silurian period; D, Devonian period; M, Mississippian period; P, Pennsylvanian period; Tr, Triassic period; E, Eocene epoch (Tertiary period). (Modified from A. C. Munyan, 1963, "Paleomagnetic Methods of Investigating Continental Drift," Polar Wandering and Continental Drift, Soc. Econ. Paleontologists and Mineralogists, Spec. Pub. 10, p. 51.)*

OBJECTIONS TO CONTINENTAL DRIFT

Although many of the reasons previously cited may lead one to feel confident that drift is a reality, there are several opposing factors. For example, land bridges are a possible means for dispersing organisms. Land bridges existed in the Tertiary Period between Asia and North America, and between North and South America. The geologic similarities we mentioned previously are but a few examples; many features cannot be correlated across the Atlantic and Indian Oceans. Even though the paleomagnetic evidence seems more concrete, the antidrifters point out that only some of the data from this source supports the drift hypothesis, not *all* of the data.

There are other aspects of the drift theory that are not yet adequately explained. The most difficult point is the mechanism invoked by Wegener to explain the breakup and movement of land masses. The objections to this mechanism will be considered in the last chapter of the book.

Another aspect of Wegener's hypothesis has also come under scrutiny. In general he regarded continents as relatively rigid masses moving over a less rigid mantle. But in explaining the formation of certain mountain ranges, he apparently reversed his thinking. Then he referred to the "frictional drag" of the continents, a concept which seems to presuppose a relatively plastic crust with respect to the mantle. Naturally the discrepancy between these two views has been questioned.

The timing of continental drift as presented by Wegener has also been questioned. If we suppose, as Wegener did, that drifting started about 200 million years before the present, we are considering only a small amount of geologic history—about 4 percent, assuming the earth to be 5 billion years old. It would seem likely that if the earth has undergone a vast rearrangement of its crust, it should have taken place rather early in its history when the earth was more unstable. However, many scientists argue that there was more than one period of continental drift and that our present arguments and observations are concerned only with the most recent and therefore the most conspicuous drift.

One of the most commonly cited evidences of drift is actually one of the poorest arguments. Although the similar shapes of the coasts of South America and Africa were one of the first features noticed by Wegener, it is difficult to understand why so much has been made of this idea. Today there are drifters who are diligently working on ways to fit the puzzle back together. It should be obvious that it cannot be done. If drift occurred, it started in the Mesozoic Era, and we cannot expect modern pieces to fit a Mesozoic puzzle. This is particularly true if we consider the amount of change that took place on continental margins during late Mesozoic and Tertiary times.

Another approach to the same problem is to match the land masses at the 500-meter contour or some other depth. Data from the continental margins of the world show many of them to be post-Cretaceous, and so these studies are not relevant to the problem. One must find the configuration of the land masses immediately after drift in order to make any contribution to fitting the pieces back together. This is quite difficult, and actually we can obtain our best reconstructions not from existing configurations of shoreline but from paleomagnetic, paleoclimatological, and structural data.

MECHANISM OF CONTINENTAL DRIFT

One of the most difficult to explain and most often questioned facets of the continental drift theory is the mechanism for breaking up Wegener's Pangaea and moving the land masses. The ancient Greeks believed that the earth was supported by the giant Atlas and that earth movements were caused by his movements. Wegener's theory suggested that most of the land masses were moving westward and toward the equator. He explained the movement of continents toward the equator by the pole-fleeing force of Eötvös, which is the resultant force produced by the difference between the force of gravity and the buoyancy of a land mass. Virtually everyone objected that such a force was inadequate in magnitude either for movement of the continents or for formation of the folded mountain ranges.

Wegener believed the westward movement was due to tidal forces, that is, the attraction between the sun, moon, and earth. Rotation of the earth causes tides to move from east to west. Wegener supposed that the tidal effect was dragging the entire crust in a westward path. Here, as above, the magnitude of the force is far too small to account for drift. Also this idea was later proved to be theoretically incorrect.

Escaping- and captured-moon hypotheses

An American geologist, F. B. Taylor, advocated drift primarily to explain the widespread Tertiary fold mountains. He called upon a unique mechanism to trigger the movement of continents. This was the capturing of the moon by the earth which increased or initiated the earth's rotation. As a result, the continents began to rotate slightly and slide toward the equator. This movement caused wrinkling of the crust in the form of the Alps, Andes, Himalayas, Rockies, and so on (Fig. 3.5). Such a theory presupposes that the moon did not become our satellite until the late Cretaceous. This theory also suggests that the age of the moon is less than that of the earth. Age dating of moon samples collected during Apollo missions demonstrates that the moon was formed at essentially the same time as the earth. Taylor's idea did not explain similar but older mountain ranges like the Appalachians or Urals.

Another drift mechanism associated with the moon has been suggested. According to this theory, because of centrifugal force, a large part of what is now the Pacific Ocean was thrown into space and retained as a satellite: the moon. The resultant "scar" was partially healed by the drift of continents, particularly by the movement of North and South America to the Pacific area in response to the imbalance created in the crust. The theory is not taken seriously by many.

DuToit's mechanism

A. L. DuToit of Johannesburg University is probably second only to Wegener as an enthusiastic proponent of, and author on, the subject of continental

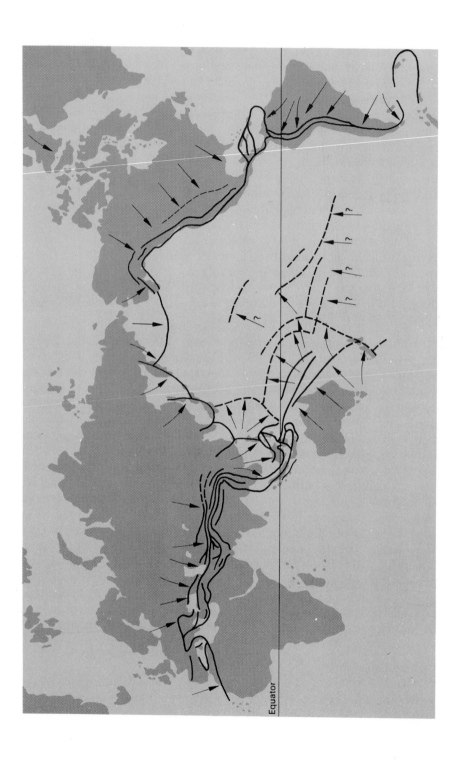

Equator

◀ FIG. **3.5** *Tertiary mountain range systems (heavy lines) with arrows showing direction of crustal movement. (After F. B. Taylor, 1928, "Sliding Continents and Tidal and Rotational Forces,"* Theory of Continental Drift, A Symposium, Amer. Assoc. Petrol. Geol., p. 165.)

drift. In his 1937 book, *Our Wandering Continents,* DuToit proposed what was at the time the most plausible theory for the mechanism of continental drift.

Geosynclines, elongate troughs which accumulate great thicknesses of sediment, commonly develop on the margin of a continent. As sediment accumulates the geosyncline slowly sinks under the great mass. In doing so the land mass is tilted and begins to slide oceanward. This is accompanied by tension cracks in the land mass which become injected with molten magma aiding the split of the continent (Fig. 3.6). The "push apart" is provided by thermal energy (magma) and the sliding of the continent due to gravity.

DuToit's theory lacked evidence and did not explain geosynclines in the middle of present continents or those that developed later along the seaward edge of the moving land masses.

Mantle convection

At one time or another, each of us has observed what takes place when a liquid is heated in a vessel. As the result of heat applied to the base of the

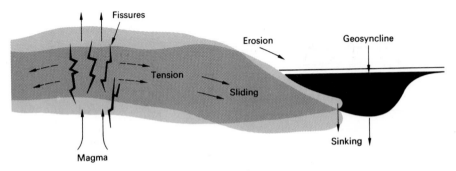

FIG. **3.6** *DuToit's mechanism for continental drift. (After Takeuchi, et al., 1967, p. 67.)*

FIG. **3.7** *Convection of liquid in a vessel showing simplified pattern of circulation.*

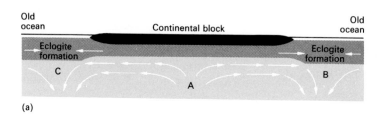

(a)

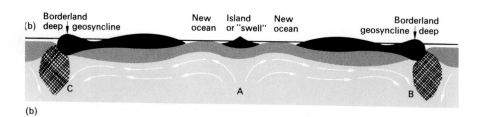

(b)

FIG. **3.8** *Mechanism for continental drift if we assume convection in the earth's mantle. (a) Upwelling causes tension which leads to (b) a fracture and lateral movement of continental blocks. (After A. Holmes, 1965,* Principles of Physical Geology *(2nd edition), New York: Ronald Press, p. 1001.)*

vessel, there is thermal convection (Fig. 3.7). D. T. Griggs of the University of California at Los Angeles was the first person to postulate that a similar type of convection was taking place in the earth's mantle. The source of heat is the radioactivity of unstable elements in the earth. Heat causes the mantle to expand and become less dense, and then convection of heat begins. The heat flow is slow but nevertheless seems to be present.

Arthur Holmes of Edinburgh University suggested that such heat transfer might take place in a solid mantle. He believed that the upwelling of the convection current and its divergence caused the sialic crust to split and move apart (Fig. 3.8). The land masses would then move laterally, as if on a conveyor belt, with the convection current. Modern proponents of the drift theory explain trenches as places where currents are moving downward and the oceanic ridges as places of current upwellings.

This idea brings the subject of continental drift in direct contact with oceanography. We should be able to decide whether convection current exists in the earth on the basis of heat flow data taken from oceanic ridges, trenches, and intermediate areas. Such data are being collected by oceanographers and are generally in support of the existence of such thermal currents. Oceanic ridges, which are volcanically and tectonically active, yield readings much above those taken on the ocean floor at considerable distances from the ridges. Similarly, the trenches reveal heat flows as low as 10 percent of the average. These data provide strong support for the existence of convection currents in the mantle, but not necessarily for their ability to transport continental masses thousands of kilometers. The presence of a layer of low-density, plastic, and spongy rock in the upper mantle, called the asthenosphere ("wave-guide layer" of some authors) lends support to the conveyor-belt concept.

Although most scientists are now inclined to support the belief in convection cells, many prominent scholars discount their connection with drift. One of them is V. V. Beloussov, a Russian geophysicist who is perhaps the most active of the antidrifters. He believes that drift did not occur and that the heat flow data should be used to explain other effects.

Gravity tectonics

The ideas proposed by those who choose convection cells to move the sea floor and continental crust all involve a general pushing motion. The spreading sea-floor conveyor is driven by an energy source along the oceanic ridges where upwelling occurs. An alternative hypothesis proposes that the pull of gravity is the mechanism for the conveyor. This pulling takes place along oceanic trenches, where the relatively old and therefore cool and dense oceanic crust descends. The descent of the oceanic crust can be traced to a depth of several hundred kilometers by earthquake foci. Some researchers believe that the dense crust descends and pulls the crust apart at the position of the oceanic ridges. Such a rift allows fresh basaltic material to rise to the surface and "feed" the conveyor belt. Others contend that the upwelling of new basaltic material at the ridges pushes the crust, thus causing it to descend in the form of trenches. The dispute has the tone of the familiar chicken-and-egg argument.

A similar suggestion, recently proposed, involves the slight tilting of the oceanic crust. In the theoretical model devised for such a mechanism, the crust slides over a viscous material which would be the upper mantle. With a tilt of only 1 in 3000, the crust could slide from the ridges toward trenches at rates comparable to those calculated for sea-floor spreading.

SUMMARY

The concept of drifting continents has run numerous cycles of support and nonsupport during past decades. Most of the opinions have resulted from the type of inconclusive evidence expressed in this chapter. It is expected that at this point the reader is left somewhat up in the air insofar as being able to make a rational decision on the validity of continental drift based on the information contained herein. Actually this represents the state of knowledge on this important topic until the late 1960s. It is no wonder, therefore, that extreme differences of opinion existed.

Discussion of the most recent evidence for drifting continents is deferred to the final chapter of this book. The rationale for such a physical separation of these two obviously closely related chapters is to allow the reader to gain a proper background before finally considering the subject of global tectonics. It is therefore appropriate that such a discussion follow the chapters on marine geology and geophysics. Why then is the present chapter placed in the first section on ocean basins?—in order to set the ocean basins and land masses in their proper framework and thus provide a rather comprehensive treatment of the first section of the book.

SELECTED REFERENCES

Dietz, R. S., and J. C. Holden, 1970, "The Breakup of Pangaea," *Scientific American,* October 1970.

DuToit, A. L., 1937, *Our Wandering Continents,* Edinburgh: Oliver and Boyd. A lengthy and interesting discussion of various aspects of continental drift by one of the early proponents of the theory.

Heirtzler, J. R., 1968, "Sea-floor Spreading," *Scientific American,* June 1968.

Holmes, Arthur, 1965, *Principles of Physical Geology* (2nd edition), New York: Ronald Press, Chapters 26–28, 31. Excellent, well-written, and well-illustrated material aimed at the beginning student.

Hurley, P. M., 1968, "The Confirmation of Continental'Drift," *Scientific American*, April 1968. General article that includes data in support of continental drift in addition to those presented by T. J. Wilson (see below). The title is somewhat strongly worded.

Knopoff, L., 1967, "Thermal Convection in the Earth's Mantle," *The Earth's Mantle*, T. F. Gashell (ed.), New York: Academic Press, pp. 171–196. Good general discussion of the mechanism most commonly credited with causing continents to drift.

Marvin, U. B., 1973, *Continental Drift, The Evaluation of Concept*, Washington, D.C.: Smithsonian Institution Press. Good general book on the subject with a very comprehensive treatment of the historical aspects of the concept.

Menard, H. W., 1965, "Sea Floor Relief and Mantle Convection," *Physics and Chemistry of the Earth*, Vol. 6, pp. 315–364. Excellent treatment of the relationship of convection to sea-floor topographic features.

Munyan, A. C. (ed.), 1963, *Polar Wandering and Continental Drift*, Soc. Econ. Paleontologists and Mineralogists, Spec. Pub. 10, Tulsa, Okla.: Soc. Econ. Paleontologists and Mineralogists. One of the more recent symposia on continental drift, with specialized papers that take both sides of the argument.

Smith, C. H., and Theodor Sorgenfrei (eds.), 1965, *The Upper Mantle Symposium—New Delhi; Part I—Physical Processes in the Upper Mantle and Their Influence on the Crust*, Copenhagen: Det Berlingske Bogtrykkeri. One of the many symposia devoted exclusively to continental drift. Arguments are presented on both sides of the question.

Sullivan, Walter, 1974, *Continents in Motion*, New York: McGraw-Hill. Excellent general treatment of the new global tectonics.

Takeuchi, Hitoshi, *et al.*, 1967, *Debate About the Earth*, San Franciso: Freeman, Cooper. Excellent general book on continental drift, stressing the geophysical aspects but written for the nongeophysicist.

Vening-Meinesz, F. A., 1964, *The Earth's Crust and Mantle, Developments in Solid Earth Geophysics*, New York: American Elsevier, Vol. 1. Rather specialized book written by a pioneer in marine geophysics.

Wegener, Alfred, 1966, *The Origin of Continents and Oceans* (translation of the 1915 original), New York: Dover. The original development and treatment of the continental theory which stimulated such great controversy among scientists.

Wilson, J. T., 1963, "Continental Drift," *Scientific American*, April 1963. Very good but somewhat biased general article on evidence for continental drift.

PHYSICAL OCEANOGRAPHY

PART **II**

PHYSICAL PROPERTIES OF SEAWATER 4

Probably everyone is aware that water is quite an amazing substance and that it is necessary for life. Water is the only substance that occurs in nature in all three states: gas, liquid, and solid. It is commonly called the *universal solvent* and is one of the few naturally occurring liquids that is inorganic.

The fact that water is the most abundant compound on the earth's surface is common knowledge; however, it is very difficult to appreciate the vast amount present. The outer few miles of the earth's crust contain about three times as much water by volume as all its other substances combined. Water is six times more abundant than feldspar, which is the most abundant mineral in the crust of the earth.

The high degree of transparency of water is perhaps one of its most important attributes. This allows penetration of light, which supports life through the photosynthesis process. This transparency also enables organisms with visual organs to see so that they may search for food and shelter.

As water changes from a liquid to a solid state, it increases in volume, contrary to the behavior of most compounds. Almost everyone has had a bottle of liquid break upon freezing or has seen the results of frostheaving in soil.

Water has great capacity for and conductance of heat. It is the best liquid heat conductor other than mercury, which is a metal. With the exception of ammonia, water has the greatest heat capacity of any fluid.

These are some of the properties that make water such a useful and valuable substance. As we become more aware of our environmental problems, the shortage and quality of water is often near the top of the list. Commonly, when scientists consider the possibility of life on other celestial bodies, the first question concerns the presence of water. The following discussion considers some of the important properties of water that affect the ocean environment.

TEMPERATURE

Any type of oceanographic investigation is at least partly concerned with the temperature of the water. Temperature changes may cause great variation in the properties of seawater and correspondingly in the life it supports. The nature of the great water masses within each of the oceans and seas depends to a large extent on the temperature characteristics of the water. For example, if we raise the temperature of a mass of water, the volume will increase due to the excitation of atoms in the water molecules and the dissolving power will be increased. Most of us have seen experiments showing that hot water dissolves more salt than cold water does.

A graph showing the relation of temperature of maximum density to freezing point is shown in Fig. 4.1. Pure water is at its maximum density at 3.94°C, or as usually considered, 4°C. From this point to freezing, water becomes less dense, although only slightly so. The temperature of maximum density is depressed as the water becomes more saline and **specific gravity** is increased. Also correspondingly depressed, but at a lesser rate, is the freezing temperature of water. At a specific gravity of 1.02000, both the freezing point

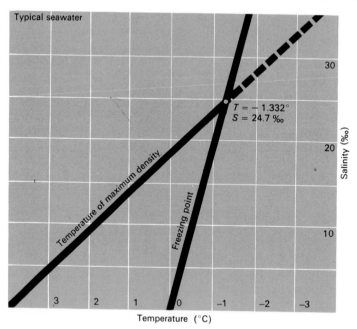

FIG. 4.1 *Relationship between temperature of maximum density and freezing temperature as affected by salinity. (After H. J. McClellan, 1965,* Elements of Physical Oceanography, *New York: Pergamon Press, p. 21.)*

and the temperature of maximum density have the same value: $-1.332°C$. However, because the specific gravity of seawater is usually more than 1.02000, freezing usually takes place before the water has cooled enough to reach maximum density.

The ability of water to conduct heat increases with temperature and also with pressure. Pure water, however, is a better heat conductor than seawater. **Specific heat**, the calories required to increase temperature of 1 gram of a substance 1°C, also changes from fresh water to seawater. It requires about 0.067 of a calorie less to heat a gram of seawater 1°C than to heat a gram of fresh water.

The temperature of seawater ranges from about $-2°C$ to 30°C in nature, but most of the water in the oceans is too deep to be influenced by climatic conditions and has a narrow range of $-1°C$ to 4°C. As expected, surface waters in the low latitudes are warm and the temperature becomes colder as the latitude increases. This simple pattern is considerably altered, however, by the major current systems of the world. Current modifications are discussed in the following chapter.

Measuring temperature

Temperature may be measured by a variety of means—some crude and some rather sophisticated. For many studies, it is only necessary to collect a sample of surface water and measure its temperature with a standard mercury thermometer. Although a simple procedure, certain precautions must be exercised, such as not taking the water sample from the rear of the ship where the temperature may be anomalously high. Similarly, the container used for collecting water should not be a metal container that has been on the deck in the hot sun. Obviously this method is not adequate for measuring minor temperature fluctuations, and it is applicable only to surface water.

It is also possible to obtain surface-temperature data from an airplane using infrared sensors. This procedure is based on Stefan's Law, which is based on the principle that the rate of heat radiation is proportional to the fourth power of the absolute temperature. By measuring the radiation given off, it is possible to obtain surface-temperature values for a large area in a short period of time. The values obtained are for the surface water only; in fact, on a very calm day with little mixing of water, there might be a thin skin of warm water not representative of the near-surface mass itself.

Most physical oceanographers are concerned with the overall temperature distribution, both vertical and horizontal. As a result, they need thermometers that are quite sensitive, usually to 0.02°C, and that will give temperature readings at depth. The most commonly used device is a **reversing thermometer**. This is a special type of mercury thermometer that is attached to the water-sampling device. As the sampling bottle is inverted to trap a water sample, the thermometer is inverted. This fixes the temperature reading by causing a break in the mercury column at the constriction (Fig. 4.2). Gravity causes the mercury to flow to the end of the tube, thus recording the temperature at the depth of reversal. There are vacuum pockets around the bulb of the thermometer to keep it from giving anomalous readings due to pressure increases at depth.

Actually, the combination of a **protected** and an **unprotected thermometer** in the sampling bottle (Fig. 4.3) is quite a valuable instrument as a pressure sensor as well as for recording temperature. The unprotected ther-

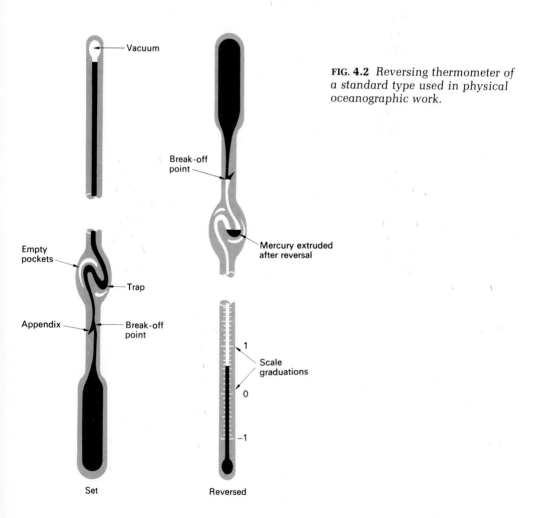

FIG. 4.2 *Reversing thermometer of a standard type used in physical oceanographic work.*

FIG. 4.3 Rack of Niskin bottles in a shipboard laboratory. Note the presence of protected and unprotected thermometers. (Photo by Keith Harris.)

FIG. 4.4 (a) Protected and (b) unprotected thermometers.

(a)   (b)

mometer is similar to that used in atmospheric temperature measurements. The protected thermometer has a second glass casing that is also filled with mercury (Fig. 4.4). Upon reversal, both thermometers record the temperature, but the temperatures actually recorded are different. The unprotected thermometer shows a higher temperature because pressure compresses the tube and causes the mercury to rise up in the tube more than in the protected thermometer. Standardization of the two thermometers will therefore permit the determination of the relationship of the pressure and temperature differential. From this information, the depth can easily be calculated because pressure and depth have a direct relationship (approximately 1 atm pressure per 10 m depth).

The reversing thermometer in combination with a sampling bottle is probably the most widely used piece of apparatus in physical oceanographic data collection. It is not limited by depth of water or intense pressure. One undesirable requisite for its use, however, is a stationary ship. The bottles are fastened at predetermined positions along a line which is weighted and lowered in vertical position. Once the bottles are in place at known depths, a messenger in the form of a weight is released on the line. This releases the first bottle, allowing it to reverse due to gravity, collect a water sample, and record the temperature. At the same time, a messenger is released by the reversing bottle to trip the next one, and so on down the line (Fig. 4.5). In this way, it is possible to obtain water samples and temperatures accurate to 0.02°C at any depth desired and in considerable numbers through a vertical column.

As was previously mentioned, the water sampling bottle apparatus requires a still ship. Often it is desirable to obtain continuous data for vertical temperature distribution, although not from considerable depths. In such cases a **bathythermograph (BT)** is employed. This is a rather small, torpedo-shaped piece of apparatus (Fig. 4.6) that is towed while the ship is under way. Although the instrument is restricted to depths of less than 300 meters, it is quite useful in that it records the temperature and depth continuously. This information is recorded on a smoked-glass slide about 2.5 by 5.0 centimeters in size. The slide is placed on a grid scale and viewed under magnification to read the temperature-depth trace.

The mechanisms for recording data with a bathythermograph are not complicated. Depth is determined by a small bellows that reacts to pressure. Temperature is measured by a thermometer which is composed of about 25 meters of coiled tubing with a pivoting Bourdon tube (Fig. 4.7). The end of the tube is fixed with a pointer which marks the trace of temperature of the smoked slide.

The apparatus has obvious limitations, depth being the most important. Also it is not designed for detailed work; consequently temperature accuracy

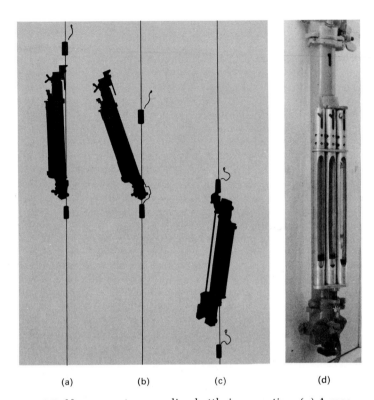

(a)   (b)   (c)   (d)

FIG. 4.5 Nansen water-sampling bottle in operation. (a) A messenger is sent down to the bottle, which is attached to a cable; (b) the messenger releases the bottle, which (c) pivots and closes, collecting a water sample and reversing the thermometer to record the water temperature. (d) Photograph of a Nansen bottle.

FIG. 4.6 Bathythermograph.

Temperature element

Pressure element

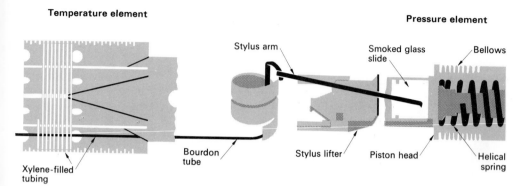

Stylus arm

Smoked glass slide

Bellows

Bourdon tube

Stylus lifter

Piston head

Helical spring

Xylene-filled tubing

FIG. **4.7** *Sketch of a bathythermograph (BT) showing the mechanism for obtaining depth-temperature profiles. (After Gerhard Neumann and W. J. Pierson, Jr., 1966,* Principles of Physical Oceanography, *Englewood Cliffs: Prentice-Hall, p. 105.)*

is probably rarely greater than a few tenths of a degree. Its most important use is in connection with the thermocline. This is discussed later when density of water is considered.

Although the reversing thermometer and the bathythermograph are classical and still widely used instruments for obtaining water temperatures,

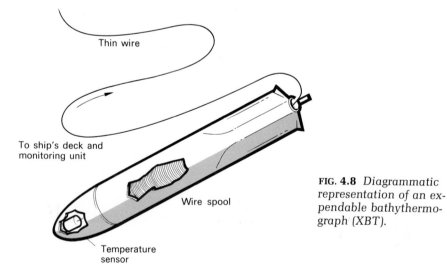

Thin wire

To ship's deck and monitoring unit

Wire spool

Temperature sensor

FIG. **4.8** *Diagrammatic representation of an expendable bathythermograph (XBT).*

modern electronics has provided new techniques. By suspending long strings of thermistor (heat-sensitive) beads underwater, it is possible to obtain nearly continuous temperature data by electronically monitoring such sensors. These devices can be fixed to floats and left unattended, and information is automatically recorded. It is possible to detect short-term fluctuations in temperature with depth using this type of apparatus.

Thermistors are utilized in a modern and less-restricted form of the bathythermograph called the **expendable bathythermograph (XBT)**. This instrument is shaped like a bomb and contains a thermistor attached to a coil of thin copper wire (Fig. 4.8) which is attached to a monitoring device on the ship. The thermistor output provides temperature information, whereas depth is calculated from the constant fall velocity and the period since release. Similar XBT instruments are dropped from airplanes and the temperature data are transmitted by radio signal. This use of the XBT has a depth restriction of a few hundred meters; however, those used from ships with data transmitted by wire can provide temperature data to considerable depths.

Temperature distribution

In ocean basins, there is a rather wide range in temperature in both the vertical and horizontal directions. As one might expect, there is a tendency for surface temperatures to be zoned along lines of latitude (Figs. 4.9 and 4.10). Although such a temperature distribution is prevalent, there are some consistent variations to this general pattern in each ocean basin. Along eastern boundaries of the basins, for example, the temperatures are anomalously low due primarily to upwelling of cold water. The west coast of the United States is a good example. Surface currents also tend to distort the isotherms. The Gulf Stream carries warm surface waters up the Atlantic coast of the United States and warms the New England states, and the Kuroshio current performs the same task for the Japanese Islands (inside cover).

The temperature of the upper 50 to 200 meters of the oceans is similar to that of the surface and ranges with latitude. From there to depths of 500 to 1000 meters, the temperature decreases rapidly, and below 1000 meters the water is almost isothermal. At great depths there is little temperature variation with latitude; it is all cold (near 0°C).

DENSITY

Density of water is undoubtedly better known than that of any other substance. We consider the density as unity for pure water at 4°C (temperature of maximum density). However, water has a rather wide range of density and its variation is extremely important in the study of oceans, as it is a major factor in water circulation and in the distribution of some organisms.

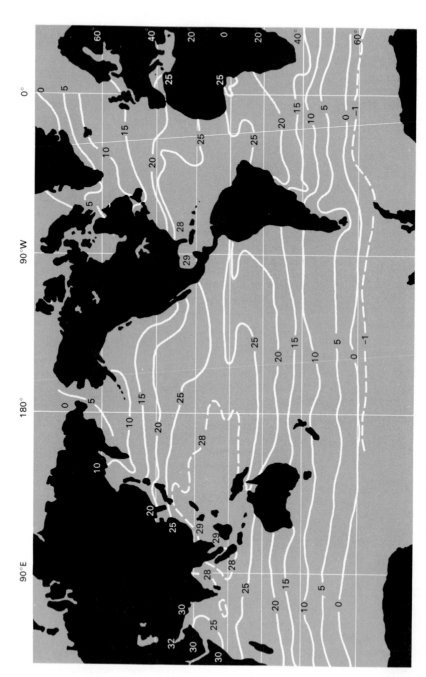

FIG. 4.9 *Surface temperature (°C) of the oceans for the month of August. (After G. L. Pickard, 1964, Descriptive Physical Oceanography, New York: Pergamon Press, p. 34.)*

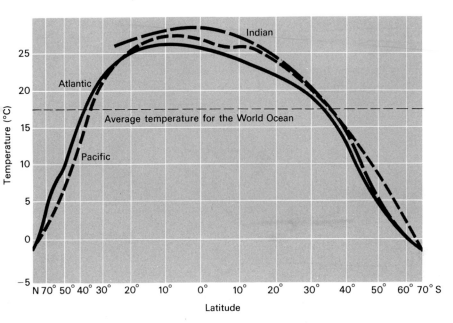

FIG. 4.10 *Mean distribution of surface temperature for three oceans with respect to latitude. (Modified from W. A. Anikouchine and R. W. Sternberg, 1973, The World Ocean, Englewood Cliffs: Prentice-Hall, p. 84.)*

In the open ocean, water density commonly falls within a range of 1.02400 to 1.03000 g/cm³, a range which may seem insignificant. Such is not the case, however, as differences of only a few ten-thousandths of a g/cm³ may be reflected in currents and water masses. Therefore, oceanographers always carry density determinations to the fifth decimal place. It seems awkward to refer to absolute density when the first two digits are constant, i.e., 1.0, and so on. As a result, a shorthand form for density is used and referred to as σ_{STP}, where STP refers to salinity, temperature, and pressure, which are the factors affecting seawater density.

$$\sigma_{STP} = (\text{density} - 1) \times 10^3/\text{cm}^3.$$

Thus for a seawater with a density of 1.02500 g/cm³, we would describe the density with the whole number 25.0. The density notation, also called sigma-tee (σ_t), is used when only temperature and salinity variations are considered.

Density increases as salinity and pressure increase, but decreases with increase in temperature. Salinity and temperature distributions are variable, being affected by geography and climate, and their influence on seawater density is more pronounced than that of pressure. Pressure varies in a predictable manner with depth. Although it increases with depth it has little influence on water density because water is difficult to compress. Water with salinity of 30‰ at 30°C has a density of 1.0180 (18.0) at the surface, but at a depth of 8 kilometers the density is 1.0570 (57.0). Pressure also tends to depress the temperature of maximum density of water. At the surface, it is 3.94°C, but it goes down to 0°C at a depth of 6 kilometers.

Density distribution

Seawater density is quite important for submarine navigation, particularly with respect to the **thermocline.** This is the depth range in which there is maximum change in temperature with depth. As a result warmer, less dense water lies on top of colder, more dense water. Such a change in density may be very abrupt or may take place over many meters. The seasonal thermocline is generally in the upper 200 meters of water, while a permanent thermocline exists from 200 to about 1200 meters. Density is important to submarines because they must be neutrally buoyant when submerged in order to navigate underwater. That means that the density of the submarine must be equal to that of the surrounding water for the vessel to maintain a particular desired depth. To accomplish this, a submarine is equipped with ballast tanks that can be filled with water to provide a desired density for the ship.

The submarine often rides on the thermocline when it is submerged. Because of the relatively great density change at the thermocline, a submarine can float on the denser, colder water while maintaining the density of the warmer, less dense water. Obviously, care must be taken to get the ship in proper trim, that is, to make the ship neutrally buoyant at the depth desired. If a miscalculation is made in the amount of water taken on the ballast or in the density of water at the desired depth, it could be disastrous. On descent a submarine will be compressed to some extent due to the increase in pressure with depth. If the ship does not come into trim as predicted, the pressure will further compress the ship, making it more dense and causing it to continue to sink. Ballast water must then be released to keep the ship from descending to the ocean floor.

Distribution of density in a vertical direction is characterized by a stable stratification. In a horizontal direction, density differences can exist only with the presence of currents, and as a result there is a strong relationship between currents and density distribution, particularly at depth. Deep ocean water is very cold and has a high density; a logical supposition is that cold,

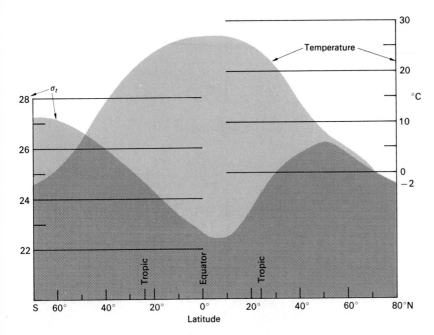

FIG. 4.11 *Generalized variation in temperature and density (σ_t) at the ocean surface with respect to latitude. (After Pickard, 1964, p. 29.)*

dense water forms in the high latitudes and sinks, filling the deep ocean basins.

Surface density (sigma-tee) ranges from about 22 to 27. It is lowest near the equator and increases progressively to about 50° or 60°. At higher latitudes it decreases slightly (Fig. 4.11).

SOUND IN SEAWATER

At one time or another, most of us have put our heads underwater and struck two rocks together. The rather sharp clicking noise is quite different from that caused by the same process in air. This is largely the result of the greater velocity of sound in water than in air. Sound travels 333 meters/second in air and averages 1445 meters/second in seawater. There is a range of about 100 meters/second in the ocean, as the velocity of sound is also affected by salinity, temperature, and pressure. In the open ocean salinity is essentially uniform; therefore, temperature and pressure (depth) are the major factors determining the speed of sound. The upper layer of water is rather isother-

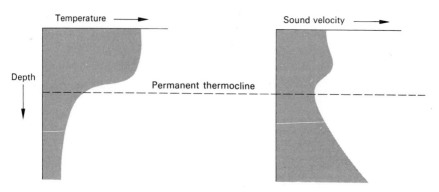

FIG. 4.12 *Relationship of temperature and sound velocity to depth.*

mal or shows a slight temperature decrease with depth. The thermocline marks a pronounced temperature change, below which there is nearly isothermal water to the ocean floor. The result is a slight increase in sound velocity in the upper layer with increasing depth (Fig. 4.12). At the thermocline, the rapid temperature decrease causes a decrease in sound velocity, reaching a minimum at this level. Below this zone is a gradual increase in sound velocity with increasing depth due to isothermal water (Fig. 4.12).

This principle is of great importance in the use of sound for detection and for depth determinations such as SONAR (Sound Navigation And Ranging). In isothermal water, sound gradually increases in velocity downward so that waves are refracted upward as they travel from the source (Fig. 4.13). However, in the natural situation there is considerable change in temperature with depth, which causes the minimum sound velocity to be located in the thermocline zone. As a result, there is a shadow zone (Fig. 4.14) where no sound waves are traveling. This zone is caused by the velocity minimum. When a submarine lies at this depth, other submarines at the same depth cannot detect it because of the way in which sound waves are refracted (Fig. 4.14). If calm, warm conditions exist, there may be a secondary thermocline very close to the surface. Such conditions would trap the sound between the two thermoclines, giving a shallow sound channel (Fig. 4.15).

Most depth-recording devices are based on the movement of sound waves in water and the ability of sound waves to bounce back off the bottom. The type of bottom surface and its constituent materials are an important factor in obtaining good data from sound reflections. A smooth, hard bottom presents the best type of surface, but in the open ocean such bottom surfaces are rare. A soft bottom tends to absorb sound and poor reflections are

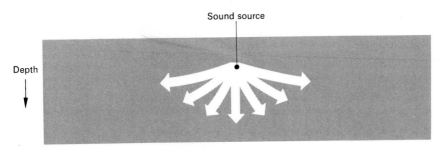

FIG. 4.13 *Refraction of sound waves in water, caused by increasing velocity downward.*

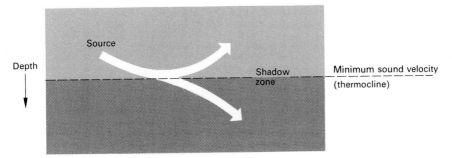

FIG. 4.14 *Position of a shadow zone, which is caused by the decreasing velocity of sound away from the thermocline in both vertical directions.*

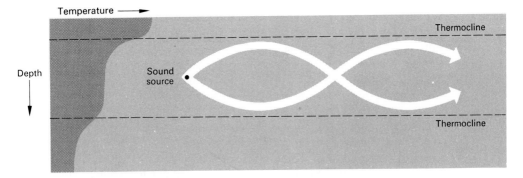

FIG. 4.15 *Sound channel between two thermoclines.*

obtained. Also irregularities cause scattering of sound waves. Therefore some corrections must be made and the type of bottom noted when depth data are presented. In this way the reader will be able to weigh the accuracy of the data.

LIGHT AND COLOR

The ability of water to transmit light is probably its most important attribute. Without it photosynthesis would not be possible and, as a result, no life could exist in the sea. In addition the sun, which provides the light, serves as a source of energy for evaporation, circulation, and heat in the sea. Water's transparency is currently of importance to man's exploration of the sea, including such techniques as photography and underwater television.

If there is no cloud cover, the position of the sun determines the amount of light that is transmitted through the water with respect to that which is reflected. Table 4.1 shows that if the angle of incidence is less than 60°, only a small percentage of sunlight is reflected, and that the percentage reflected increases rapidly at angles above 60°.

TABLE 4.1 *Reflection of light from a smooth water surface.*

Angle of incidence	Percent reflection
0°	2.0
20°	2.1
40°	2.5
60°	6.0
70°	13.4
80°	34.8
90°	100.0

It is commonly necessary to determine the amount of light penetration, particularly in studies concerning photosynthesis. The simplest way to do this is with a **Secchi disk**. This is a white porcelain disk measuring 30 centimeters (about 12 inches) in diameter. A line is fastened at the center and the disk is lowered until it disappears from sight. This depth is called the depth of visibility and may be as much as 50 meters in the Pacific Ocean. Note that the light rays are really traveling twice the depth in order for the observer to view the disk. Consequently, a diver could see approximately as well at a depth of 100 meters as the observer on a ship can see the disk at 50 meters. The Secchi disk is limited in its application due to inconsistency in observations and comparison of data. Each observer will undoubtedly obtain somewhat different results because of individual variations in sight. The roughness of the water surface, elevation of the sun, and cloud cover will also be factors. Nevertheless, data seem to be comparable if averages of many readings are taken.

In recent years the Secchi disk has become almost obsolete due to the development of photoelectric cells. A light-receptive surface is lowered to a given depth. This surface is monitored on board ship by a meter giving the number of lumens of light being received. With such an instrument, it is possible to determine the continuous distribution of light intensity with respect to depth.

The color of seawater in the open ocean is in the blue-green portion of the spectrum. If a bottle of this seawater is held up to the light, it is colorless. How then does the great volume of the sea obtain its color? If organisms and impurities are temporarily eliminated as possibilities, it must be due to the way various visible light rays are affected by transmission through water. Most important is absorption of light rays. The wavelengths of light in the visible spectrum range from about 0.4 micron to 0.8 micron, with the shortest of the visible rays being violet and the longest being red. The degree to which rays are absorbed by a substance depends on their wavelength, but in water it is not a straight-line relationship. Long, red rays are absorbed most, whereas blue and green rays are absorbed least. Thus a wavelength of about 0.5 micron is absorbed least, with increasing absorption in both shorter and longer rays (Table 4.2). Because of the relative absorption of various wavelengths, ocean water appears blue to green. The complete absence of light with depth is due to total absorption and scattering of light rays.

Locally, many factors may affect the color of the sea. Equatorial waters are deep blue, whereas in high latitudes the color is blue-green. This difference is primarily the result of greater biological production in higher latitudes.

On a small scale, such things as plankton blooms might give the sea an anomalous color for short periods of time. It may be red, such as the famous "red tides," or green. In shallow areas of carbonate production, water turns milky white, due probably to fish activity stirring up the bottom. Organic material may give water a brownish tint, especially near deltaic areas where

TABLE 4.2 *Absorption coefficients of pure water for light of different wavelengths.*

Microns	Color	Absorption coefficient
0.4	violet	0.072
0.5	↑	0.016
0.6		0.125
0.7	↓	0.840
0.8	red	2.400

there is a high influx of such detritus. Where silt and clay are abundant, turbulence may cause these particles to become suspended, changing the color of water to gray or brown. Color changes due to organic and inorganic suspended particles also reduce the penetration of light through seawater.

SELECTED REFERENCES

Groen, P., 1967, *The Waters of the Sea*, London: Van Nostrand, Chapter 2. Very good general discussion of seawater, including sampling and analytical methodology.

Hill, M. N. (ed.), 1963, *The Sea—Ideas and Observations*, New York: Wiley, Vol. 1, Chapters 1, 8–14. Detailed consideration of each physical property of seawater. A general background is necessary for good comprehension.

King, Thomson, 1953, *Water—Miracle of Nature*, New York: Macmillan. A popular treatment of water in all its forms and what it means to life. Covers the entire subject but suffers from total absence of illustrations.

Kuenen, P. H., 1955, *Realms of Water: Some Aspects of Its Cycle in Nature*, New York: Wiley. A general treatment of water written for the science student.

McClellan, H. J., 1965, *Elements of Physical Oceanography*, New York: Pergamon Press, Chapters 4–6. A thorough but highly mathematical treatment of temperature, density, pressure, and salinity.

National Research Council, 1958, *Conference on Physical and Chemical Properties of Sea Water*, Easton, Maryland, Washington, D.C.: National Academy of Sciences—National Research Council Pub. No. 600.

Neumann, Gerhard, and W. J. Pierson, Jr., 1966, *Principles of Physical Oceanography*, Englewood Cliffs: Prentice-Hall, Chapter 3. A comprehensive but highly mathematical treatment of this subject. Probably the best treatment for a student of physical oceanography.

Sverdrup, H. U., M. W. Johnson, and R. H. Fleming, 1942, *The Oceans—Their Physics, Chemistry, and Marine Biology*, Englewood Cliffs: Prentice-Hall, Chapters 3–5. Very extensive and easily understood material on water properties. Suffers from being somewhat out of date.

von Arx, W. S., 1962, *An Introduction to Physical Oceanography*, Reading, Mass.: Addison-Wesley, Chapter 5. Brief but fairly well-written treatment of seawater properties.

Williams, Jerome, 1973, *Oceanographic Instrumentation*, Annapolis: Naval Institute Press. Several excellent chapters dealing with physical properties of water and their measurement.

ATMOSPHERIC CIRCULATION 5

Virtually all water motion results either directly or indirectly from the complex interactions of the atmosphere with the oceans. It is therefore appropriate to briefly discuss atmospheric circulation before considering water motion, that is, currents and waves. It is not intended that this chapter provide the reader with anything beyond some fundamental principles which will enable proper comprehension of the two succeeding chapters. For in-depth discussions of this subject, the references at the end of the chapter should be consulted.

IDEAL CIRCULATION

Before considering the complexities of atmospheric circulation as it exists on our rotating earth with its rather randomly distributed land masses, it is appropriate to discuss circulation under some simplified and idealized conditions. For purposes of discussion, the earth will be considered at rest and completely enveloped with water; that is, land masses are absent.

Movement of the atmosphere, or wind, results from changes in temperature which cause air to expand when heated or contract when cooled. Because of spatial variations in the temperature of the air, temperature gradients are created. These cause air to move. The sun provides the original source of energy that generates these temperature differences. If we then consider the variation in the sun's energy which falls on the nonrotating, uniform surface of our ideal earth, it is apparent that the area of greatest heating of the atmosphere will occur in the equatorial zone and the area of least heating will occur at the poles. This difference in radiation creates temperature gradients, and the expanding warm air of the equatorial zone will rise. Likewise the relatively cold, and therefore contracting, air of the polar regions will descend because it is heavier. This combination will result in a convection-cell type of circulation on the idealized and nonrotating earth, with the upper atmosphere moving toward the poles and the lower atmosphere moving toward the equator (Fig 5.1). This simplified model

North pole

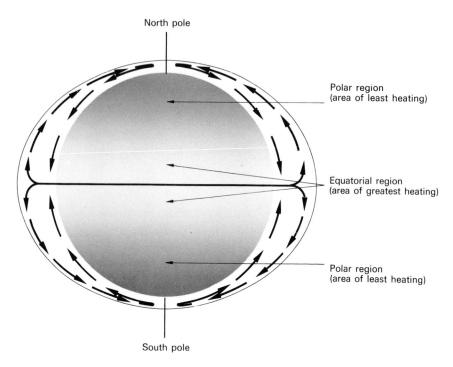

Polar region
(area of least heating)

Equatorial region
(area of greatest heating)

Polar region
(area of least heating)

South pole

FIG. 5.1 *Ideal atmospheric circulation for a uniform, non-rotating, nonrevolving earth. (After N. Bowditch, 1962, American Practical Navigator, U.S. Navy H. O. Pub. No. 9, p. 795.)*

would have the equatorial zone characterized by low atmospheric pressure, cloudy conditions, and high precipitation, whereas the polar zones would experience high atmospheric pressure and little precipitation.

CORIOLIS EFFECT

In order to understand the true conditions and patterns taken by atmospheric circulation, the rotation of the earth must be considered. It is apparent from our position on earth with respect to other celestial bodies that the earth is indeed rotating. The fact that the earth is spherical in shape coupled with its rotating movement causes the need for modification of Newton's Second Law of Motion, which states that

$$F = Ma.$$

Because objects moving on the rotating earth or within its gravitational field are moving in an apparently curved rather than straight path, some adjustment must be made to Newton's Second Law.

Although this effect was recognized by La Place in the eighteenth century, it was first quantified by the French scientist Gaspard Gustave de Coriolis in 1844. The phenomenon is commonly regarded as the Coriolis force; however, it is only an apparent force with respect to the observer. Simply stated, from the observer's point of view, an object moving in any direction over the earth's surface is deflected to the right in the Northern Hemisphere and to the left in the Southern Hemisphere. This deflection is really the apparent change in the path of the object with respect to an observer stationed on the rotating earth.

A simple example would be an airplane flying from Chicago to New Orleans, essentially a southerly direction. The navigator, however, must correct for the earth's rotation in planning the flight. Once the plane takes off, it is not under the influence of the rotating earth. Therefore, if compensation is not made the earth would rotate, west to east, under the plane while it is in flight, and the plane would land near Houston, Texas (Fig 5.2). To eliminate the Coriolis effect, the pilot must constantly veer to the left. Similar compensation must be made when shooting missiles, long-range cannon, and so on.

According to the principle of Coriolis, the deflection is apparent in any direction. At first, this does not seem consistent with the earth's rotation if we consider the return flight from New Orleans to Chicago. But since the earth is a sphere, its rotational velocity changes considerably from the equator to high latitudes (Fig 5.3). Keeping this in mind, it is evident that an object moving from south to north would also be deflected to the right in the Northern Hemisphere. At the equator, there is no Coriolis effect when the path of the equator is followed, and the effect increases toward both poles.

Although not as obvious as deflections in paths of objects moving through air, there is a similar effect on trains, automobiles, and water moving over the earth. A train imparts greater pressure on the right-hand rail in the Northern Hemisphere. The amount of wear is so small, however, that it is nearly impossible to distinguish from that on the left side. The same tendency is present in an automobile moving along a highway, but the great amount of friction between tires and pavement masks the effect. Studies have been made on riverbank erosion and although the results are inconclusive, they indicate more rapid erosion on the right bank than the left in the Northern Hemisphere. Driftwood seems to collect in greater quantities on the right bank.

The mathematical expression developed by G. G. Coriolis is

$$C = 2V\Omega \sin \phi,$$

where C is the Coriolis effect, V the velocity of the particle, Ω a constant

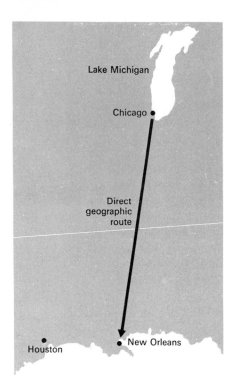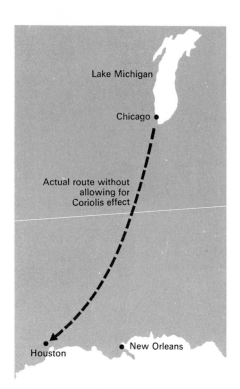

FIG. **5.2** *The Coriolis effect causes apparent deflection of an airplane traveling from Chicago to New Orleans. The plane leaves Chicago at 1:00 P.M. and is scheduled to arrive in New Orleans at 2:00 P.M. If the plane flies a due south course, it will land near Houston, Texas, because of the earth's west-to-east rotation during the one-hour flight.*

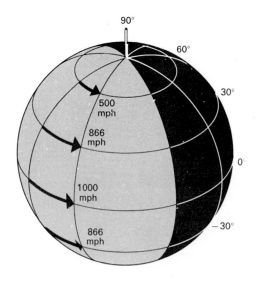

FIG. **5.3** *Tangential velocities at various latitudes on the earth's surface. (After W. S. von Arx, 1962, An Introduction to Physical Oceanography, Reading, Mass.: Addison-Wesley, p. 88.)*

(angular velocity of earth's rotation), and ϕ the latitude. Commonly, $2\Omega \sin \phi$ is abbreviated as f, because it is used in so many motion equations.

GENERAL ATMOSPHERIC CIRCULATION

If the principles of the Coriolis effect on the rotating, spherical earth are then applied to the simplified model described in the first section of this chapter, certain complexities occur. The equatorial air is deflected to the right in the Northern Hemisphere and to the left in the Southern Hemisphere as it moves toward the poles. This causes an overall decrease in the poleward component of air movement and results in accumulation of air, therefore high-pressure zones, near 30° latitude. The high-pressure condition causes air to sink and thus develop the so-called **horse latitudes** (Fig. 5.4). Air in the lower atmosphere moves both poleward and toward the equator, causing permanent circulation cells. The low-pressure zone near the equator is known as the **doldrums** because of the relatively calm conditions that persist.

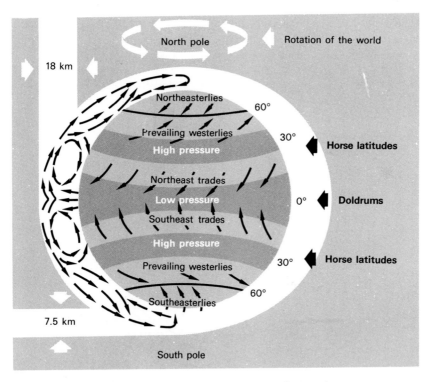

FIG. 5.4 *Generalized diagram of atmospheric circulation showing major circulation patterns. (After Bowditch, 1962, p. 797.)*

In addition to the above regions of little surface wind activity, there are also belts of persistent winds. In the low latitudes there are the **trade winds** which blow from the belts of high pressure toward the equatorial low-pressure region. The earth's rotation causes these winds to be deflected to the west (Fig. 5.4). In the midlatitudes, the poleward decrease in atmospheric pressure gives rise to the **prevailing westerlies.** These winds are also influenced by the Coriolis effect and are deflected to the east (right) in the Northern Hemisphere and to the west (left) in the Southern Hemisphere (Fig. 5.4). At the poles the low temperatures cause atmospheric pressure to remain relatively high with respect to the adjacent areas. As a result circulation is away from the poles, forming the **northeasterlies** in the Northern Hemisphere and the **southeasterlies** in the Southern Hemisphere.

LARGE-SCALE MODIFICATIONS CAUSED BY LAND MASSES
Further complication of our model is achieved by considering the continental land masses. By adding land masses to the model, two steps are taken in the development of the model. First of all, the land masses cause surface winds to tend toward forming closed cells over the ocean (Fig. 5.5). If land masses were similar in size and shape and uniform in their global distribution, a regular and uniform pattern would result. However, as was discussed

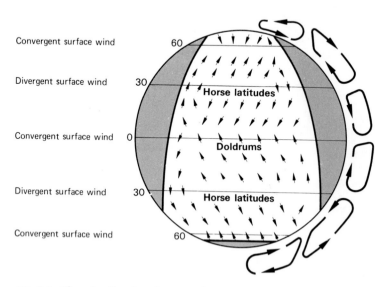

Convergent surface wind

Divergent surface wind

Convergent surface wind

Divergent surface wind

Convergent surface wind

60

30

0

30

60

Horse latitudes

Doldrums

Horse latitudes

FIG. 5.5 *Closed cells of surface winds over the oceans due to the presence of land masses. (After Anikouchine and Sternberg, 1973, p. 92.)*

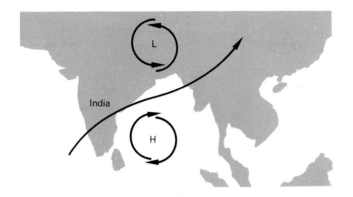

FIG. **5.6** *The summer monsoon situation in southeast Asia. (After Bowditch, 1962, p. 800.)*

in Chapter 1, there is considerable difference between the land masses in the Northern Hemisphere and those in the Southern Hemisphere. As a result of this, the surface cells are displaced northward so that the meteorological equator is some 5 to 10 degrees north of the geographic equator.

Another important alteration caused by land masses occurs in the polar regions. In the Artic the winds blow over a water body surrounded by land. This causes winds in the Arctic Ocean to be variable and of low intensity. On the other hand, the southern polar region is occupied by a land mass surrounded by water. As a result circulation is enhanced and the high-pressure condition is stronger than in the Arctic. The result is strong and persistent surface winds.

An important influence of land masses is pronounced seasonal temperature changes. In summer the land masses are quite warm in contrast to the oceans, whereas in winter the opposite is true. Therefore, during the summer low pressures prevail over the land; a high-pressure system which moves toward a land mass is weakened or broken up, whereas a low-pressure system is intensified. The opposite effect is produced during the winter.

A rather unique and extreme example of this situation takes place in the Indian Ocean where the **monsoons** of southeast Asia arise. In summer, the temperature contrast creates extreme storm conditions accompanied by intense rainfall as the result of winds moving between a high-pressure system over the ocean and a low-pressure system over the land mass (Fig 5.6).

DAILY MODIFICATIONS
Along coastal areas, there are daily or diurnal changes in atmospheric circulation. Typically the land is warmer than the ocean during the daytime; as a result there is a steady onshore breeze. At night the land is cooler than the ocean, causing an offshore breeze.

SELECTED REFERENCES

Blair, T. A., and R. C. Fite, 1965, *Weather Elements* (5th edition), Englewood Cliffs: Prentice-Hall. Rather comprehensive and readable introduction to meteorological principles.

Donn, W. L., 1951, *Meteorology, With Marine Applications* (2nd edition), New York: McGraw-Hill. Although rather old, this book is an excellent, well-illustrated text in meteorology for the marine scientist.

Stewart, R. W., 1969, "The Atmosphere and the Ocean," *Scientific American,* Vol. 121, pp. 76–86. Brief and general article on atmospheric and oceanic interrelationships.

OCEANIC CIRCULATION 6

Circulation of ocean waters is a tremendously complex phenomenon. There is considerable variation in factors which control currents in both the vertical and the horizontal directions, and such variation presents problems in observing and recording circulation data. The discussion of oceanic currents presented here is made much simpler than its treatment in most texts in order to omit the mathematics of physical oceanography books. The reader is referred to the reference list at the end of the chapter, where excellent and thorough treatments of oceanic circulation are listed.

Radiation from the sun is the ultimate factor in the circulation of ocean waters. The sun's radiation is responsible for atmospheric circulation, which is closely linked to that of the hydrosphere.

Most ocean water movement can be included under one of two headings: **wind-driven** or **thermohaline** circulation. Wind-driven currents, also referred to as surface currents, are directly influenced by atmospheric circulation and are found only in surface and near-surface waters. Thermohaline currents are the result of density differences in water and are deep and slow-moving compared to surface currents. The term "thermohaline" refers to the temperature and salinity of water, which control density and, therefore, also control deep-water circulation. In addition to the above mentioned differences in location and rate, these two types of currents also differ in direction. Surface or wind-driven currents move in a generally horizontal manner, whereas thermohaline currents move vertically or at least have a vertical component to their movement.

SURFACE CURRENTS

It does not require keen observation by a scientific mind to notice the effect of air moving over a motionless body of water. Most of us have been in a boat or at the shore when a gust of wind blew over a still pond or lake and have seen the resulting ripples build into small waves. This phenomenon is a result of friction between the moving air and the motionless water. Not only ripples and waves are caused by this friction, but also the current, or net horizontal movement of water, which is not so easily seen. The stress implemented by wind on the water involves a transfer of energy from the atmosphere to the hydrosphere. The energy causes water motion which may be turbulent (water particles moving in a variety of directions) or laminar (water particles moving parallel to one another), depending on wind velocity. Because such currents are generated across an interface between fluids of different state (gas and liquid), the current will decrease from the surface downward and will be absent at shallow depths. A depth of 100 meters or so is usually considered the practical lower limit for wind-driven currents.

Patterns displayed by surface currents, therefore, should reflect the prevailing wind patterns of the globe. In both the Northern and Southern Hemispheres, there are prevailing westerlies and trade winds. The trade winds move from the east diagonally toward the equator in the low latitudes, whereas westerlies are from the west diagonally away from the equator in the midlatitudes (Fig 6.1). The result is a system of large, somewhat elliptical cells of circulation commonly referred to as **gyres**. It is evident from Fig. 6.1

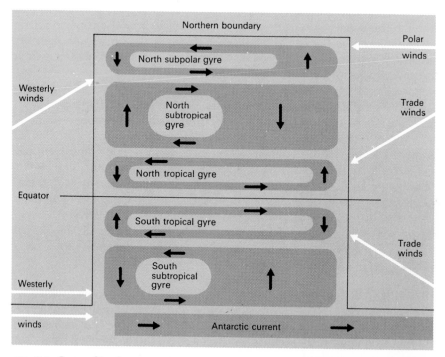

FIG. 6.1 *Generalized current patterns in a typical ocean basin, showing the influence of prevailing winds. (After J. Williams, 1962, Oceanography, An Introduction to the Marine Sciences, Boston: Little, Brown, p. 136.)*

that most ocean currents depend on prevailing winds. If, however, we look at a world map with its current patterns (inside cover), it is also evident that wind is not the sole factor in oceanic circulation. Land masses also play an important role, and if we move the continents around in our minds, we can imagine current patterns quite different from those present today.

Actually the land masses in their present distribution contribute to a rather uniform system of currents, so that Fig. 6.1 applies to the Atlantic, Pacific, or Indian Ocean. In the last case, of course, only the Southern Hemisphere is affected. The Northern and Southern Hemispheres display current patterns which are alike in shape and distribution but opposite in direction. An exception to this generalization is found in the high latitudes, largely because of a difference in land distribution. In the Northern Hemisphere is the landless area known as the Arctic Sea; in the Southern Hemisphere the same latitudinal belt is occupied by Antarctica. Thus the west wind drift, where water moves from west to east circumscribing the globe, exists only in the Southern Hemisphere (Fig. 6.1 and inside cover).

The movement of the earth itself also affects oceanic and atmospheric circulation. This relationship is complicated because we are dealing with liquid movement over a solid rotating sphere. Again referring to Fig. 6.1 and the inside cover, we see that there is some displacement of the gyres, so that within a given gyre the current is squeezed together on the west side and spread out on the east side. This phenomenon is the result of the earth's rotation. The rotating solid earth moves under the liquid oceans and the result is this concentration of current because the water is more or less pushed toward the land mass on the west side of the ocean while the reverse takes place on the east side of the oceans. This phenomenon is not reversed from one hemisphere to the other but occurs in the same manner from pole to pole because it is due only to the west-to-east rotation of the earth.

The Coriolis effect has a direct influence on the currents of the ocean. Looking at the surface currents of the world (Fig. 6.1), we see that the major gyres in the Northern Hemisphere travel clockwise, and those in the Southern Hemisphere in the opposite direction. A common misconception is that the Coriolis influence causes water to drain from tubs and bowls in the above-described manner. This is not true, primarily because the vessels and the amount of water are far too small for Coriolis effect to be a factor. Size is, therefore, a consideration in studying the Coriolis effect, and the huge oceans, along with the atmosphere, exhibit the considerable influence exerted by the rotating earth even when friction with it is minimal.

In addition to the obvious areal effects of the Coriolis influence, there is also a vertical effect. As mentioned above, the earth's rotation causes a concentration of major currents on the west side of ocean basins. Because of this rotation under the fluid ocean, the sea surface is really not level but slopes upward to the right as the primary gyres move in the Northern

Hemisphere. Thus sea level is higher in the center and on the west side of an ocean basin and achieves a maximum in midlatitudes (Fig. 6.2). Around the Antarctic continent there is a slope upward toward the land mass. Note also the minimal slope near the equator, where Coriolis effect is small.

Some local examples of the Coriolis effect are evident in and on margins of the oceans. Local topographic irregularities such as seamounts or guyots cause currents to be deflected in a predictable manner, in accordance with the Coriolis principle. Currents are deflected around the right side of topographic highs in the Northern Hemisphere, but to the left side of depressions. The opposite is true in the Southern Hemisphere.

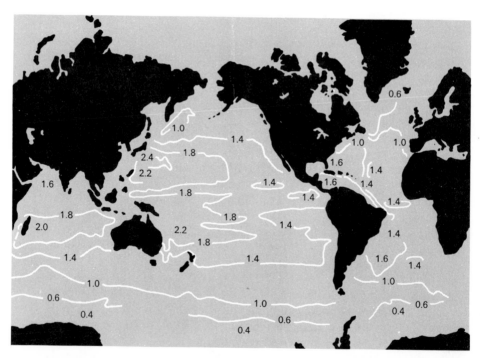

FIG. 6.2 *The uneven surface of the sea caused by the earth's rotation. (After H. Stommel, 1964, "Summary Charts of the Mean Dynamic Topography and Current Field at the Surface of the Ocean, and Related Functions of Mean Wind-Stress," Studies in Oceanography, Seattle: University of Washington Press, p. 54.)*

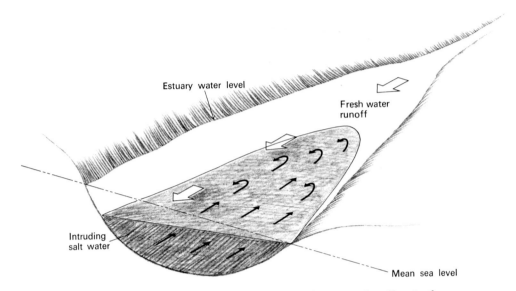

Estuary water level

Fresh water
runoff

Intruding
salt water

Mean sea level

FIG. 6.3 *The Coriolis effect in estuaries. This example illustrates the effect in the Northern Hemisphere; the opposite would be the case in the Southern Hemisphere.*

An **estuary** is a bay on the ocean margin where there is mixing of fresh and marine water caused by runoff from the land. In many estuaries there is a stratification of the heavier salt water under the lighter fresh water. The boundary, however, is rarely horizontal but somewhat tilted in a predictable fashion. These estuaries are subjected to tidal currents (see Chapter 8) that cause a current in and out of the estuary. Due to the Coriolis effect, the boundary between salt water and fresh water slopes down and to the right (Fig. 6.3) as there is a net flow out of the estuary because of runoff.

Ekman spiral
Surface currents receive essentially all of their energy due to friction of the wind moving over the water surface. Observations of the net drift of icebergs in high latitudes in the Northern Hemisphere showed that the mean direction of drift is as much as 40 degrees to the right of the wind direction. In 1902 V. W. Ekman, a Norwegian physicist, explained this phenomenon by considering forces of the winds, the earth's rotation (Coriolis effect), and friction. Assuming no acceleration, he found that surface water is transported

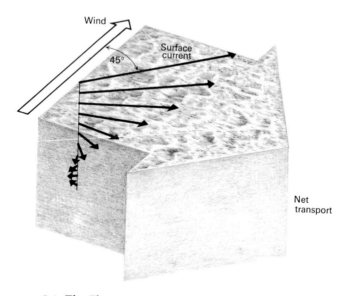

FIG. 6.4 *The Ekman spiral showing deflection and change in velocity of currents and depth for the Northern Hemisphere. (After H. U. Sverdrup, 1958, Oceanography for Meteorologists, Englewood Cliffs: Prentice-Hall, p. 125.)*

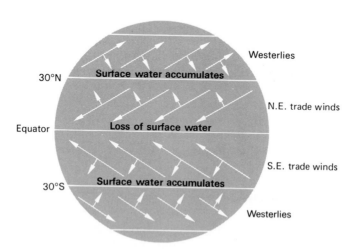

FIG. 6.5 *Movement of surface water resulting from the Ekman transport due to the surface winds over the sea. Short arrows represent the direction of Ekman transport. (After Anikouchine and Sternberg, 1973, p. 103.)*

45 degrees to the right of the wind direction in the Northern Hemisphere and 45 degrees to the left in the Southern Hemisphere. As depth increases, the direction of water transport deviates to the right in the Northern Hemisphere relative to the overlying water in the same fashion as the surface water is transported by the wind. Friction causes each succeeding layer to decrease in speed. The diagrammatic representation of this relationship is called the **Ekman spiral** (Fig. 6.4).

At depth there is a reversal of water transport, with the current moving in the opposite direction of the wind but at a very low magnitude: about 4 percent of the speed of the surface current. This depth is commonly referred to as the depth of frictional resistance and is at approximately 100 meters. One of the most important aspects of Ekman's theoretical work is that the *net* transport of the water column moved by the wind is at 90° angles to the wind direction, to the right in the Northern Hemisphere and to the left in the Southern Hemisphere. The Ekman spiral has not been conclusively observed in the oceans because of the assumptions imposed; however, evidence does support the theory. The best indications are the tendency for ice to drift to the right in the Northern Hemisphere and the phenomenon of upwelling (p. 103).

The Ekman transport is largely responsible for the "hills" of water described previously (Fig. 6.2). Surface water is concentrated near 30° latitude by the combination of the westerlies and trade winds and the resulting Ekman transport (Fig. 6.5). Placement of the land masses in their proper position thus yields "hills" of water near the center of the gyres but displaced to the west. A side or cross-sectional view of this phenomenon shows how zones of convergence become "hills" and zones of divergence become "troughs" due to the Ekman transport (Fig. 6.6).

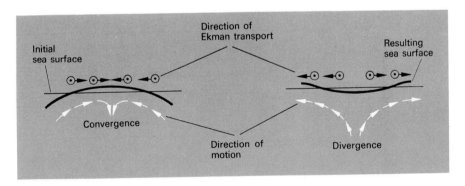

FIG. 6.6 *The shape of the sea surface and water motion associated with a convergence and divergence. The symbol ⊙ indicates a wind moving toward the reader. (After Anikouchine and Sternberg, 1973, p. 103.)*

Geostrophic motion

Fluid motion may be unaccelerated and frictionless, in which case it is **geostrophic.** Oceanographers assume that the major surface currents of the world are geostrophic. In the discussion of the Coriolis effect and Ekman spiral, it was pointed out that the net movement of water is nearly at right angles to the wind direction. The major current gyres therefore have some tendency to pile up low-density water near the center. These "hills," as they are called, are shown in Fig. 6.2.

The "hills" provide a gentle slope away from the center of the ocean down which water tends to move due to gravity. As the water moves, it is deflected by the Coriolis effect at right angles to it. This causes the water to change direction until the two components (Coriolis effect and pressure gradient) balance each other. This then is a geostrophic or earth-turned motion in which the water movement is essentially parallel to imaginary contours on the "hill" of water (Fig. 6.7).

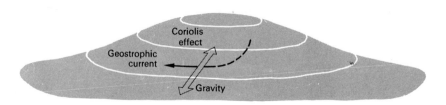

FIG. 6.7 *Geostrophic current where gravity and the Coriolis effect balance each other. (After von Arx, 1962, p. 96.)*

Although the above assumption is theoretical, empirical data compare well with predicted geostrophic motion. Actually, the water flows slightly downhill and eventually reaches the "bottom" because of its viscosity. Some energy, however, must be expended to keep water flowing, and so a deviation from the ideal exists.

THERMOHALINE CIRCULATION

Deep-water or thermohaline circulation is a response to density variations caused by salinity and temperature changes in the oceans. It was recognized quite early in oceanic investigations that deep ocean water was cold even at the equator. It was also known that atmospheric circulation was ultimately controlled by heat differences produced by radiation from the sun. Logically, it was thought, the oceans ought to follow an analogous pattern, with heavy, cold, polar water moving toward the equator and the warm, light, equatorial water moving poleward (Fig. 6.8).

This general concept was presented by Alexander von Humboldt in 1814. He pointed out that such density circulation would cause the entire ocean to be constantly in motion. Although the theory made sense with respect to temperature, it contradicted observations of density changes caused by salinity. The fact that evaporation was greatest near the equator would cause salinity, and therefore density, to increase at the low latitudes and act in opposition to density changes caused by temperature. However, Humboldt believed that such salinity-related density changes were minor and would not alter the predicted pattern.

Actually Humboldt's ideas were far too general and ignored many factors which affect deep-water circulation. More recent and detailed analyses of data show that the deep waters of the oceans move in a great complexity of directions at various depths and in different parts of the ocean basins.

Motion in deep waters cannot usually be determined by current meters or tracing floats, because the movement is quite slow in most places and is at a considerable depth. It is determined from temperature, salinity, and dissolved oxygen content of water samples. Systematic plots of the distribution of these properties provide us with inferences on water motion. Density is also used to study deep-water motion. It is possible to obtain pressure distribution from density distribution and then predict the location and direction of currents that would balance the pressure gradients. Of course, the Coriolis effect must be considered. The pressure distribution is plotted

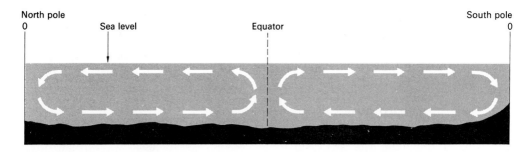

FIG. 6.8 *Simple convection type of deep-sea circulation advocated by von Humboldt. Water rises near the equator, where it is warmed, and sinks near the poles, where it is cooled.*

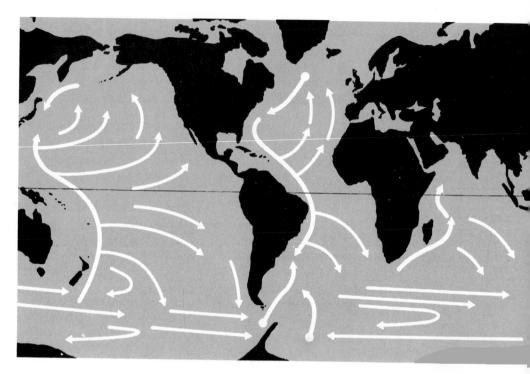

FIG. 6.9 *General circulation patterns of deep ocean water. (After H. Stommel, 1958, "A Survey of Ocean Current Theory," Deep Sea Research 5, 82.)*

areally in the form of **isobars**, lines of equal pressure, much like the technique used by meteorologists.

Today most oceanographers accept the deep sea circulation patterns set forth by Henry Stommel. A quick glance at Stommel's map (Fig. 6.9) reveals a resemblance to the surface circulation patterns. The primary sources of deep water are in the North Atlantic and in the Weddell Sea adjacent to Antarctica. These areas are represented by the large dots on the map. Arrows indicate general direction of flow and the heavy lines mark the strong, concentrated deep currents. Note that they are on the western side of the oceans, due to the earth's rotation. The deep drift around Antarctica serves as a path for interoceanic circulation.

Stommel's map is a generalized areal view. By looking at a cross section through the Atlantic Ocean, we can see the various major currents which are operating in deep waters (Fig. 6.10). It is evident from the diagram, also somewhat generalized, that there are at least three major currents operating at intermediate, deep, and bottom levels of the ocean.

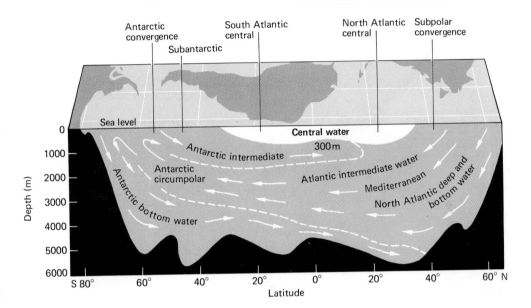

FIG. 6.10 *Generalized cross section of the deep-water circulation in the Atlantic Ocean.*

LOCAL CIRCULATION

So far we have considered oceanic circulation on a large scale. There are also local currents along coastal areas that are of considerable importance for the fishing industry, coastal engineers, and as recreation possibilities. These currents, like other types of oceanic circulation, also draw their ultimate source of energy from the sun's radiation. Surface winds are a major factor, as are waves and bottom topography.

Upwelling and sinking

Along some coastal areas there are what appear to be anomalous water temperature conditions. These are caused by the sinking or upwelling of water in response to coastal winds and show the Coriolis effect according to the Ekman spiral. For example, consider a coastal area in the Northern Hemisphere where surface winds blow from north to south on the west side of a land mass. This wind causes upwelling because of the net transport of surface water to the west which is replaced by deeper, colder water (Fig.

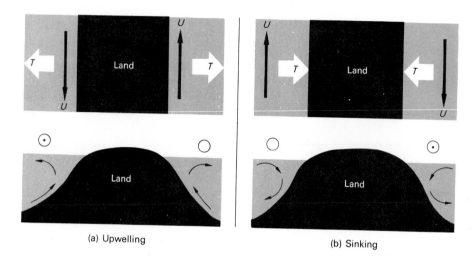

(a) Upwelling   (b) Sinking

FIG. 6.11 *Ekman upwelling and sinking along coastlines in the Northern Hemisphere. U stands for wind direction, T for water transport; ⊙ means wind blowing toward the reader and ○ means wind blowing away. (Partially after von Arx, 1962, p. 115.)*

6.11). The same situation in the Southern Hemisphere produces the opposite transport and the sinking of coastal waters.

In the process of upwelling, the light surface water is carried offshore, causing deeper, colder water to rise to the surface. Such is the case off the coasts of California, Peru, and northeastern Africa. For example, the southerly flowing California Current (inside front cover) is deflected to the right (west), giving it an offshore component. Deeper, colder waters move toward the surface in response to the warm surface waters being carried in an offshore direction (Fig. 6.11). The phenomenon of upwelling is extremely important in the productivity of the ocean in that it circulates great quantities of nutrient materials.

Sinking takes place in response to warm surface water moving onshore and down, forcing lighter, warmer water to greater-than-normal depths. These phenomena take place in the upper few hundred meters of water; upwelling of bottom water is rare. Local upwelling and sinking conditions affect the position of fish and other organisms that move with the temperature of the water.

Wave-induced currents

Waves in shallow water cause transport of water which may be of considerable importance. These currents are restricted to shallow, nearshore areas where waves are breaking, and they are quite variable in both velocity and direction. Because it is fundamental to understand waves in order to comprehend the generation of these currents, their discussion is included in the following chapter on waves.

CURRENT-MEASURING DEVICES AND TECHNIQUES

Some of the principles of oceanic circulation have been discussed above, but it is also necessary to understand how the data are collected that ultimately provide us with maps and charts of ocean currents. There has been considerable development of apparatus and techniques for this purpose; however, many of the early and rather primitive techniques are still often used. The type of instrument or method utilized is determined by what information we wish to collect and where we must collect it. The scientist has to keep his procedures in proportion to his goals.

Current measurements can be accomplished in two general ways: by direct methods or by indirect methods. In the former case, actual water motion is measured, whereas in the latter, water motion is calculated and predicted from indirect measurements of such factors as density, radiocarbon concentration, and electromagnetic properties. Our discussion here is restricted to representative standard methods.

Direct current measurement

Although all types of direct current measurement have the same objective—to obtain data directly from movement of water—they can be conveniently grouped in two categories: **Eulerian** methods and **Lagrangian** methods. The flow of water past a fixed geographical point is measured with Eulerian methods, whereas Lagrangian methods involve tracing and monitoring the movement of floats or tagged particles.

In measuring currents, we are interested in their velocity, direction, and geographic location. All these are important; however, some techniques do not provide all three. Velocity of ocean currents ranges considerably from meters/second in surface currents to kilometers/year in deep waters. It is appropriate here to introduce the term **knot** as it is commonly used to express currents. The term refers to nautical miles/hour, and several of its equivalents are listed in Appendix D. Direction is expressed as a compass direction either in degrees, with north as 0°, or as a geographic compass bearing. Location is probably the most important part of this or any other oceanographic data. All the oceanographic data in the world are of no value if we do not know their precise location. Locations are determined by astronomical calculations and are expressed by latitude and longitude coordinates. The utilization of satellites has largely replaced celestial navigation, however, because it enables us to ascertain locations more accurately and more quickly.

Eulerian methods.   There is a variety of methods which can be used to measure a current as it passes a fixed point. One of the greatest difficulties is in establishing that fixed point. Most data are collected from a ship, and even anchored ships have some motion. Some measurements can be recorded from bottom-based instruments, but these also have drawbacks.

Principles which are commonly utilized are rotation of a free-moving propeller, torque on an arrested propeller, and pressure exerted on a monitored surface. Probably most current meters use the propeller rotation principle. They come in numerous forms, but can be characterized by a few.

One of the first widely used types was the Ekman meter, named after its inventor, V. W. Ekman, who designed it in 1905 (Fig. 6.12). This instrument has the advantage of recording both direction and velocity. A large vane, much like a weather vane, orients the meter with the flow of water. A

Starting and
stopping mechanism

Orienting vane

Paddle wheel
impellor

Revolution
counters

Crate and
compass

Weight

FIG. 6.12 *Ekman current meter. (After von Arx, 1962, p. 218.)*

messenger frees the propeller to turn and record the velocity of water movement.

Another commonly used current meter which is similar in some respects to the Ekman meter is monitored from a vessel. Many of these meters are used primarily for velocity and have no means of recording direction. The rotation of the propeller creates an electronic signal which can be converted to the speed of the current and read or recorded on the shipboard monitor. This device eliminates the need for messengers to start and stop the propeller. It is discouraging to retrieve a meter from considerable depth only to find that a messenger did not work properly and as a result the meter did not function.

A third type of current meter can be left unattended for relatively long periods of time during which data are recorded. These rather complicated electronic devices can be installed on the bottom, at the surface, or at intermediate depths. Both direction and velocity can be recorded either continuously or at certain time intervals.

One example of this type of recording current meter uses a simple home-movie camera, a wrist watch, and a bulbous-type compass. Both direction and speed of the current are indicated by the compass and recorded on film at predetermined intervals. Current speed is shown by the tilt of the bulbous compass which reflects the inclination of the suspended fins of the meter. The Savonius rotor-type current meter (Fig. 6.13) is rapidly becoming

FIG. 6.13 *Savonius-type current meter rigged to a tripod for emplacement on the sea floor. (Photo by M. A. H. Marsden.)*

the most widely used Eulerian system. The meter utilizes an S-type rotor rather than a propeller. This has two significant advantages: The response is unlikely to be influenced by vertical water motion, and it has a very low threshold velocity.

Lagrangian methods. There are some advantages and some disadvantages to tracing the movement of water through space and time. A wider variety of gadgets and techniques is available than for Eulerian methods. As with Eulerian methods, modern electronics has provided a large measure of sophistication to Lagrangian techniques, which were originally primitive.

Drift bottles were probably the first means of tracking oceanic currents. Bottles are stoppered and contain some ballast to give them a density only slightly less than water; then they are set adrift from a particular place in the sea. A card is placed in the bottle asking the finder to fill in the location and date where the bottle was found and return it by mail. Plastic envelopes dropped from a plane are used in the same manner. This technique is quite simple and does provide some data; however, there is a major drawback in that we know the location of the release, the retrieval points, and the time involved, but not the path the bottle followed. Even the time information is not of much value, because many bottles are recovered on beaches where they may have rested for long periods.

Another Lagrangian method involves the release of dyes or contaminants; their movements are determined either by visual tracing or resampling and analysis of water. In the case of dyes (usually fluorescent types) there is the advantage of visual tracing; however, the rate of mixing with ocean water is high, and after a short time the colors are difficult to trace. Contaminants in the form of radioactive tracers, chemical wastes, or selected ions may also be used. These cannot be traced visually and consequently many water samples must be collected and analyzed in order to detect movement of water. Both dyes and contaminants are best suited to quite local, qualitative current studies.

Radio buoys are becoming one of the most widely used Lagrangian instruments. These buoys are tracked by direction-finding receivers, thus providing accurate data on the actual path of water movement. The buoy transmits a signal when triggered by an interrogating signal.

Another way that surface currents can be estimated is by the actual drift of a ship. Here the motion of water carries the ship with it; this also should be regarded as a qualitative technique.

Drogues may be used to trace currents near the surface. A cross made of plastic, wood, cloth, or other relatively impermeable material, or a parachute, is suspended at a known depth from a small buoy (Fig 6.14). As the water moves, the resistance presented by the cross causes it to move at the same

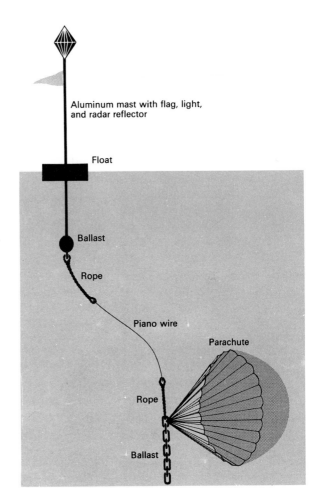

FIG. 6.14 *Simple drogue for current measurement.*

velocity as the water. Commonly a flag, light, or radar reflector is attached to the buoy to permit constant tracing. Care should be taken in the installation of flags and reflectors because they might provide wind resistance and cause the drogue to move at a faster rate than the water.

Neutral buoyancy floats are relatively new to Lagrangian current studies, but they are widely used for intermediate and deep currents. The Swallow float, named after its English inventor, consists of a tube of aluminum alloy several feet long. It is constructed so that there is no compression even at extreme depths. Ballast is added to make the apparatus neutrally buoyant at the desired depth. This is a critical part of the operation, because a small difference in density may represent hundreds of meters in depth. Swallow floats are traced by monitoring acoustical signals which are emitted by a battery-operated transmitter. These signals can be picked up at distances of several miles, and the ship follows along, keeping a record of the float's movement.

In small areas, floats such as partially filled balls or weighted, upright shafts can be traced. However, sighting is a major problem which limits the area studies. Both visual methods are best used in shallow, nearshore areas.

Indirect current measurement

Some carbon dioxide is dissolved in water and all carbon dioxide contains radioactive carbon (^{14}C) in a known ratio with normal ^{12}C. At high latitudes where the cold surface water sinks to become deep water, the radioactive carbon begins to decay. Since there is no means of replenishing the ^{14}C unless it is in contact with the air, the amount of ^{14}C in the water provides a yardstick by which we can determine rates of deep-water movement. Usually several liters of seawater are collected and analyzed for radioactive carbon. From this measure, we can determine how long it has been since the water was in contact with the atmosphere. Assuming that it originated in the high latitudes, in accordance with Stommel's theory on deep-sea circulation, *and* assuming that it has not been recycled from other deep-water systems, we can determine its rate of movement. This technique is being used extensively, especially for analyzing bottom water. It is not uncommon for deep waters to have been isolated from the atmosphere for hundreds of years.

The principle of geostrophic motion discussed earlier also provides us with an indirect means of determining deep-water movement. The method is based on pressure gradients calculated from water properties such as temperature, salinity, and hydrostatic pressure. Samples and temperatures are collected at various depths using water sampling bottles (Chapter 4). The vertical pressure gradient is calculated at each sampling station and profiles are constructed (Fig. 6.15) so that pressure anomalies can be seen diagrammatically. It is possible to compute the horizontal component of the pressure gradient and the geostrophic motion; however, the pressure differentials (and therefore the geostrophic motion) are relative unless we select an arbitrary reference surface (Fig. 6.14). This is commonly called the **dynamic height** method.

It should be noted that this method assumes unaccelerated and frictionless flow, a pressure field that is in steady-state, and simultaneous collection of samples. The latter may be critical, but usually the movement being measured is so slow that we can safely make the assumption.

Modern electronics has also provided a technique for indirect current measurement: the geomagnetic electrokinetograph (GEK). The principle on which it is based is the measurement of electric forces in seawater which are caused by the earth's magnetic field. Seawater is a good conductor of elec-

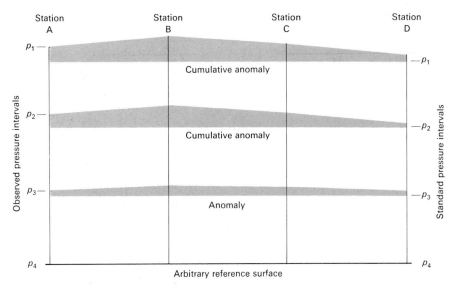

FIG. 6.15 *Generalized diagram of pressure anomalies, which are used to determine circulation by the dynamic height method. (After von Arx, 1962, p. 249.)*

tricity, and its movement in a magnetic field (the earth's) generates an electromagnetic force (emf) which varies with the speed of the water (the conductor) and is normal to the movement and magnetic field. The great advantage of the GEK is that it is towed behind a ship while it is under way (Fig. 6.16). Of course, it must be towed far enough astern to eliminate the interference of the steel ship in the geomagnetic field.

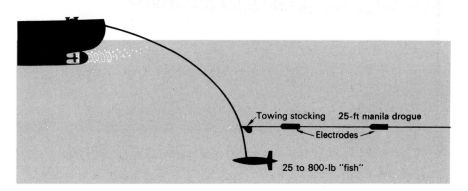

FIG. 6.16 *GEK apparatus being towed behind a ship.*

WATER MASSES

Though there is an apparent homogeneity to the oceans, there are subtle differences in water characteristics which cause certain volumes of quite homogeneous properties to move as a distinct mass and retain their properties for great distances without appreciable mixing. These **water masses** are defined by salinity and temperature, density or trace-element characteristics. They obtain their characteristics at or near the surface but may eventually move to the ocean floor.

Stratification of water masses according to density may yield as many as five distinct layers; the surface and upper layers are under the influence of wind, while intermediate, deep, and bottom layers move in response to density differences. A profile section of the Atlantic Ocean (Fig. 6.10) shows the movement of some of these layers.

Perhaps the best example of how a water mass retains its distinctiveness is illustrated by the Mediterranean Sea water as it enters the Atlantic via the Strait of Gibraltar. Compared to the Atlantic, the Mediterranean is warm and saline and the Gibraltar sill serves as a partial barrier, restricting circulation and mixing. When heavy Mediterranean water does flow over the sill, it moves several hundred kilometers into the Atlantic at a depth of about 1000 meters (Fig. 6.17). The Mediterranean water stabilizes at this depth because, although more saline than the Atlantic water, it is much warmer and therefore less dense than the deep Atlantic water.

Distinct surface expression of water masses is shown in Fig. 6.18. This map can easily be related to the surface current pattern (inside cover) and

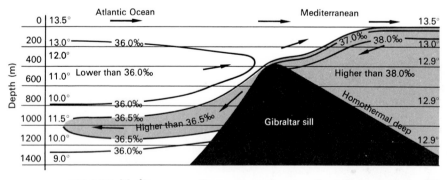

FIG. 6.17 *Mediterranean Sea water as it enters the Atlantic over the Gibraltar sill. Shaded portion represents Mediterranean water; temperatures are in °C. (After Ph. H. Kuenen, 1963, Realms of Water, New York: Wiley, p. 67.)*

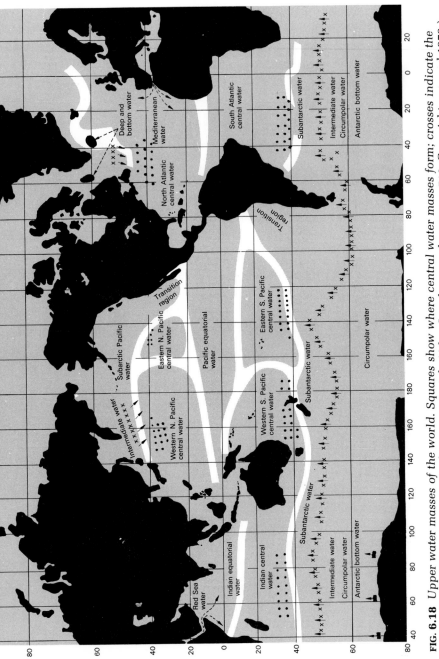

FIG. 6.18 *Upper water masses of the world. Squares show where central water masses form; crosses indicate the regions in which the polar intermediate masses sink. (After Sverdrup, et al., 1942, p. 710. Copyright renewed 1970.)*

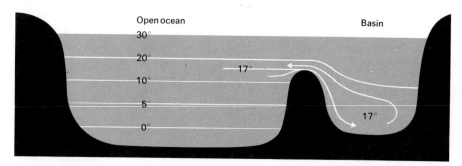

FIG. **6.19** *Generalized cross section showing how a basin is filled with warm ocean water when the barrier is present. (After Kuenen, 1963, p. 69.)*

also to the deep-water circulation map of Stommel (Fig. 6.9). There is no deep cold water forming in the northern Pacific Ocean that corresponds to the North Atlantic deep water, because the Bering Strait prevents the cold Arctic water from moving freely toward the Pacific.

Physical barriers play a major role in determining water characteristics of a basin or of different water masses within a basin. The examples of the Mediterranean Sea and northern Pacific Ocean have already been mentioned. Both involve the effect of mediterraneans on oceanic water. The reverse situation may also be brought about by a physical barrier if a restricted basin receives its water characteristics from the ocean. A sill prevents the cold dense water from entering, and the basin, even at depth, becomes filled with relatively warm water that spills over the top (Fig. 6.19). Such is the case in fjords and other similar topographic relationships.

SELECTED REFERENCES

Bascom, Willard, 1959, "Ocean Waves," *Scientific American*, August 1959. General but well-written article typical of the level found in *Scientific American*.

Chapin, Henry, and F. G. W. Smith, 1962, *The Ocean River*, New York: Charles Scribner's Sons. Popular but comprehensive treatment of the Gulf Stream written by two of its most avid investigators.

Defant, Albert, 1961, *Physical Oceanography*, New York: Pergamon Press, Vol. 1, Chapters 12–21. Exhaustive and highly mathematical treatment of oceanic circulation.

Fomin, L. M., 1964, *The Dynamic Method in Oceanography*, New York: American Elsevier. Comprehensive treatment of a specific method for determining sea-water circulation. This text should be attempted only by advanced students and specialists in physical oceanography.

Gaskell, T. F., 1972, *The Gulf Stream*, London: Cassell and Company. A popular treatment of one of the most fascinating subjects in oceanography.

Groen, P., 1967, *The Waters of the Sea*, London: Van Nostrand, Chapters 6 and 7. Very good general discussion of oceanic circulation without mathematics.

Hill, M. N. (ed.), 1963, *The Sea—Ideas and Observations*, New York: Wiley, Vol. 2, Chapters 10–14. Excellent collection of papers on specific provinces of ocean circulation, with considerable detail.

Neumann, Gerhard, 1968, *Ocean Currents*, New York: American Elsevier. Comprehensive and mathematical treatment of the subject.

Neumann, Gerhard, and W. J. Pierson, Jr., 1966, *Principles of Physical Oceanography*, Englewood Cliffs: Prentice-Hall, Chapters 7, 8, 13, and 14. Detailed and sophisticated mathematical treatment of the subject with a striking absence of field methodology.

Proudman, J., 1953, *Dynamical Oceanography*, New York: Wiley, Chapters 1–10. Quite advanced treatment of ocean dynamics, including currents, tides, and waves.

Robinson, A. R. (ed.), 1963, *Wind-Driven Ocean Circulation*, New York: Blaisdell. Detailed mathematical treatment of upper-ocean circulation only.

Smith, F. G. W., 1973, *The Seas in Motion*, New York; Thomas Y. Crowell. An excellent, nonmathematical treatment of all aspects of water motion. Well illustrated and comprehensive.

Stommel, Henry, 1955, "The Anatomy of the Atlantic," *Scientific American*, January 1955.

Stommel, Henry, 1957, "A Survey of Ocean Current Theory," *Deep Sea Research* **4**, 149–184. Excellent review article of ocean circulation theories by one of the foremost researchers in the subject.

Stommel, Henry, 1958, *The Gulf Stream*, London: Cambridge University Press. Detailed treatment of the Gulf Stream in more sophisticated and quantitative fashion than that by Chapin and Smith (above).

Sverdrup, H. U., M. W. Johnson, and R. H. Fleming, 1942, *The Oceans—Their Physics, Chemistry, and Marine Biology*, Englewood Cliffs: Prentice-Hall, Chapters 11–13 and 15. Principles of ocean circulation in various basins. Many new data have been made available since this book was written.

von Arx, W. S., 1962, *An Introduction to Physical Oceanography*, Reading, Mass.: Addison-Wesley, Chapters 6–11. Excellent and very thorough treatment of ocean circulation. Somewhat mathematical and designed for the serious student of physical oceanography.

Wiegel, R. L., 1964, *Oceanographical Engineering*, Englewood Cliffs: Prentice-Hall, Chapters 13 and 14. Fairly good discussion of currents with strong emphasis on nearshore circulation.

WAVES 7

The previous chapter dealt with the movement of large water masses in virtually any direction. Waves also involve water motion, but it is largely confined to the surface. A wave may be described as a surface disturbance of a fluid medium. Although we normally consider this definition as complete, there are also waves that form at depth, on the interface between water masses.

The generation and propagation of waves may be accomplished in two ways, both of which put some stress on the water surface. They are (1) change in air pressure caused, for example, by the wind acting directly on the surface, and (2) any significant disturbance of the container holding the water, that is, the earth's crust. The former accounts for the vast majority of waves.

SURFACE WAVES
Any description of waves must be greatly simplified from what is actually observed in nature or produced in a laboratory. Diagrammatically and ideally, waves are sinusoidal (Fig. 7.1) and they contain several components that are incorporated into various formulas to express wave motion (see standard texts on physical oceanography). In the simplified diagram, λ or L represents the wavelength, the distance between two corresponding points on succes-

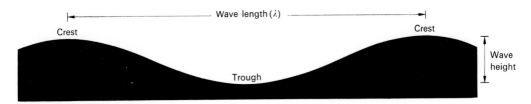

FIG. 7.1 *Ideal sinusoidal wave showing various components.*

117

sive waves. The wave height is expressed as H; it is the maximum relief between the crest or top of a wave and its trough or bottom. The wave period T is another measure commonly used in the study of waves; it is the time required for any successive corresponding points of two waves to pass a fixed point. Such a measurement is much easier to determine than wavelength because of the constant motion of waves.

Wind blowing over a water surface imparts energy to the fluid and causes a disturbance on even apparently motionless water. At first, small ripples or capillary waves are formed, and with time these may build up to larger gravity waves. **Capillary waves** are less than 1.73 centimeters (about 1 inch) in wavelength. Their shape is largely controlled by surface tension of the fluid and approximates a sinusoidal form but with slightly V-shaped troughs. **Gravity waves** are longer than 1.73 centimeters and are controlled largely by gravitational forces. The actual transfer of energy from the wind to the water surface is not yet fully understood.

The size of waves is a function of three factors:

$$H, T = f\,(W, F, D),$$

where W is wind velocity, F is fetch or distance of the water surface over which the wind blows, and D is the length of time the wind blows. There is apparently no maximum height that wind-generated waves may reach; they have been observed over 30 meters in height in oceans. In lakes the fetch is a limiting factor in wave size.

Wind-generated waves may conveniently be assigned to three categories: sea, swell, and surf. Each of these can be readily distinguished from the others and is formed under its own unique conditions. The term **sea** (Fig. 7.2) covers most wind-generated waves and refers to waves under the direct influence of wind. Wave patterns and shapes of this kind are often quite complex. They have a rather peaked crest and a rounded trough. Seamen make estimates of wind speed by the type of sea. In fact, there is a scale for wind speed and the corresponding sea characteristics. This is the Beaufort scale, developed by the English admiral Sir Francis Beaufort in the early eighteenth century (Table 7.1).

Swell is the name applied to waves that are not under the direct influence of wind, either because the wind has ceased or because the waves have moved beyond the area of active wind. These waves exhibit a rather simple pattern, with rounded crests and troughs approaching a sinusoidal shape (Fig. 7.3). Swells are characterized by a smooth, undulating water surface. They may exist simultaneously with a sea-type wave configuration.

Anyone who has been to the coast has observed and listened to the surf. **Surf,** unlike the other two types of waves, is restricted to shallow water. It is

FIG. 7.2 *Complex pattern of sea waves with rather peaked crests. Note the general lack of continuity of the crests.*

FIG. 7.3 *Long and low swell waves with a smaller set of sea waves superimposed along the Big Sur Coast, California.*

TABLE 7.1 *Beaufort wind scale and the state of the sea.*

Beaufort number	Descriptive term	Speed		Appearance of the sea
		knots	m/sec	
0	Calm	—	—	Like a mirror
1	Light air	1	0.5	Ripples with the appearance of scales are formed, but without foam crests
2	Light breeze	3	1.5	Small wavelets, still short but more pronounced; crests have a glassy appearance and do not break
3	Gentle breeze	5	2.5	Large wavelets; crests begin to break; foam of glassy appearance, perhaps scattered white horses
4	Moderate breeze	7½	3.7	Small waves, become longer; fairly frequent white horses
5	Fresh breeze	10	5.0	Moderate waves, taking a more pronounced long form; many white horses are formed (chance of some spray)
6	Strong breeze	12½	6.2	Large waves begin to form; the white foam crests are more extensive everywhere (probably some spray)
7	Near gale	15	7.5	Sea heaps up and white foam from breaking waves begins to be blown in streaks along the direction of the wind
8	Gale	18	9.0	Moderately high waves of greater length; edges of crests begin to break into spindrift; the foam is blown in well-marked streaks along the direction of the wind

the steepening and eventual breaking of the wave form (Fig. 7.4) originally produced as either sea or swell. The break in the wave may have various characteristics. It may be a **spilling breaker** (Fig. 7.5), in which the break is gradual over some distance and the water appears to be spilling over the side

TABLE 7.1 (*continued*)

Beaufort number	Descriptive term	Speed knots	m/sec	Appearance of the sea
9	Strong gale	21	10.5	High waves; dense streaks of foam along the direction of the wind; crests of waves begin to topple, tumble and roll over; spray may affect visibility
10	Storm	24	12.0	Very high waves with over-hanging crests; the resulting foam, in great patches, is blown in dense white streaks along the direction of the wind; on the whole the surface of the sea takes a white appearance; the tumbling of the sea becomes heavy and shocklike; visibility affected
11	Violent storm	27	13.5	Exceptionally high waves (small and medium-sized ships might be for a time lost to view behind the waves); the sea is completely covered with long white patches of foam lying along the direction of the wind; everywhere the edges of the wave crests are blown into froth; visibility affected
12	Hurricane	—	—	The air is filled with foam and spray; sea completely white with driving spray; visibility very seriously affected.

of a container. **Plunging breakers** steepen and curl over, eventually breaking with a crash of water (Fig. 7.6). These are the typical surfing waves of the Pacific Ocean. **Surging breakers** steepen and rush up the beach face (Fig. 7.7), rather than spilling or plunging.

FIG. 7.4 *Surf along the coast displaying a broad zone of nearly continuously breaking waves.*

FIG. 7.5 *Spilling breaker.*

FIG. **7.6** Plunging breaker.

FIG. **7.7** Surging breaker.

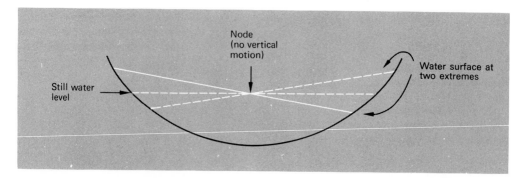

FIG. 7.8 *Standing wave in a container with the length of the container equal to one-half a wave length.*

The above wave types are collectively referred to as **running waves** because the wave form is moving across the water surface. **Standing waves** also exist. This wave form moves up and down and does not advance horizontally. A simple and commonly observed example can be seen whenever one carries a cup, pail, or dishpan full of water. The oscillation of water in the container is really a standing wave (Fig. 7.8), with the crest and trough alternately moving from one side of the container to the other. The vessel is therefore a half wavelength in diameter. Similar waves can often be seen next to a seawall or jetty, from which they are reflected. The wave is reflected from the wall and then reinforced by an advancing wave, thus producing a standing wave (Fig. 7.9).

Water motion in waves

The actual movement of water particles is not at all as it appears to the casual observer. Anyone who has been to a beach has seen waves moving toward the shoreline and eventually breaking. The water in general, however, is not moving in this direction. A simple experiment confirms our conclusion: Any object floating on the surface is repeatedly passed by as each successive wave moves toward shore. The floating object moves in the apparent up-and-down direction; actually it is a small circular path which corresponds to the movement of individual water particles. The progression of waves toward the shore is the result of the water surface changing shape, and this change proceeds longitudinally. The water itself is moving in small orbital paths (Fig. 7.10).

At the crest of a wave, the water particles move in the direction of propagation (the direction the wave form travels), whereas in the trough it is the reverse. This motion can easily be studied in a laboratory wave tank. Although theoretically there is orbital motion only, there is in reality a small amount of net movement in the direction of propagation.

FIG. 7.9 *Reinforced wave caused by reflection at a seawall.*

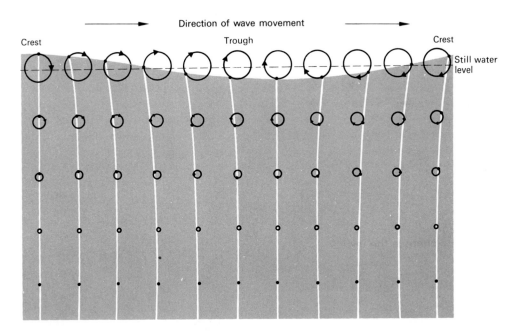

FIG. 7.10 *Generalized cross section of a wave showing the distribution of orbital motion.*

The diameter of an orbit for a surface water particle is equal to the wave height, with the still-water level bisecting the orbit horizontally (Fig. 7.10). By using neutrally buoyant spheres or drops of oil it is possible to observe the orbital motion below the surface. There is an obvious decrease in orbit diameter from the surface downward; orbit diameter is halved with each descent of one-ninth of a wave length. For example, at a depth of $\frac{1}{3} L$ ($\frac{3}{9}$), orbits will have a diameter of one-eighth wave height H. At a depth equal to half the wave length, water motion is almost undetectable, as the orbits are $\frac{1}{23}$ their original diameter.

An impressive demonstration of this principle can be experienced by a SCUBA diver as he ascends toward the water surface. First he is aware of a subtle motion of the water when he enters the lower part of the wave. This motion increases toward the surface, until eventually the diver is being moved in an orbit with the water particles. It is not uncommon to become seasick because of this underwater movement, just as one might on shipboard.

The term **wave base** has been applied to the maximum depth at which bottom material is disturbed by wave motion. Generally, a depth of about half the wave length is considered to equal wave base. However, this is a quite simplified generality. The length of time during which motion takes place and the type of bottom material are also critical factors. In some places, water motion in waves has been found to extend to depths of a few hundred meters.

Wave shape

The generalized shape of the sinusoidal wave shown in Fig. 7.1 is not common in surface waves. Their shape more closely approximates the trochoidal profile (Fig. 7.11), with crests peaked and troughs smoothly rounded. The still-water level of such a profile is slightly below one-half the wave height.

There is a limit to the peakedness or steepness of the crests. This was first theorized by J. H. Mitchell in 1893 when he showed a minimum inclusive angle of 120° and a wave height equal to $\frac{1}{7}$ the wave length (Fig. 7.12). If these limits are exceeded, the wave breaks.

Shallow-water waves

The most significant change in surface waves takes place as waves enter shallow water and approach the shore. For this discussion, shallow water is defined as any depth less than half the wave length. Near such a depth, water motion (orbits) begins to "feel bottom." This causes the wave to slow down and also to steepen because of the interference of the bottom with orbiting water particles; circular orbits are squashed to ovals (Fig. 7.13). At the bottom, water motion is essentially back and forth with no vertical compo-

FIG. 7.11 *Trochoidal wave, the shape assumed by most waves. Note that more than half of the wave height is above the mean water level.*

FIG. 7.12 *Theoretical limits of surface wave shape. (After G. Dietrich, 1963, General Oceanography, New York: Wiley Interscience, p. 362.)*

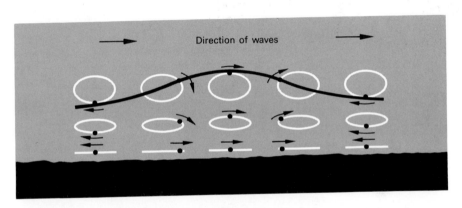

FIG. 7.13 *Water motion as waves are affected by the bottom, causing particles to move in elliptical paths whose short diameters decrease with depth.*

nent. As the wave enters shallower water, this effect takes place in larger orbits, until eventually the surface-water orbits are moving faster than the velocity of propagation, and the wave steepens beyond its limits and breaks.

Along a rather gently sloping nearshore area, we might observe waves breaking more than once as they approach shore (Fig. 7.14). This is a result of irregular bottom topography and illustrates the fact that waves still retain some energy after breaking and may break again. As a shallow area is reached, the wave breaks; then beyond the shallow area the water is somewhat deeper again and allows orbital motion without great hindrance until shallower water is reached (Fig. 7.15).

FIG. **7.14** *Two distinct lines of breakers.*

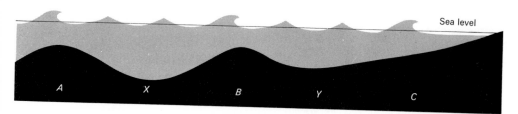

FIG. **7.15** *Profile of the nearshore zone illustrating how topography can be responsible for waves breaking at more than one location. Waves may break at A, B, and C, but not at X or Y because of bottom topography. This is the reason that bands of breakers form parallel to the shore in many areas with a gently sloping bottom.*

Wave reflection, diffraction, and refraction

In some ways, surface waves act much like light rays. They can be reflected, diffracted, and refracted. Waves that meet vertical walls may be **reflected** with little energy loss. This reflection causes the formation of standing waves, which were mentioned earlier in this chapter. At the wall or other barrier, water motion is only up and down. Even though the striking waves are quite large, there is little energy imparted to the jetty or seawall. Such reflection is not very common, however, because it is necessary that the crests of the waves be parallel to the barrier.

Diffraction can be observed in sound, light, and also surface waves. It can be simply thought of as "traveling around barriers," as sound does when it comes from an adjacent room, even though the observer is out of direct line with the source. The same phenomenon takes place as waves pass a barrier, such as one end of a jetty or seawall (Fig. 7.16). A boat placed at point X behind the breakwater will feel wave activity even though it is behind the

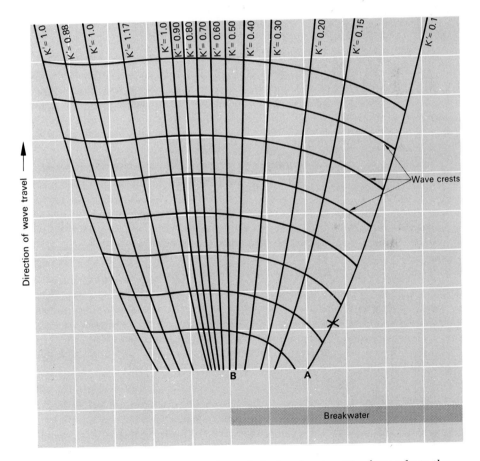

FIG. 7.16 *Diffraction as waves pass the end of a breakwater. K' values refer to the proportion of the unaffected wave height. (After F. P. Shepard, 1972, Submarine Geology (3rd edition), New York: Harper and Row, p. 51.)*

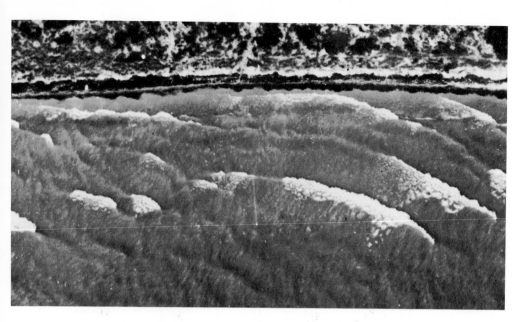

FIG. 7.17 *Wave refraction near the shore.*

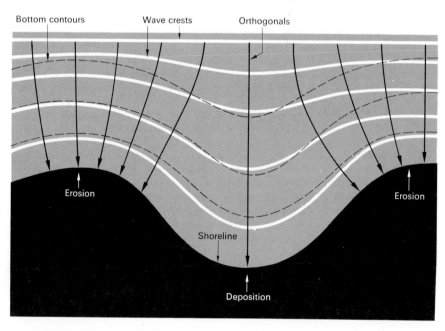

FIG. 7.18 *Wave refraction pattern for an irregular shoreline with a uniformly sloping bottom.*

breakwater. The barrier causes some wave energy to move laterally along the crests, producing smaller waves in a shadow behind it. Along line *A* the wave height is 0.1 times that of unaffected waves; along line *B* it is 0.5 times the unaffected wave height, and so on. Wave diffraction is not restricted to shallow water.

Water **refraction** is the bending of wave fronts. Knowledge of refraction is a very important phase of military strategy and coastal engineering. Recall that in our discussion of shallow-water waves, it was pointed out that the velocity of waves decreases as they are impeded by the bottom. Waves tend to become compressed in shallow water because the length decreases in response to this bottom friction.

The refraction of waves is caused by one part of a wave reaching shallow water before another. As a result, that part where the bottom is slowing the wave down is proceeding toward shore at a slower rate than the unaffected part of the wave. Such refraction is really the rule in most coastal areas, and the waves are usually bent so that they nearly parallel the shore (Fig. 7.17).

A simple but typical example is shown in Fig. 7.18. The shore line is irregular with bays and headlands, and the bottom slopes uniformly toward

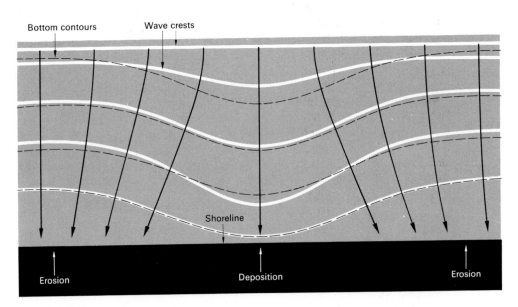

FIG. 7.19 *Wave refraction pattern for a straight shoreline with an irregular bottom topography.*

deep water. An approaching wave "feels bottom" adjacent to the headland area first, because the water is shallow there. The part of the wave not affected by the bottom is moving faster, and as the wave comes closer to shore it tends to parallel the shore. A somewhat different wave pattern will form if the shore line is straight but the bottom topography is irregular (Fig. 7.19). Wave crests tend to bend around islands also.

It is common practice to construct lines called **orthogonals** or wave rays on wave refraction diagrams. These are lines of equal energy constructed perpendicular to any wave crest. The result is a convergence of orthogonals on headlands and thus a convergence of wave energy. Orthogonals diverge in bays and energy is spread. Consequently, headlands become eroded and embayments tend to be filled in with erosion products from the headlands.

There is change in wave height as well as direction during refraction. Where orthogonals spread energy, the wave height decreases, and the opposite takes place in converging areas. This is due to concentration or depletion of energy per unit area along the wave crest.

WAVE-GENERATED CURRENTS

Wave-drift current

Our earlier discussion of wind-driven surface waves described the water particle motion as circular. This is basically correct; however, these orbits are not closed and there is a small current generated due primarily to friction caused by the wind. This is commonly called **wave-drift current**, and it is caused by the small amount of forward movement associated with each orbit of a surface water particle (Fig. 7.20).

A previous example was a floating object with waves passing it, indicating no apparent lateral movement of the object, hence no net lateral movement of the water. However, if we observed this floating object for some length of time, we would see it move eventually in the direction of wave propagation. This movement is slow, and it is not due to surface winds blowing the object. It is possible to predict this current velocity from wave characteristics in deep water but not in shallow water where waves are steepened or breaking.

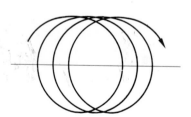

FIG. 7.20 *Net lateral movement of a water particle during orbital motion. This causes a slight wave-drift current.*

Longshore currents

The discussion of wave-drift currents shows that there is some transport of water due to wave activity. As the wave enters shallow water and is slowed by bottom interference, this forward transport of water increases but is then halted abruptly by the beach. Wave-refraction diagrams show that there is a tendency for waves to parallel the shore because of velocity changes during refraction. Rarely, however, are waves completely parallel to the shoreline as they break and strike the beach. There is a component of energy which causes water to move parallel to the shore and away from the acute angle made by the breaking wave as it approaches the beach (Fig. 7.21). This is a **longshore current**, which exists primarily between the breakers and the shoreline.

The velocity of longshore currents is a function of breaker height and wave angle. During storm conditions these currents may be moving in excess of one meter/second. Because of their mode of generation, longshore currents are variable in both direction and speed as well as being ephemeral. On many coasts, such currents may follow predictable patterns directly related to meteorological conditions. This is particularly true where cyclonic or frontal systems approach and pass over a coastal area. For example, consider a coast oriented essentially north-south in the midlatitudes. Low-pressure systems (cyclones), with their counterclockwise winds, move west to east. As such a weather system approaches the coast, it generates winds from the southwest which cause waves to approach the coast from that direction. These waves approach the coast and break, causing longshore currents to move toward the north along the coast (Fig. 7.22). As the storm (cyclone) approaches, waves get larger and currents become swifter. After the center of the storm passes the coast, the wind direction reverses because the coast is now subjected to winds from the northwest on the trailing side of the storm. This causes a subsequent change in wave orientation and longshore currents are reversed, moving to the south along the coast (Fig. 7.22). This pattern is cyclic and somewhat predictable with the location and intensity of the storm controlling the size of waves and the speed of the longshore currents.

Rip currents

As wave-drift currents and breaking waves continue to move toward shore, water moves laterally along the shore, but locally it is also moving normal to and away from the shore. This type of water transport is called a **rip current**. Such currents are local and are fairly narrow (Fig. 7.24). They are caused by a buildup of water due to its shoreward movement. Rip currents commonly form where there is a path of low resistance, such as a topographic low in the nearshore sandbars (Fig. 7.23). This type of current is probably what is experienced by those who claim to be caught by "undertow." Many ocean

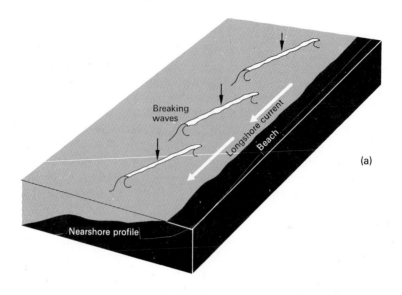

(a)

(b)

FIG. 7.21 *Generation of a longshore current due to wave action in the nearshore zone, as shown (a) diagrammatically and (b) photographically. The photograph shows refraction of waves as they approach the coast and pass through decreasing water depth.*

Breaking waves

Longshore current

Beach

Nearshore profile

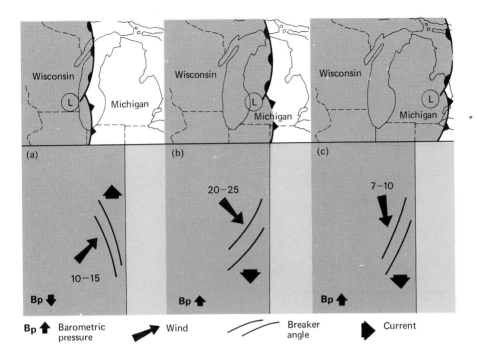

FIG. 7.22 *Diagram showing approach and passage of a cyclonic system along the Lake Michigan coast, with associated changes in barometric pressure, wind speed and direction, wave approach, and longshore current.*

FIG. 7.23 *Broad beach along the Oregon coast showing large and deep rip currents oriented obliquely to the shore. (Photo by W. T. Fox.)*

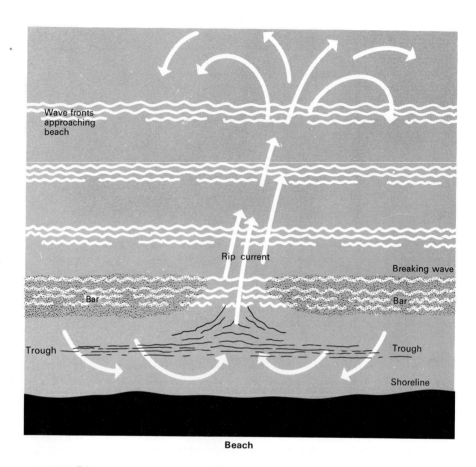

Wave fronts
approaching
beach

Rip current

Breaking wave

Bar

Bar

Trough

Trough

Shoreline

Beach

FIG. 7.24 *Rip current system caused by water piling up behind nearshore sandbars.*

beaches used to have signs warning swimmers of the "dangerous undertow." A downward bottom current such as that implied by the term "undertow" does not exist. Rip currents may be quite strong, but because they are narrow one can swim a few strokes parallel to shore and be out of the current's influence. Commonly they are recognizable by bubble trains or plumes of suspended sediment moving seaward along the surface normal to the shore.

LARGE AND CATASTROPHIC SURFACE WAVES
The waves discussed so far are short; that is, their period may be measured in terms of a few seconds. There are also surface waves of much longer wave

length and periods of up to several minutes. Waves of an intermediate size in the spectrum are rare.

Storm waves (storm surges, or storm tides)

Onshore winds associated with severe storms or hurricanes may cause the water level to rise substantially, sometimes more than five meters. Usually this high water is assisted in its destruction by high waves caused by storm winds. What is taking place is a piling up of water along the coast due to the inability of the water to flow back to normal sea level as fast as it is transported toward shore by high winds and large waves. Although this phenomenon takes place along rather smooth, open coasts, it is most significant in estuaries or other coastal bays where water is funneled into the embayment and temporarily trapped.

Atmospheric low-pressure systems associated with a storm may produce a slight hill of water that reinforces the high water level. An unusually high water level is also likely if such a storm surge is coincident with a high tide. These storm surges are quite destructive, particularly along coastal plain areas where an increase of a few meters in water level can inundate large areas of land.

Seiches

Bays, lakes, or local enclosed portions of the ocean may experience sudden high water that is repeated in periods of several minutes or a few hours. Such a wave phenomenon is termed a **seiche**. It is long with respect to wind-generated waves and generally has a small amplitude. The period depends on the wave length and the limits of the basin. The long waves are commonly reflected from one side of the basin to another and consequently may occur at one place with some repetition.

Seiches can originate in two ways. First, an abrupt halt of storm winds allows storm waves to lower, and the water seeks to regain its equilibrium. In a small basin this might bring about seiches, as the water which was piled up on one side of the basin seeks its own level. A second way is by significant differences in barometric pressure over different areas of the basin. Quick change from this pressure differential to a uniform pressure will produce the same effect: a seiche.

Seiches are generally not of major consequence, though they may come without warning and can reach heights of a few meters as they approach the shore. Seiches are not uncommon in the Great Lakes, where they are the result of local squalls and regional storms. For instance, on Lake Michigan a storm at Chicago may build up a storm wave. If that storm abates quickly, the eastern side of the lake might receive a seiche quite unexpectedly, and that wave might be reflected back toward Milwaukee on the opposite side of the lake (Fig. 7.25).

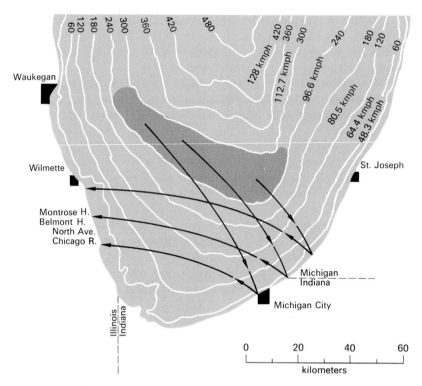

FIG. 7.25 *Reflection of a seiche wave in southern Lake Michigan on June 26, 1954. The dark shading indicates the place where the seiche originated. (After M. Ewing, et al., 1954, "An Explanation of the Lake Michigan Wave of 26 June 1954," Science **120**, 685.)*

Seismic sea waves (tsunamis)

Unquestionably the most destructive of all wave types is the **tsunami**, commonly called a tidal wave by the layman. This latter term is incorrect, as these waves are not in any way related to tides. The true origin of the wave is related to submarine earthquakes or other catastrophic events in or on the earth's submarine crust.

Broadly speaking, an event which causes the shape of the ocean's container to be altered is the triggering mechanism for a tsunami. Such an event could be an abrupt displacement of the sea floor as movement takes place along a fault (Fig. 7.26). This would cause the bottom of the ocean to drop locally and would cause a distortion of the sea surface. As the water seeks to return to normal sea level, waves are sent out from the area of disturbance. These waves are tsunamis. Most of them occur in the Pacific Ocean because of the relationship to seismic activity. A similar effect can be produced by a submarine landslide or perhaps a volcanic eruption. Most of the latter do not produce tsunamis, however, because they evidently do not generate enough wave energy.

Tsunamis are quite harmless in the open sea; only when they enter shallow water do they become sizable and destructive. In the open ocean their wave height is usually less than a meter and the wave length is tens or hundreds of kilometers. Consequently they might go unnoticed by ships at sea. The velocity that tsunamis reach as they move across the open ocean is tremendous, more than 400 knots. Fortunately, the Pacific Ocean is so large that people on the surrounding land masses can generally be warned in time to evacuate the nearshore areas. Presently there is a Pacific-wide warning system which includes automatic alarm sounds when large earthquakes are felt. Data from the seismic stations are analyzed to determine the location of the disturbance and the likelihood of a tsunami.

The great buildup of the tsunami as it enters shallow water is due to the long wave length and the slope and configuration of the ocean bottom and shoreline. Wave height also depends on the wave shape in deep water and on diffraction. The speed of the wave causes net transport of a large mass of water which builds up to a height of several meters in some cases; however, in others there is little change in water level even at the coast. We can estimate, within some limits, the height that waves will reach at the shore, but much more research is necessary before we can be confident of our predictions. Sometimes the first wave is the largest, in which case there

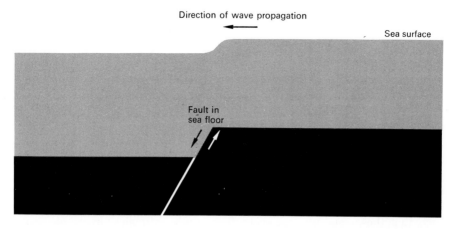

FIG. 7.26 *Submarine faulting as a cause of a tsunami or seismic sea wave.*

seems to be a dampening effect on later waves. But in other cases the largest wave may be the third, fourth, or fifth.

Perhaps the highest of all recorded tsunamis is that which occurred on April 1, 1946, originating from an earthquake off the Aleutian Islands. A lighthouse on nearby Unimak Island was completely destroyed. The foundation of the lighthouse was about 15 meters above sea level and a destroyed radio tower above it was over 32 meters above sea level. The same tsunami caused considerable damage in Hawaii and elsewhere in the Pacific.

MEASURING SURFACE WAVES

Like most data in oceanography, wave measurements have undergone substantial sophistication during the past 20 to 25 years. Until that time virtually all data on wave characteristics were obtained by direct or indirect visual methods involving a variety of sighting methods or photographic techniques. Electronic wave recorders are now widely used and provide scientists with much more quantitative information.

Visual techniques

In measuring waves, we are most often interested in the wave length, wave height, and period. Fairly simple techniques are available for determining all three, although the reader should keep in mind the difficulties of measuring these parameters on a continually moving wave.

Wave height or breaker height can be measured easily in shallow water by using a graduated stick. The observer stands in the water at the place where measurements are to be taken and watches the minimum and maximum depth as waves pass the stick (Fig. 7.27). The difference is equal to the wave height. A similar technique is to observe the elevation at which crests and troughs pass by pier pilings, breakwaters, or other fixed reference points. A calibrated scale can be fastened to the pier and the crests and troughs observed as they pass. This is a crude technique and not very accurate, but often it is all that is necessary. The place of observation is limited by water depth, presence of reference levels, and wave activity (one could not collect these data under rough storm conditions).

A slight modification of the above is to use a weighted drogue attached to a graduated pole and place it below the zone of wave action (Fig. 7.28). This provides a stable reference point to determine the trough and crest as they pass the pole, and it can be used in deep water.

Wave height is frequently measured at sea by sighting from the ship along wave crests to the horizon. Knowing the height of the observer above the water line on the ship gives the wave height. Rarely is the ship's position such that these readings are more than crude estimates. In rough water particularly, the ship pitches back and forth so that the readings are of little value.

FIG. 7.27 *Man measuring wave heights in a breaker zone with a graduated pole.*

FIG. 7.28 *Floating pole with a weight attached for measuring deep-water waves.*

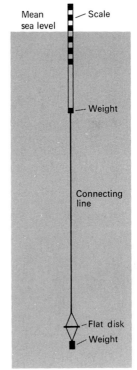

Mean sea level

Scale

Weight

Connecting line

Flat disk

Weight

Wave period can be measured by using most of the same techniques that have been described for measuring wave height. It is easy to time the passage of wave crests as they pass by the poles or a breakwater. The preferred method is to determine the time for 11 crests (10 waves) to pass and then find the average period for each one.

Photography is the best practical method to measure wave length. Aerial photographs are commonly used to determine wave length and examine refraction patterns of waves as they enter shallow water (Fig. 7.17). Stereoscopic pairs of aerial photographs would enable researchers to determine wave height also; however, in a dynamic environment like the sea, such photographs have to be taken simultaneously so that wave position is the same in both photographs. A similar method has been used successfully from a ship, with two photos being taken about 5 meters apart. Stereoscopic photographs of a small area of the sea are obtained and the waves are measured from the photos.

Wave-recording devices
Without actually observing waves, it is possible to determine wave characteristics by electronic recording instruments. There are four types of these instruments: tide gauges, underwater pressure gauges, reverse echo-sounding gauges, and acceleration gauges.

Tide gauges record changes in sea level associated with diurnal tidal changes (see next chapter). They also record the up-and-down movement of waves as they pass by the measuring station. A pen records the up-and-down movement directly on a revolving drum. On the chart, wave height (Fig. 7.29) is superimposed on tidal fluctuations.

Waves can also be monitored by a pressure-sensing device which is placed on the nearshore bottom. As wave crests and troughs alternately pass

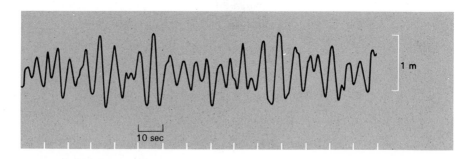

FIG. 7.29 *Wave record showing height and period. (After Groen, 1967, p. 149.)*

over the sensor, the changing height of the water column causes a change in pressure on the sensor. The signal is transmitted to a recorder on shore which provides graphic representation of the waves.

A step resistance wave staff provides similar data. The staff contains numerous, closely spaced, vertical terminals. As the wave passes by the staff, various terminals are alternately submerged and exposed, which causes changes in the resistance along the terminals on the staff. These data are also recorded graphically and provide a continuous record of the waves.

By placing an echo-sounding transmitter and receiver on the bottom, it is possible to record wave height. Like the above techniques, it is monitored on shore. The principle is the same as using an echo-sounder to determine depth, except that the sound is transmitted from the bottom, reflected off the water surface back to the bottom, and picked up there by the receiver. The time for the sound to make its trip is a function of the amount of water between the instrument and the surface. There are some problems with this method, in that the sound does not travel in a narrow vertical path but spreads out in a cone above the transmitter. As a result, small waves with a wave length less than the width of the top of the cone will not be measured. The same principle is used by submarines with an echo-sounder directed toward the surface.

As an object moves up and down with a wave, there is an acceleration which can be recorded and transformed into wave-height data. Floating buoys are used, or the ship itself is used. In either case the acceleration on the object is recorded. This technique is advantageous because it does not require a fixed location and can be used in deep water.

INTERNAL WAVES

There are distinct interfaces in the ocean between water of differing densities. The surface that separates these water masses may be in motion, much like the surface between the ocean and the atmosphere. This motion is generally recognized by the distribution of water temperature gradients (Fig. 7.30).

The water motion is much like that of surface waves. Below the interface is an orbital motion in the direction of propagation, while above the interface similar orbital motion moves in the opposite direction. Orbits decrease in size in both directions from the interface. Internal waves are slow-moving in comparison with surface waves and have a generally sinusoidal shape. The reason for this is the ease with which the interface is distorted, due to the small density difference across the interface compared to the air-water interface on the surface. Usually the period of internal waves ranges between a few minutes and a day, indicating their long wave length. Amplitudes range up to 100 meters with velocities between a few centimeters/second and one meter/second.

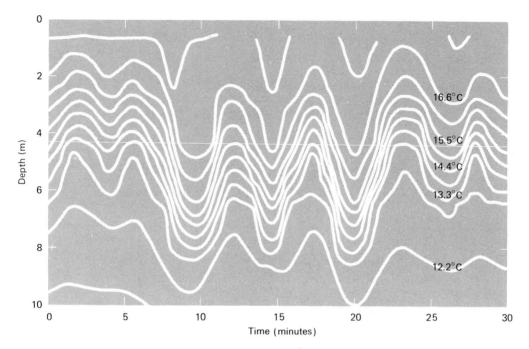

FIG. 7.30 *Distribution of temperature with depth, showing location of internal waves. (After O. S. Lee, 1961, "Effect of an Internal Wave on Sound in the Ocean,"* Jour. Acoust. Soc. Am. **33**, p. 678.)

In some instances there is surface expression of internal waves in the form of slicks. These are parallel wave crests that are located about midway between the crest and trough of the internal wave.

Internal waves may be generated by a wide variety of phenomena including ships, storms, tidal action, or combinations of these.

SELECTED REFERENCES

Bascom, Willard, 1964, *Waves and Beaches—The Dynamics of the Ocean Surface*, Garden City, N.Y.: Doubleday, Chapters 2–8. Very excellent general book on waves which is written for the general public or beginning student in oceanography. Illustrations are good but they and the examples used are restricted to the Pacific Coast of the United States.

Defant, Albert, 1961, "Physical Oceanography," New York: Pergamon Press, Vol. 2, Chapters 1–6, 16.

Groen, P., 1967, *The Waters of the Sea*, London: Van Nostrand, Chapter 4. Well-written nonmathematical treatment of various wave types.

Hidy, G. M., 1971, *The Waves, The Nature of Sea Motion*, New York: Van Nostrand Reinhold. Concise treatment of oceanic circulation including both currents and waves.

Hill, M. N. (ed.), 1963, *The Sea—Ideas and Observations*, New York: Wiley, Vol. 1, Chapters 15–22. Detailed and in-depth articles on various aspects and types of water waves. A general background is necessary for good comprehension.

Kinsman, Blair, 1965, *Wind Waves—Their Generation and Propagation on the Ocean Surface*, Englewood Cliffs: Prentice-Hall. Most complete treatment of the subject available, but it is designed for the specialist. Highly mathematical.

Neumann, Gerhard, and W. J. Pierson, Jr., 1966, *Principles of Physical Oceanography*, Englewood Cliffs: Prentice-Hall, Chapters 9, 10, and 12. Well-written and thorough treatment of waves for the student of physical oceanography.

Phillips, O. M., 1966, *The Dynamics of the Upper Ocean*, Cambridge: Cambridge University Press. Very extensive and mathematical treatment of all types of waves except tsunamis.

Proudman, J., 1953, *Dynamical Oceanography*, New York: Wiley, Chapters 15 and 16. Fundamental mathematics of waves as well as other ocean dynamics.

Russell, R. C. H., and D. H. Macmillan, 1952, *Waves and Tides*, New York: Hutchinson's Scientific and Technical Publications (Part I). General treatment of waves with a minimum of mathematics.

Smith, F. G. W., 1973, *The Seas in Motion*, New York: Thomas Y. Crowell. Seven chapters devoted to waves in a fashion which is easily understood by the layman.

TIDES 8

Anyone who has spent even a few hours at an ocean beach has observed the slow change in sea level that we call the **tide**. At certain times of the day, a sunbather must move his beach towel or else be inundated by the rising water. Such sea-level changes are significant, regular, and predictable.

The obvious water-level changes caused by tides were noticed and recorded by many of the ancients in the Mediterranean area: Herodotus (450 B.C.), Aristotle (350 B.C.), Pytheas (325 B.C.), and Strabo (54 B.C.). These scholars also correctly related tidal cycles to the moon. However, not until Sir Isaac Newton (1642–1727) formulated his universal law of gravitational attraction was the tidal phenomenon adequately explained.

Tides are actually waves, in that they are a form of water-surface disturbance. They are by far the longest waves in the ocean, with periods of more than 12 hours; their wave length is thousands of kilometers and their wave height ranges from almost zero to more than 18 meters. Tides occur in all water bodies; however, as in the case of the Coriolis effect, small basins do not exhibit tides that are important or practical to measure.

TIDAL FORCES

Newton's Law of Universal Gravitation is important in oceanography because it serves as the basis for any discussion of the earth's tides. The law states that any two objects are attracted to each other by a force that is dependent upon both their masses and the distance between them. Expressed symbolically,

$$F = G \frac{M_1 M_2}{d^2},$$

where F is the gravitational force, G the gravitational constant, M_1 and M_2 the respective masses of the bodies, and d the distance between the centers of the two bodies.

This is a universal law; that is, it applies to any two bodies, a knife and a fork in a table setting or two people sitting on opposite sides of the table. The earth, therefore, is attracted to other celestial bodies in accordance with this law. Only two celestial bodies exert appreciable force on the earth; they are the moon and the sun. Although the moon is small, its effect on tides is greater than that of the sun. The sun is 389 times farther away than the moon, and its mass is 26×10^6 times greater. The gravitational pull of the sun is 100 times more than that of the moon. However, the *difference* between the gravitational attraction on one side of the earth and that on the other is the important factor for tides, and in this respect the moon predominates. The tidal affect of the sun is only 46.6 percent that of the moon.

In addition there is a second force to consider in explaining tides, namely, **centrifugal force**. This force results from the movement of a given mass along a curved path. Here the force on the earth in its orbit in the earth-moon system is of concern. This force, expressed as

$$CF = \frac{MV^2}{r},$$

is directed away from the center of this orbit. M is the mass of the orbiting body, V is the velocity, and r is the radius of curvature of its orbital path.

Tides are caused by water distortions resulting from the combined effects of the above two forces in the system of the earth, sun, and moon. Water, being a fluid, is readily "pulled out of shape," causing bulges in sea level (Fig. 8.1). This distortion is actually the net result of the attraction of the sun, moon, and earth, along with the centrifugal force between the earth and moon. The earth and moon are moving in a balanced system in which the

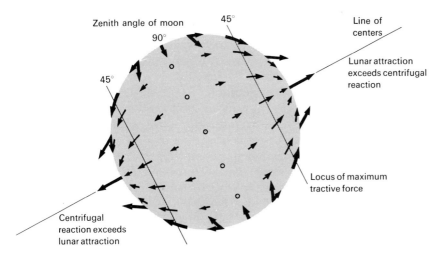

FIG. **8.1** *Tide-producing forces and resulting vectors of attraction. (After von Arx, 1962, p. 47.)*

gravitational attraction is balanced by centrifugal force caused by their revolution with respect to the center of this system. On the surface of the earth, there is an imbalance between the centrifugal and attractive forces. On the side toward the moon, the attractive force exceeds the centrifugal force, and on the side away from the moon the opposite is true. Consequently, there are forces acting in opposite directions along a line drawn through the center of the earth toward the moon. The forces are called **tractive**, in that they act tangentially to the earth's surface, pulling the water in the direction of the bulges.

The movement of the moon with respect to the earth causes variation in the magnitude of these forces and likewise in the magnitude of the tides that are produced. During the new moon and again during the full moon position, the moon, sun, and earth are aligned and the forces are maximal, yielding tides of highest magnitude called **spring tides** (Fig. 8.2a). During the first quarter and third quarter, the sun and moon are acting in opposition to each other and minimal tides called **neap tides** are produced (Fig. 8.2b). Note that each of these tides occurs every two weeks and not with the season, as the name seems to suggest.

The lunar day is 24 hours and 50 minutes in duration. That is, the moon passes a given location on earth once in that period of time. If we assume a water-covered earth, then a given point would pass under two highs and two lows each tidal day (24 hours and 50 minutes). The highs correspond to wave crests and the lows to troughs. There are two wavelengths in the complete circumference of the earth, each having a period of 12 hours and 25 minutes (one-half a lunar day).

Figure 8.2 shows tide-producing forces in the plane of the equator and also at an angle to it. The difference reflects the change in the moon's position with respect to the equator, which may be as much as 28.5° to the north or the south. As a result, the crests of tide waves (not to be confused with tsunamis) move in response to the moon's movement, producing wide variation in tidal range at a given location.

From the above principles we would expect to observe maximum tidal range at any location when the moon is directly above or below that location. In fact, that is not the case. The earth is rotating at great speed (up to 1600 kilometers/hour) relative to the moon. In order for tidal bulges to coincide exactly with the moon's position, they would have to keep up with this movement. This does not happen, nor is the earth's crust frictionless with respect to the ocean, which would mean that the earth would rotate under a motionless envelope of water. There is a balance between the pull of the moon and the tendency for the bulge to rotate with the earth (Fig. 8.3). As a result the tidal crest actually precedes the moon in its path around the earth.

Thus far our discussion of tides has assumed a uniform envelope of water encasing a smooth sphere (the earth). Obviously this is an oversim-

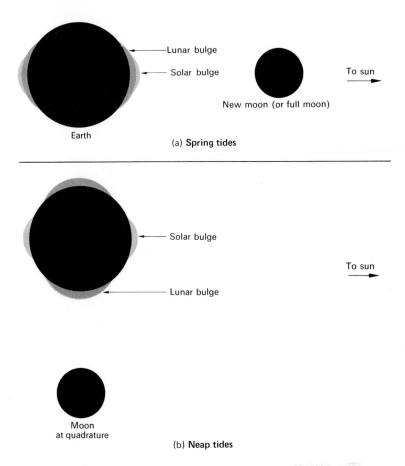

Lunar bulge

Solar bulge

To sun

New moon (or full moon)

Earth

(a) **Spring tides**

Solar bulge

To sun

Lunar bulge

Moon
at quadrature

(b) **Neap tides**

FIG. 8.2 *(a) Sun, moon, and earth aligned to produce spring tides. (b) Sun and moon acting at right angles to each other to produce neap tides.*

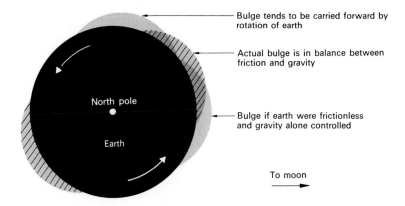

Bulge tends to be carried forward by rotation of earth

Actual bulge is in balance between friction and gravity

North pole

Bulge if earth were frictionless and gravity alone controlled

Earth

To moon

FIG. 8.3 *Position of the tidal bulges caused by equilibrium between friction and the moon's attractive force. (After W. Bascom, 1964, Waves and Beaches, Garden City, N.Y.: Doubleday, p. 87.)*

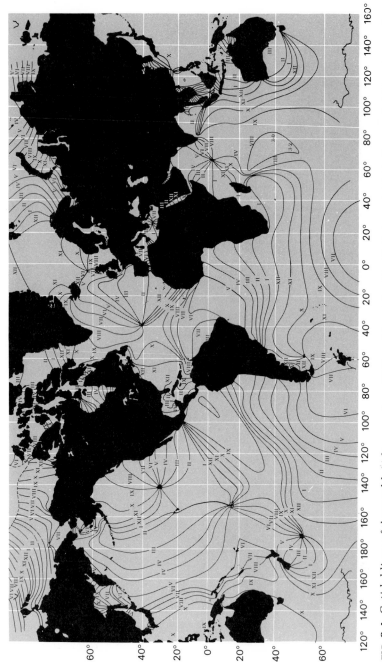

FIG. 8.4 *Cotidal lines of the world. (After von Arx, 1962, p. 51; original by H. Poincaré, 1910, Leçons de Mécanique Céleste, Gauther-Crofts, vol. 3.)*

plification; the earth is not a regularly shaped sphere, and water covers only 71 percent of it. In dealing with the movement of tides we must consider many of the same principles that apply to tsunamis and other large surface waves. Velocity and height of tide waves are controlled in part by water depth, but by far the most complicating influence is the distribution of land.

Tide waves, although influenced by many of the same factors as other waves, differ in that they are formed by external forces, the moon and sun. As a result, they cannot move due to their own internal energy and are consequently called **forced waves**.

Complications caused by the land-water distribution on the globe cause considerable variation in the arrival of crests (high tide) or troughs (low tide) of the tide waves. For instance, if the moon is directly above the equator we might expect that high tide would occur at the same time along the longitudinal meridians. However, bottom topography and particularly land masses cause considerable alteration of this simple pattern.

It is possible to map the distribution of simultaneous high tides by drawing lines connecting these locations. These lines, shown in Fig. 8.4, are called **cotidal lines**. Roman numerals on cotidal lines are time differences in lunar hours (one twenty-fourth of a lunar day). This map illustrates the complexity of tide-wave movement. The areas of connection between basins allow waves from one, the Antarctic area for instance, to enter an ocean (Atlantic), causing a further complication. That is, the crest of the Antarctic area enters the Atlantic, which because of its confined area has its own tide. Irregularities along the coasts cause tidal crests to be reflected and increase the complexity of the cotidal map.

Another feature of the cotidal map is the numerous points from which several cotidal lines radiate. These are called **amphidromic points**. They are caused by turning points in the tidal crests due to the effect of the earth's rotation on the horizontally moving currents (Fig. 8.5). The best example of this movement is in estuaries or embayments where there are tidal currents. During high or flood tide, water level is increasing in a direction away from open water, and during low or ebb tide, water is returning to open water (Fig. 8.6). This causes a tilting of the water surface in the direction of water movement, with the nodals or nodal lines marking the axis of tilting. In addition, the earth's rotation and consequent Coriolis effect also cause a slope on this surface. During flood tide it slopes up to the right and likewise during ebb tide, thereby causing a tilt on either side of the estuary or embayment. The circulation of water in the estuary is therefore around a central point where there is little or no motion; this is the amphidromic point (Fig. 8.7). This same phenomenon occurs in the oceans and in partially enclosed portions of oceanic basins. The entire North Atlantic basin responds in a similar manner, with an amphidrome near the center, and in the North Sea there are three such points.

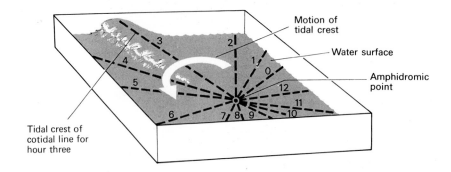

FIG. 8.5 *Motion of the water surface in an exaggerated, hypothetical, amphidromic tidal system in the Northern Hemisphere. (After Anikouchine and Sternberg, 1973, p. 156.)*

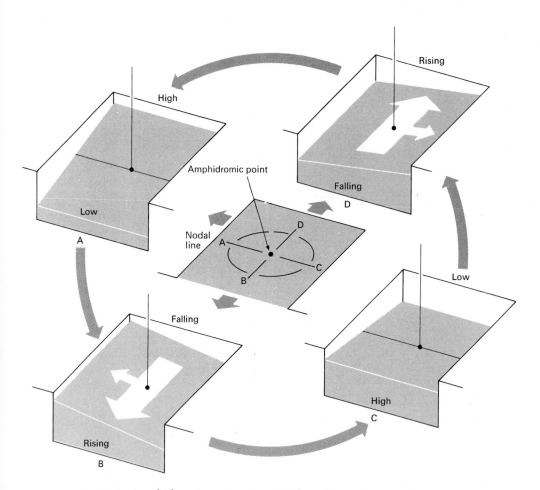

FIG. 8.6 *Amphidromic motion in a Northern Hemisphere embayment or estuary, showing various positions of the water surface. (After von Arx, 1962, p. 57.)*

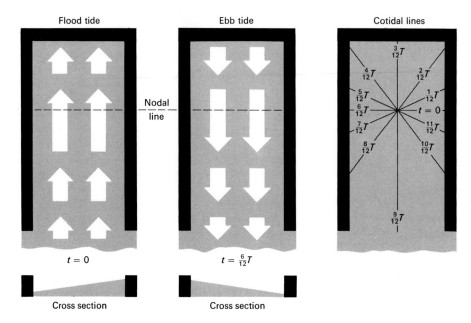

FIG. 8.7 *Map view of amphidromic motion, with times in twelfths of a tidal period. (After Groen, 1967, p. 203.)*

OBSERVATION AND PREDICTION OF TIDES

At any given place, tides occur either semidiurnally, when there are two high and two low tides every 24 hours, diurnally, when there is one of each, or they may be mixed within the lunar cycle. The semidiurnal type is most widespread, with diurnal tides generally being restricted to areas like the Gulf of Mexico at times when the sun's declination is zero (equinoxes) and the moon's nearly zero. One of the highs becomes suppressed, and the result is a diurnal tide with one high and one low tide (Fig. 8.8).

The most common semidiurnal fluctuation in sea level is such that the highs and lows are not of equal magnitude (Fig. 8.9). This is due to the earth being tilted on its axis so that, when the moon is over the tropics, the bulges are not distributed evenly with respect to latitude (Fig. 8.10). As the earth rotates about its axis, a given parallel of latitude passes under a high tide and a low high tide. A typical semidiurnal curve would reflect this (Fig. 8.8). Actually, all possible combinations of high and low tides may be present in the semidiurnal tide.

The magnitude of tides varies with time as the positions of the moon and sun change with respect to the earth. Another and more significant cause for differences in tidal range is the coastal configuration and distribution of land. At sea, where water is deep and there are no land masses, the tide wave is

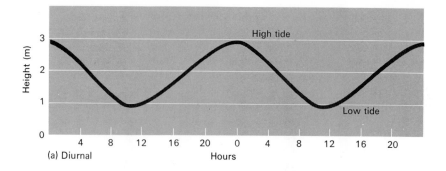

(a) Diurnal — Hours

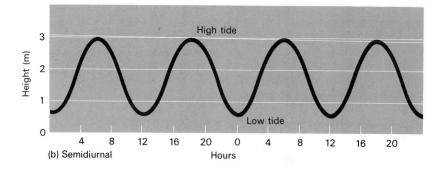

(b) Semidiurnal — Hours

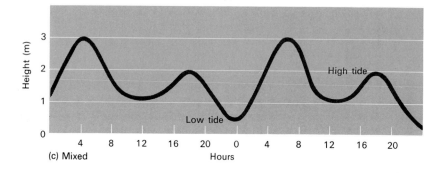

(c) Mixed — Hours

FIG. 8.8 *Generalized illustrations of (a) diurnal, (b) semidiurnal, and (c) mixed types of tides.*

much like a tsunami; it has small amplitude. As the tidal crest moves toward a land mass, however, its amplitude can increase many times. It is affected by land and shallow water much as a tsunami is, which explains why tsunamis have been called tidal waves.

Oceanic islands experience tides which have a range of only a meter or less. Coastal areas may have tides of this magnitude, but if conditions are right tidal range may be as much as 18 meters. In general, areas of irregular

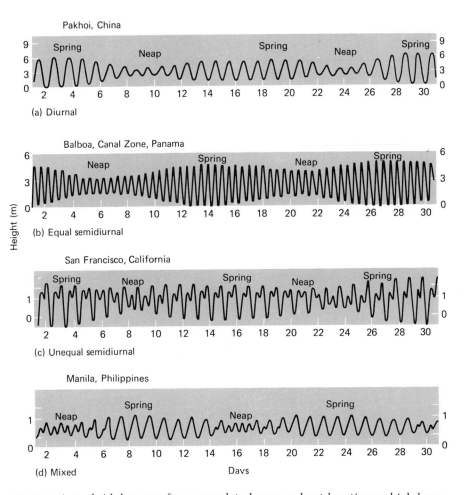

Pakhoi, China

(a) Diurnal

Balboa, Canal Zone, Panama

(b) Equal semidiurnal

San Francisco, California

(c) Unequal semidiurnal

Manila, Philippines

(d) Mixed

Days

Height (m)

FIG. 8.9 *Actual tidal curves for a complete lunar cycle at locations which have diurnal, semidiurnal, and mixed tides.*

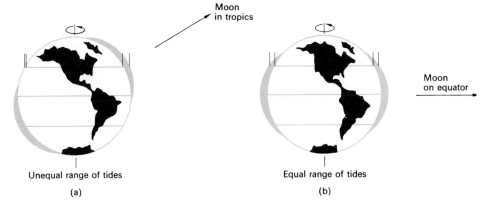

Moon in tropics

Moon on equator

Unequal range of tides

(a)

Equal range of tides

(b)

FIG. 8.10 *Tilted earth showing bulges indicating crests of tides. As the earth rotates, a point on the parallels of latitude may show (a) two different tidal ranges or (b) two equal ranges.*

(a)

(b)

FIG. 8.11 *Photographs of an estuary along the New Hampshire coast showing (a) high-tide and (b) low-tide conditions. (Photos courtesy of J. C. Kraft.)*

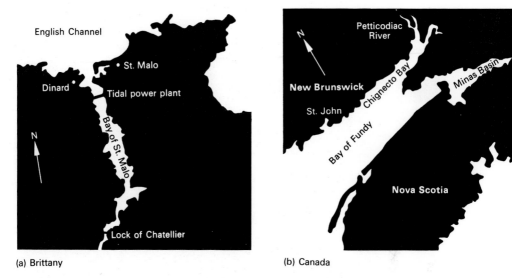

(a) Brittany   (b) Canada

FIG. 8.12 *Outline maps of two of the world's most famous tidal areas where tides may exceed 15 meters. (a) The Bay of St. Malo on the coast of Brittany and (b) the Bay of Fundy in Canada form typical bays where water piles up during high tide.*

coast line with embayments narrowing toward land (Fig. 8.11) provide the largest tidal amplitudes. Maximum tides occur on the coast of Brittany in France (Bay of St. Malo) and in the Bay of Fundy in Canada (Fig. 8.12).

Cape Cod illustrates the great differences in tidal amplitude that may exist between two geographically adjacent areas. On Nantucket Sound, the range is less than one meter, whereas across the peninsula in Massachusetts Bay, the range may be nearly five meters, as at Provincetown. Water is trapped and piles up north of the peninsula, but along the southern coast it moves southwesterly, parallel to the coast.

Tides are present in all bodies of water, even a birdbath. However, unless the basin is quite large the fluctuation of water level is not noticeable. The Mediterranean has tides of 30 centimeters, the Black Sea 10 centimeters, and Lake Michigan 3 centimeters. There are also tides in the solid earth, but they are considerably smaller than in the oceans.

Predicting and recording tides
Factors which control tides are the position of the moon and sun, the earth's rotation, the configuration of the coastline, and latitude. All these factors are

FIG. 8.13 *Tide-predicting machine of the U.S. Coast and Geodetic Survey (now ESSA). (After Dietrich, 1963, p. 407.)*

either constant (latitude, coastline*) or regularly changing (position of moon and sun, earth's rotation), and so we are able to predict the tides. A large number of observations at many different locations enables us to determine both short-term variations (within a month) as well as longer variations (throughout the year) which are caused by the arrangement of the celestial bodies involved. This enables us to predict future tides or determine past tides. These values are quite accurate (±2 centimeters) if atmospheric pressure and winds are eliminated from the calculations.

A computerlike apparatus called a tide-predicting machine will provide predictions when the correct data are fed into it. These data consist of ten different tidal species, which are values dependent on astronomical relationships. The first machine of this kind was developed by Lord Kelvin in 1872. Modern tide-predicting machines (Fig. 8.13) are much like that of Kelvin but they operate more rapidly. The machine pictured can predict tides for any place over a period of one year in less than a working day. The predictions are published each year for various locations around the world.

Before tides can be predicted, it is necessary to observe and record tidal fluctuations in sea level. There are hundreds of such tide recording stations on the coast of the United States alone. The devices are relatively simple, consisting of a float and counterweight on either end of a line which moves over a pulley. A recording device is attached to the line and makes a continuous trace on a revolving drum as the float moves up and down (Fig. 8.14). This produces a trace of tidal fluctuations in sea level and wave motion. Often it is desirable to eliminate the wave trace from the graph. This can be accomplished by sealing the pipe which houses the float and drilling a small hole near the bottom. This change prevents the float from moving with each wave but allows the water level in the tube to fluctuate with the tide.

Periodic readings can be made on graduated poles fastened to a pier piling or seawall, or on poles secured to the bottom. Although waves may be a nuisance, this type of tidal data may be all that is necessary.

In the previous chapter storm tides, or wind tides, were discussed. Frequently it is of interest to determine the actual magnitude of such phenomena. Because storm tides represent changes in water level, they will be contained within the record of a tide gauge. Lunar tidal changes have cycles of essentially 12 or 24 hours, whereas storm tides rise and fall more slowly, perhaps over at least a few days. By taking the actual water-level record and subjecting it to harmonic analysis, it is possible to extract the wind tides (storm tides) (Fig. 8.15) by equating them to the residual fluctuations after the lunar cycles, with their known period, have been extracted.

---

* Although the coastline is not fixed in time, it changes so slowly that for these purposes it can be considered fixed.

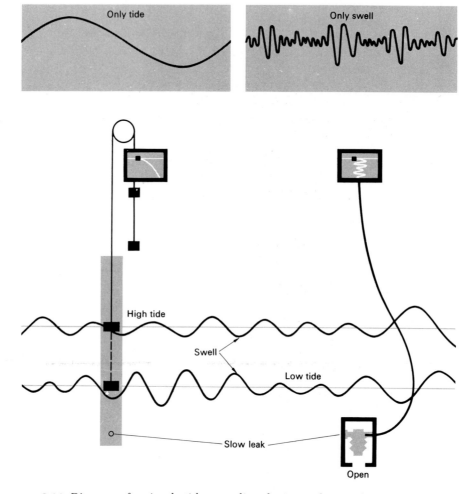

FIG. 8.14 *Diagram of a simple tide-recording device and typical traces of tidal curves. (After Bascom, 1964, p. 136.)*

TIDAL BORE

Some estuaries, or more commonly river mouths, experience a wall of water coming in with the tide. Such a phenomenon is a **tidal bore**. This wall of water is well defined and usually less than a meter in height; however tidal bores of several meters have been observed on the Chien Tang Kiang River in China and on the Amazon River.

Tidal bore is caused by the steepening of the tide wave as it enters shallow water, such as an estuary or river mouth. This steepening is due to

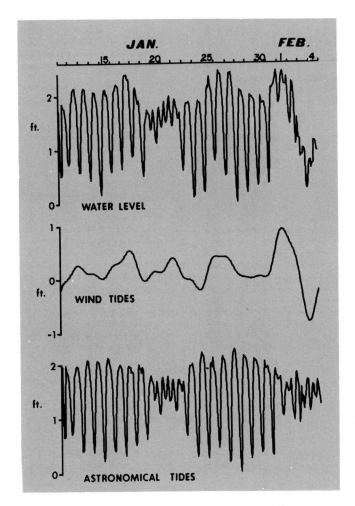

FIG. 8.15 *Curves showing recorded water-level fluctuations, wind tides or storm surge, and astronomical tides. The sum of the wind and astronomical tides yields the water-level fluctuations.*

the same factors that cause wind-driven gravity waves to steepen. In rivers this advancing tide wave is opposed by discharge from the river, which further steepens and defines the tidal advance. The speed with which the bore travels depends on river depth and the height of the bore. Deep water and a high bore yield maximum velocity. The velocity of these bores may exceed six meters/second, and they may be recognizable for hundreds of kilometers upstream of the river mouth.

SELECTED REFERENCES

Bascom, Willard, 1964, *Waves and Beaches*, Garden City, N.Y.: Doubleday, Chapter 5. General article on causes of tides with some information on tide-recording devices.

Clancy, E. P., 1968, *The Tides, Pulses of the Earth*, Garden City, N.Y.: Doubleday. Well-written general book exclusively on tides.

Defant, Albert, 1960, *Ebb and Flow—The Tides of the Earth, Air and Water*, Ann Arbor: University of Michigan Press.

Godin, Gabriel, 1972, *The Analysis of Tides*, Toronto: University of Toronto Press. A quantitative treatment of tides designed for the advanced student of mathematics and physics.

Groen, P., 1967, *The Waters of the Sea*, London: Van Nostrand, Chapter 5. The best general explanation available for tides and tidal theory.

Hill, M. N. (ed.), 1963, *The Sea—Ideas and Observations*, New York: Wiley, Vol. 1, Chapter 23. Good, highly quantitative treatment of tidal theory.

Macmillan, D. H., 1968, *Tides*, New York: American Elsevier. More mathematical than Clancy's treatment, but most students should be able to comprehend the text.

McLellan, H. J., 1965, *Elements of Physical Oceanography*, London: Pergamon, Chapter 15. Principles of tidal circulation, but lacking discussion of tidal measurement and prediction.

Neumann, Gerhard, and W. J. Pierson, Jr., 1966, *Principles of Physical Oceanography*, Englewood Cliffs: Prentice-Hall, Chapter 11. Highly mathematical treatment of tidal theory.

Russell, R. C. H., and D. H. Macmillan, 1952, *Waves and Tides*, London: Hutchinson's Scientific and Technical Publications, Part II, Chatpers 1–12. Fairly extensive discussion of tides without much mathematics.

Tricker, R. A. R., 1965, *Bores, Breakers, Waves and Wakes, An Introduction to the Study of Waves on Water*, New York: American Elsevier, Chapters 1–5. Very good treatment of tidal theory as well as wave theory. Excellent color photographs provide a nice supplement.

von Arx, W. S., 1962, *Introduction to Physical Oceanography*, Reading, Mass.: Addison-Wesley, Chapter 3. Very well-written and understandable short explanation of tides and tidal circulation.

CHEMICAL OCEANOGRAPHY

PART **III**

COMPOSITION OF SEAWATER 9

Marine chemistry is an extremely complicated portion of the ocean sciences. In considering the chemistry of seawater we must be concerned with not only the descriptive aspect, that is, the observable composition, but also the origin of the constituents, how long they remain in seawater, and where they go when they leave. Interactions of organisms, how they affect and are affected by seawater, and oceanic sediments are also important to the study of marine chemistry. Our discussion will keep the details of chemical reactions and equilibria to a minimum while stressing the principles involved. Those interested in a more detailed consideration should consult the several excellent references on marine chemistry listed at the end of this and succeeding chapters.

Chemical and geological oceanography have experienced great growth during the past two decades. The percent increase of personnel in these areas was much greater during this time than in physical and biological oceanography, the older and more classical areas of ocean study. As a consequence, the level of sophistication and the quantity of research, literature, and emphasis have undergone corresponding increase. As in many fields, however, some of the original procedures, studies, and instruments are still in wide use.

ORIGIN OF THE HYDROSPHERE

It seems likely that the origin of the atmosphere and hydrosphere of the earth bears some relationship to the origin of the lithosphere. This relationship, along with the evolution of the atmosphere and hydrosphere, must be considered from what we can see in the geological record of the earth's history and from what is known about the elements and compounds involved.

A variety of opinions have been expressed concerning the age and development of the hydrosphere. Some scientists maintain that the oceans have been fairly constant in volume since Early Precambrian times, whereas others propose a more recent development of deep oceanic areas and have suggested dates as late as the end of the Paleozoic for the age of the oceans.

Primeval hydrosphere and atmosphere

The generally accepted theory on the origin of the atmosphere and the hydrosphere is based largely on the work of W. W. Rubey. If we compare the abundance of certain gases in the earth with their cosmic abundance, it indicates that the atmosphere and hydrosphere are probably secondary in origin. Dissolved cations in seawater cannot be accounted for solely through the weathering of rocks. They probably were also provided by volcanic and hot-spring activity. These and other "excess volatiles" such as H_2O, CO_2, Cl, N, S, and so on are far too abundant in the hydrosphere to have been produced by weathering alone. Consequently, the degassing of the earth's interior is considered by most scientists to have been the best mechanism for producing these materials.

The rate at which this degassing has taken place is as yet undetermined. Rubey described two possibilities: One involves rapid degassing followed by a reaction between the primordial ocean and the earth's crust, whereas the other depends on slow degassing with the hydrosphere evolving over most of the earth's history. Most writers believe that the primordial atmosphere initially contained abundant methane (CH_4) and ammonia (NH_3), but that due to their chemical instability, carbon dioxide (CO_2) and nitrogen (N_2) rapidly evolved. Free oxygen (O_2) was probably nearly absent, and this atmosphere was in a reduced state.

Two prominent geochemists, F. T. Mackenzie and R. M. Garrels, have supported Rubey's idea of combining igneous rocks with excess volatiles to produce sedimentary rocks, the atmosphere, and the hydrosphere. In addition, their work indicates that there have been long periods during which the chemical nature of seawater changed. The initial seawater was acidic. From about 3.5 billion years to 1.5 billion years before present, there was a gradual change to the present, slightly basic condition. This is supported by numerous trends in the rock record such as the abundance and nature of iron formations, limestones, and fossil algae. Since 1.5 billion years before present, the oceans have essentially maintained their present chemical composition.

CONSTANCY OF COMPOSITION

The chemical composition of seawater has a rather wide range, even though most of the volume is chemically similar. **Salinity** is a measure of the total amount of dissolved solids in seawater by weight when the carbonate is converted to oxide, the bromide and iodide are converted to chloride, the organic matter is oxidized, and the remainder is dried to 480°C: This is the formal definition. Commonly salinity is called the dissolved salts in seawater. It is measured in parts per thousand (‰) and is usually between 33‰ and 38‰, but it may be a great deal higher or lower locally. Normal salinity is considered to be 35‰. Details of the various aspects of salinity will be considered in the next chapter. It is defined here because its definition is critical to an understanding of how seawater is analyzed.

The first refined analyses of seawater samples, and still some of the most detailed, were carried out by Dittmar in 1884. He investigated samples collected by the *Challenger* Expedition a decade before. Crude analyses of seawater were performed for one or more constituents more than a century before Dittmar's work. During the eighteenth and nineteenth centuries, it was common opinion that the seas were chemically homogeneous and that the constituents were present in the same proportions everywhere.

Dittmar's work on 77 samples from different depths and various parts of the world is usually cited as a classic effort and the one that provided the best data for the composition of seawater. His laboratory techniques were thoroughly tested on synthetic samples, and all available methods were used to eliminate errors. Dittmar's analyses for eight major constituents showed no significant differences in their relative abundance. These results provided the necessary data for the "law of relative proportions" or the "constancy of composition" of seawater. In other words, regardless of the absolute values of the major constituents, their relative abundance is constant.

There was some criticism of Dittmar's analyses because no samples were included from the high latitudes or from the Mediterranean Sea. Also, the samples were stored in glass bottles for two years, which might have had some effect on their composition. Nevertheless, Dittmar's values almost exactly match those obtained by the best analytical techniques of modern chemistry (Table 9.1).

TABLE 9.1 *Dittmar's values for major constituents of seawater (with references to a chlorinity of 19‰). (Modified from H. Barnes, 1955, "The Analysis of Sea Water," Analyst **80**, 575.)*

	Ratio to chlorinity		
Ion	Dittmar's original values	Present best values	g/kg
Cl⁻	0.99894	0.99894	18.9799
Br⁻	0.00340	0.00340	0.0646
SO_4^{2-}	0.13880	0.13940	2.6486
HCO_3^-	0.00760	0.00735	0.1397
Ca^{2+}	0.21630*	0.02106	0.4001
Mg^{2+}	0.06801	0.06695	1.2720
K^+	0.02029	0.02000	0.3800
Na^+	0.55300	0.55560	10.5561

* Dittmar's value is for calcium plus strontium as oxide; best value today on that basis is 0.02165.

Because of these consistent ratios in seawater, it is possible to analyze major constituents of seawater by determining only one constituent. Salinity, or total dissolved solids, can be determined in the same manner. The procedure most often used is to determine the chlorinity and solve the following equation for salinity:

$$S\text{‰} = 1.80655 \text{ Cl‰}.$$

To alleviate problems of comparing analyses from different laboratories in various parts of the world, a standard seawater (eau de mer normale) was established by the Oceanographic Congress of 1902. Samples of normal seawater are obtainable from the Hydrographic Laboratory in Copenhagen so that all people can use the same standard for comparison.

ELEMENTS IN SEAWATER
Sodium, chlorine, and a few other dissolved solids such as calcium, magnesium, and potassium are the most obvious constituents of salinity, with Na^+ and Cl^- accounting for more than 85 percent of the total. Analyses have shown that most naturally occurring elements are present in seawater (see Table 9.2). The number of elements which has been positively identified in the ocean has continued to rise as our techniques have become more refined. To date more than 70 elements have been discovered in the oceans, and many researchers believe that in time all the naturally occurring elements will be found.

The residence time of an element is the average time an element is in seawater before being removed by chemical precipitation and is equal to the quantity of the substance in the ocean divided by its rate of inflow into the ocean. It is based on the assumption that oceans are in a steady state such that the amount of a particular element brought into the sea is balanced by precipitation of a like amount. This is not exactly the case because the total concentration of many elements is increasing very slowly. Residence time also assumes total mixing in the oceans during a short period of time. A considerable range of residence exists—from aluminum (Al) at 100 years to sodium (Na) at 2.6×10^8 years. Those elements with long residence times are relatively inert in the marine environment.

METHODS OF SEAWATER ANALYSIS
With the exception of sodium and chlorine, virtually all the constituents of seawater are present in small concentrations. Their scarcity demands a great deal of sensitivity in the techniques used to determine their concentration. In addition, it is necessary that all scientists standardize analytical methods for determining each of the elements, so that results from various laboratories and sample locations may be compared in a meaningful way.

The equation based on the constancy of composition which is used to determine salinity has but one variable: chlorinity. Determination of most major dissolved constituents of seawater is made by first establishing the chlorinity. It is therefore necessary to be extremely accurate in chlorinity determinations. Precision is possible for the major constituents because they occur in fixed ratios; however, trace-element concentration may vary locally and less precise results are acceptable.

Sampling and preparation

Like most data gathering, the collection of seawater samples is critical in obtaining meaningful chemical data. It is important that the sampling device fill rapidly and seal tightly, so that the sample will be truly representative of the water at the sample station and not an accumulation over a wide area. The Nansen bottle (Chapter 3) is commonly used for routine work, but it does not seal completely. A somewhat better apparatus is the Fjarlie sampler or the Niskin bottle, which is attached to a line at both ends with spring-closing hinged ends. A messenger closes the bottle with a good seal.

Corrosion and deterioration of metal-lined samplers may cause changes in water composition in an hour or so. Copper, zinc, lead, and iron in the metal linings often contaminated seawater samples. Today plastics have solved this problem for both collection and storage of seawater.

Another extremely important source of contamination in samples is suspended particulate matter and organisms. The suspended matter may be particles of mineral matter of a micron or so in diameter or it may be organic debris. In either case, ions can be adsorbed or liberated, causing a change in seawater composition. Tiny living organisms present similar problems as they take in nutrients and expel waste products. In order to remove these materials, it is necessary to filter the sample. The most commonly used filter is made of cellulose; it is available with pore sizes ranging from $10m\mu$ to 5.0μ.

If samples are to be stored before analysis, even for a fairly short time, care should be exercised to prevent any reactions with the containers. Recall that one of the objections to Dittmar's analyses was the long time that the samples were stored in glass bottles. Dense polyethylene or hard glass containers with airtight stoppers should be employed. Some low-density plastics allow moisture and gases to pass through. Regardless of what containers are used, there will be changes in composition through time, particularly in the gaseous components. If samples are to be analyzed for certain constituents rather than undergo general analysis, methods are available to fix those constituents. For instance, a special freezing and storage process will stabilize nitrate, phosphate, and silicate and stop many chemical processes that might be taking place.

TABLE 9.2 *Chemical composition of seawater. (Modified from E. D. Goldberg, 1963, "The Oceans as a Chemical System," The Sea, M. N. Hill (ed.), vol. 2, pp. 4–5.)*

Element	Concentration in mg/l (ppm)	Residence time in years
Hydrogen (H)	108,000	
Helium (He)	0.000007	
Lithium (Li)	0.18	2.0×10^7
Beryllium (Be)	0.0000006	1.5×10^2
Boron (B)	4.5	
Carbon (C)	28	
Nitrogen (N)	0.5	
Oxygen (O)	857,000	
Fluorine (F)	1.4	
Neon (Ne)	0.0001	
Sodium (Na)	10,800	2.6×10^8
Magnesium (Mg)	1350	4.5×10^7
Aluminum (Al)	0.01	1.0×10^2
Silicon (Si)	3	8.0×10^3
Phosphorus (P)	0.07	
Sulfur (S)	885	
Chlorine (Cl)	19,400	
Argon (Ar)	0.45	
Potassium (K)	416	1.1×10^7
Calcium (Ca)	422	8.0×10^6
Scandium (Sc)	0.00015	5.6×10^3
Titanium (Ti)	0.001	1.6×10^2
Vanadium (V)	0.002	1.0×10^4
Chromium (Cr)	0.0005	3.5×10^2
Manganese (Mn)	0.002	1.4×10^3
Iron (Fe)	0.003	1.4×10^2
Cobalt (Co)	0.0008	1.8×10^4
Nickel (Ni)	0.002	1.8×10^4
Copper (Cu)	0.003	5.0×10^4
Zinc (Zn)	0.005	1.8×10^5
Gallium (Ga)	0.00003	1.4×10^3
Germanium (Ge)	0.00006	7.0×10^3
Arsenic (As)	0.003	
Selenium (Se)	0.004	
Bromine (Br)	68	
Krypton (Kr)	0.0003	

TABLE 9.2 (*continued*)

Element	Concentration in mg/l (ppm)	Residence time in years
Rubidium (Rb)	0.12	2.7×10^5
Strontium (Sr)	8	1.9×10^7
Yttrium (Y)	0.00015	7.5×10^3
Zirconium (Zr)	0.000025	
Niobium (Nb)	0.00001	3.0×10^2
Molybdenum (Mo)	0.01	5.0×10^5
Technetium-Palladium (Tc-Pd)		
Silver (Ag)	0.0001	2.1×10^6
Cadmium (Cd)	0.00011	5.0×10^5
Indium (In)	0.00001	
Tin (Sn)	0.00001	5.0×10^5
Antimony (Sb)	0.005	3.5×10^5
Tellurium (Te)		
Iodine (I)	0.06	
Xenon (Xe)	0.0001	
Cesium (Cs)	0.0005	4.0×10^4
Barium (Ba)	0.03	8.4×10^4
Lanthanum (La)	0.0003	1.1×10^4
Cerium (Ce)	0.0004	6.1×10^3
Praeseodymium-Tantalum (Pr-Ta)		
Tungsten (W)	0.0001	1.0×10^3
Rhenium-Platinum (Re-Pt)		
Gold (Au)	0.000004	5.6×10^5
Mercury (Hg)	0.00005	4.2×10^4
Thallium (Tl)	<0.00001	
Lead (Pb)	0.00003	2.0×10^3
Bismuth (Bi)	0.00002	4.5×10^5
Polonium-Astatine (Po-At)		
Radon (Rn)	0.6×10^{-15}	
Francium (Fr)		
Radium (Ra)	1.0×10^{-10}	
Actinium (Ac)		
Thorium (Th)	0.00005	3.5×10^2
Protactinium (Pa)	2.0×10^{-9}	
Uranium (U)	0.003	5.0×10^5

MAJOR CONSTITUENTS OF SEAWATER

A small number of constituents comprise about 99.7 percent of the dissolved elements in seawater. These are commonly called the major constituents (Table 9.3) and are treated separately from the trace or minor constituents. The major constituents comply with the principle of constancy of composition and thus can be determined by their chlorinity ratios. In addition, special means have been developed to determine many of the major constituents; such techniques are included in the following discussions of the elements and ions.

TABLE 9.3 *Major constituents of seawater salinity. (After Sverdrup, et al., 1942; and J. P. Riley and A. Chester, 1971, Introduction to Marine Chemistry, New York: Academic Press, p. 81.)*

Ion	Chlorinity ratio g/unit Cl	g/kg of H_2O of 19‰ chlorinity
Chloride (Cl^-)	0.99840	19.3530
Sodium (Na^+)	0.55610	10.7620
Sulfate (SO_4^{2-})	0.13940	2.7090
Magnesium (Mg^{2+})	0.06680	1.2930
Calcium (Ca^{2+})	0.02125	0.4110
Potassium (K^+)	0.02060	0.3990
Bicarbonate (HCO_3^-)	0.00735	0.1420
Bromide (Br^-)	0.00348	0.0673
Boric acid (H_3BO_3)	0.00023	0.0044
Strontium (Sr^{2+})	0.00041	0.0079
Fluoride (F^-)	0.00007	0.0013

Chloride (Cl^-)

Chloride makes up more than half the total dissolved constituents in seawater. Chlorinity has been the basis for salinity determination until recently, when it was largely replaced by electrical conductivity. It is defined as the amount of halogen (chlorine, bromine, and iodine) in one kilogram of seawater. It is usually determined by titration with silver nitrate, with potassium chromate as the indicator. The silver nitrate is added until all halogens are precipitated and a red color, due to formation of silver chromate, appears. This marks the end of the titration. Formulas for the reactions are below:

$$Cl^- + Ag^+ \rightarrow AgCl, \qquad Br^- + Ag^+ \rightarrow AgBr,$$

$$I^- + Ag^+ \rightarrow AgI, \qquad CrO_4^{2-} + 2Ag^+ \rightleftharpoons Ag_2CrO_4.$$

Sodium (Na$^+$)

The most abundant cation in the sea is derived largely from weathering of sodium-rich feldspar and to some extent clay minerals, as well as other, less significant sources.

The first quantitative attempts to determine the age of the oceans involved estimating the total amount of sodium present in seawater and comparing it to the annual amount of sodium being carried to sea by runoff. An age of nearly 90 million years was calculated, but this figure was not accepted by most scientists. Residence time (Table 9.2) was not considered, and annual discharge data were not accurate.

Determinations of sodium are rarely made directly, because analysis is difficult due to the high activity of alkali metals. When analysis for sodium is undertaken, however, gravimetric methods are used.

Sulfate (SO$_4^{2-}$)

Most sulfur in seawater occurs as sulfate (SO$_4^{2-}$). It forms the third most abundant constituent dissolved in seawater, but it is considerably less than chloride and sodium. Anomalously high SO$_4^{2-}$ concentrations may be present at river mouths. Lower than normal amounts are found in seawater at high latitudes, because SO$_4^{2-}$ is preferentially incorporated into ice. Under certain chemical conditions, hydrogen sulfide (H$_2$S) is present in seawater. Such conditions prevail in oxygen-starved areas where there is organic matter and little or no circulation of water, such as in fjords or the Black Sea.

Sulfate determinations are generally made by using a gravimetric technique on barium sulfate precipitate. This method has been used since the early nineteenth century and is fairly accurate.

Magnesium (Mg^{2+})

The second most abundant cation in seawater, magnesium, is currently being studied carefully by carbonate geochemists. Magnesium and calcium often occur together in minerals (dolomite, hi-Mg calcite), and magnesium may substitute for calcium.

The ratio of magnesium to chlorinity is uniform, making its determination an easy task. Commonly, however, gravimetric or volumetric techniques are used to make direct magnesium analyses. In the former, precipitation of magnesium ammonium phosphate is used, and in the latter rather complex titrations are employed.

Calcium (Ca^{2+})

Calcium has been studied more than any other nongaseous constituent of seawater because of its great importance in tests (or shells) and skeletons of organisms, and as a mineral precipitate. Virtually all precipitated calcium is in the form of calcium carbonate or calcium sulfate, and it may result either from organic activity or from inorganic precipitation.

Locally there may be variation in the ratio of calcium to chlorinity, especially near river mouths, where calcium is more plentiful. The photic zone of the ocean contains low concentrations of calcium due to its removal by plants and animals. A detailed treatment of calcium precipitation is presented in Chapter 11.

Determination of calcium concentration is usually by gravimetric analysis of calcium oxolate. It can also be determined by titration methods.

Potassium (K$^+$)

Potassium is one of the most active cations in seawater, and this fact causes some variations in its concentration, although generally the ratio of potassium to chlorinity is constant. Potassium ions in solution may be absorbed by clays and other detrital particles. Upon entering seawater there may be further loss of potassium from solution as the mineral glauconite (clay mineral rich in potassium and iron) is formed. Certain plants, such as kelp, concentrate potassium and may contain up to 15 percent by dry weight.

Direct determinations of potassium are difficult because of its low concentration and high activity. Potassium precipitates can be measured gravimetrically, but most of the time the chlorinity ratio is used.

Bicarbonate (HCO$_3^-$)

See Chapter 11.

Bromide (Br$^-$)

Apparently bromine occurs only as the bromide ion in seawater; as yet, bromates have not been found. Although one of the lesser major constituents, the amount of bromine in seawater seems to be more than adequate for our needs. It is commercially extracted from seawater by treatment with chlorine and aniline.

Concentration of bromine in seawater is not easily determined because of the great amount of chlorine present. Originally, determinations were made by precipitating with silver nitrate and then liberating bromine by chlorine treatment and calculating the weight loss of the precipitate. A more successful technique is to oxidize bromide, treat it with potassium iodide, and then employ titration of the liberated iodine. This can be done with much greater accuracy than the former method.

Boric acid (H$_3$BO$_3$)

Although boric acid is relatively weak, the normal hydrogen ion concentrations in seawater do not dissociate it. The concentration of boron has a constant relationship to chlorinity.

Certain organisms, particularly plants, apparently concentrate boron in their tissues.

A relatively simple titration with sodium hydroxide can be used to determine boric acid concentrations, although a variety of more precise and more complicated procedures are also employed.

Strontium (Sr^{2+})
Strontium in seawater is currently being studied extensively because of its inclusion with calcium carbonate in precipitation by plants and animals. Strontium concentrations which are orders of magnitude higher than in seawater are found in some red and green algae. Calcium carbonate shells and tests contain strontium, and some show greater concentrations in Arctic areas due to apparent replacement of calcium carbonate by the strontium.

Dittmar's original analyses of calcium included strontium because they precipitate together. Modern techniques for strontium analysis are based on flame photometric techniques.

Fluoride (F^-)
Early analyses of fluoride content in seawater showed a rather wide variation in concentrations; however, better techniques indicate a constant concentration. Some fluorides have been found in boiler scale from ships and in mollusk shells.

Most methods for fluorine determination are based on colorimetric measurements. Fluorine breaks down the color of some metallic complexes with standard dyes like alizarin red. The decrease in color intensity is a measure of the fluoride content, and it can be determined quantitatively with a colorimeter.

MINOR CONSTITUENTS OF SEAWATER
The minor constituents may show local variations and hence cannot be determined by the constancy-of-composition concept, as the major constituents can. Most of the elements which are present in low concentrations are not of major importance; however, some are important to living organisms or as possible economic sources of precious elements. The following paragraphs describe the mode of occurrence of several important elements found in low concentrations in the sea.

Aluminum (Al)
Small amounts of aluminum are present in the sea, with coastal waters generally being relatively high in aluminum. This is because aluminum in the sea is derived from the colloidal clay particles that run off from the land. Some seasonal fluctuations in aluminum concentration can be related to changes in river discharge. In the open ocean, aluminum is present in quantities below the detection limits of most analytical methods.

Arsenic (As)

Although the chemical state in which arsenic occurs is not known, it is present in seawater. Many organisms concentrate arsenic, perhaps in place of phosphorus. Mollusks and arthropods in particular have a high concentration of arsenic, with some lobsters having as much as 100 parts per million.

In analysis for arsenic, special care must be taken with glass containers to ensure freedom from contamination. Also, alkaline samples will dissolve some arsenic from the glass if it is present.

Barium (Ba)

Published values of barium concentrations in seawater show considerable range. It has the lowest seawater concentration of the alkaline earth elements. Many marine organisms and a variety of marine sediments contain some barium. Its most unusual occurrence in the sea is in the form of concretions and modules which are high in barium sulfate. These were found near Ceylon and, as yet, their formation is not understood.

Copper (Cu)

Concentrations of copper show some variation in coastal areas, with high values in brackish waters near river mouths. The most studied aspect of copper in seawater is its uptake by organisms. Both plants and animals, particularly invertebrates, contain copper. Oysters have a high concentration of copper, and oyster blood cells have yielded as much as 0.5 percent copper. Higher concentrations, however, are toxic to marine life, and for this reason some antifouling paints for boats and ships contain abundant copper.

Gold (Au)

Several attempts have been made and theories have been proposed for extracting gold from seawater. Although the concentration is quite low, there is more than a million dollars' worth of gold in each cubic kilometer of seawater. Gold concentrations found by early investigators were much higher than the presently accepted figure (Table 9.2). Probably it occurs in a variety of colloidal and suspended forms in addition to the dissolved ions. This particulate gold could yield anomalously high values in specific samples.

Iodine (I)

The original discovery of the element iodine was in the sea, where it was extracted from kelp, a marine alga. Iodine has received more study than most elements in seawater because of its importance in human physiology. Marine plants and animals which concentrate iodine serve as our primary source of the element. Commercial sources are algae, but we ingest iodine in fish and other seafoods. Oysters, clams, and lobsters have the highest concentrations, but all edible fish, mollusks, and arthropods contain abundant iodine.

Iron (Fe)

Iron concentrations range from less than 1 to nearly 50 parts per million. Weathering and erosion of mafic minerals is the source of most of the iron in the sea. Much of it is carried to the sea with colloidal clays, and it eventually becomes an important constituent of marine sediments. Manganese nodules (Chapter 11) contain about 30 percent iron. Iron may be a limiting factor in plant growth, and it is found highly concentrated in plankton. A fair amount of the iron present in seawater is not dissolved and may be retained on fine filter papers.

The distribution of iron is not uniform. Nearshore waters are relatively high in iron, reflecting seasonal changes in runoff. Several systematic analyses have shown that surface waters contain a minimum of iron, with a noticeable increase with depth.

Manganese (Mn)

The interest in manganese in the marine environment has increased greatly in past years because of extensive accumulations of manganese nodules on the sea floor (Chapter 21).

Some organisms, particularly animals, concentrate manganese. It is present in at least trace quantities in most marine sediments.

Nitrogen (N)

A wide variety of nitrogen compounds are present in seawater, including nitrates, nitrites, ammonia, and organic nitrogen. Their concentration shows some local variation because of ulitization by organisms and also because of production by certain bacteria. The water temperature at time of contact with the atmosphere will also cause local variations.

Oxygen (O)

See Chapters 11 and 13.

Phosphorus (P)

Although present in small concentrations, phosphorous is one of the most important of the sea's constituents because of its presence in living organisms. Many plankton concentrate phosphorous which returns to seawater after the death and breakdown of the living tissues.

Phosphorus determinations are most often made by colorimetric methods.

Silicon (Si)

Large amounts of silicon as silica (SiO_2) are taken up from seawater by various organisms, particularly diatoms and radiolarians. Tests of both are composed primarily of SiO_2. There is considerable seasonal variation in silicon concentrations in the sea. Diatom blooms take up considerable silicon, and the concentration in seawater is thereby lowered.

Coastal areas are relatively rich in silicon due to its transport of runoff. Many scientists believe that silicon precipitates on the sea floor in the form of chert but no conclusive evidence of "silica jels" or inorganic precipitates has yet been found.

Both gravimetric and colorimetric methods are used for silicon determinations. Care should be utilized in storing samples to be tested for silicon; wax-coated or plastic containers should be used.

SELECTED REFERENCES

Goldberg, E. D., 1963, "The Oceans as a Chemical System," *The Sea*, M. N. Hill (ed.), New York: Wiley, Vol. 2, Chapter 1. Good brief discussion of seawater chemistry by one of the world's foremost geochemists.

Harvey, H. W., 1960, *The Chemistry and Fertility of the Sea*, Cambridge: Cambridge University Press, Chapters 1, 8, and 9. A somewhat dated but good treatment of marine chemistry with emphasis on character and distribution of various elements.

Hood, D. W., 1965, "Chemistry and the Oceans," *Chemical and Engineering News Special Report*, July 1, 1964. A well-written and illustrated article on oceanography with emphasis on marine chemistry. Aimed at the layman or nonoceanography science student.

Horne, R. A., 1969, *Marine Chemistry, The Structure of Water and the Chemistry of the Hydrosphere*, New York: Wiley. Comprehensive coverage of the subject but the treatment is rigorous and is not recommended for those without a chemistry major or the equivalent.

Martin, D. F., 1968 and 1970, *Marine Chemistry*, New York: Marcel Dekker, Vols. 1 and 2. Comprehensive treatment of marine chemistry that can be assimilated by anyone having a year or two of college chemistry. Volume 1 covers tools and techniques, while Volume 2 deals more with the "meat" of marine chemistry.

Riley, J. P., and R. Chester, 1971, *Introduction to Marine Chemistry*, New York: Academic Press. Excellent text on marine chemistry with considerable treatment of relationships of chemistry to other marine disciplines.

Riley, J. P., and G. Skirrow, 1975, *Chemical Oceanography* (2nd edition), Vols. 1 and 2, New York: Academic Press. Classic reference on marine chemistry. Comprehensive and in-depth treatment of the subject.

SALINITY AND pH 10

The chemical environment of the sea is just as important as the physical environment. In the marine world, the distribution of all living organisms and inorganic precipitates is largely controlled by the chemical characteristics of the water. With increasing attention and speculation focused on the sea as a possible source of new foods and economic mineral deposits, it is necessary to have a basic understanding of some of these important chemical factors. Two such factors are salinity, which was briefly introduced in the previous chapter and pH (acidity-basicity). The overall chemical makeup of seawater is little affected by variation in these two factors; however, they are significant in controlling the distribution of organisms and the precipitation of various compounds.

SALINITY

The single most noticeable distinguishing characteristic of seawater is its salty taste. The taste is derived largely from only two of the more than 70 elements present: sodium and chlorine. Actually most of the elements and their ionic forms which are present in the sea are not salts. Therefore, the term "salinity" seems somewhat inappropriate if we consider the variety of composition of seawater, but it is meaningful in describing over 99 percent of all the dissolved materials in the sea.

A restatement of the complete definition of salinity is appropriate here. The 1902 International Oceanographic Congress defined salinity as the total dissolved solids in one kilogram of seawater when all carbonate is converted to oxide, bromine and iodine are replaced by chlorine, and organic matter is oxidized to 480°C. Constancy of composition permits us to define the standard relationship between salinity and chlorinity:

$$S\text{\textperthousand} = 0.03 + 1.805 \ Cl\text{\textperthousand}.$$

This equality was derived from empirical data which showed the relationship between density, salinity, and chlorinity. These data have since become known as Knudsen's tables and have served as standards of comparison for more than half a century. In recent years the above relationship has been modified so that S = 1.80655 Cl.

Several assumptions must be made when salinity determinations are calculated using this relationship. First, it must be assumed that constancy of composition is an accurate and reliable relationship. In general this is true; however, even with the major constituents there may be some range in abundance, particularly on the continental shelves where influence of runoff is greatest. The above equation is also based on the assumption that seawaters of the same salinity have their constituents distributed in the same ratios. These assumptions are not completely true, especially for very detailed analyses. Nevertheless, in the vast majority of salinity determinations, the accuracy provided by the equation is sufficient.

Chlorinity determination

A change in the definition of chlorinity is about the only alteration which has taken place in the standard titration procedure used to determine salinity. The previous definition did not allow for variations in the atomic weight of chlorine. The revised definition is expressed in grams/kilogram and is 0.3285233 times the weight of silver equivalent to all the halogens present.

The apparatus used in the titration procedure for chlorinity is relatively simple (Fig. 10.1). The special Knudsen buret and pipette must be accurately calibrated. Some scientists have expressed dissatisfaction with chlorinity determinations which they believe were made with inaccurately constructed burets.

The titration procedure for chlorinity should be followed to the letter. Great care should be exercised to ensure clean equipment and to avoid contamination. A silver nitrate solution is prepared and standardized against a sample of known chlorinity. This is the standard seawater, obtainable from the Hydrographic Laboratory in Copenhagen. An indicator solution of potassium chromate is also prepared, although in recent years a fluorescein indicator has been widely used instead.

The actual titration involves addition of the indicator solution to the seawater sample, followed by slow addition of silver nitrate solution during constant stirring. The silver combines with the halogens (Cl^-, Br^-, I^-) until all are taken up. Then excess silver combines with chromate from the indicator solution to give a red precipitate of silver chromate. The titration is completed when this red precipitate forms, and the exact amount of silver nitrate solution used is in direct relation to the abundance of halogens in the sample. The appropriate calculations are then made to determine salinity. The complete procedure, including the proper preparation of solutions, is detailed in marine chemistry references at the end of this chapter.

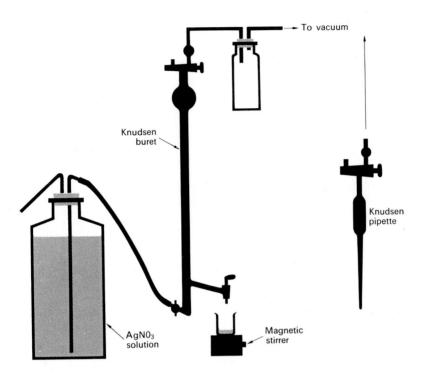

FIG. 10.1 *Knudsen buret and pipette assembly for chlorinity determinations.*

Procedures in titration analyses lend themselves to the possibility of error, but marine chemists have taken steps to reduce these errors to a minimum. The primary reasons for their extreme care are to make salinity determinations more precise and to increase reproducibility. Stirring is important in order to make sure all the halogens have an opportunity to combine with the silver. Magnetic stirrers provide a means of constant, uniform mixing. Any error in pipette reproducibility can be eliminated by carefully weighing the samples used.

Perhaps the biggest problem is determining the endpoint of the titration by the color indicator; but the endpoint can be standardized by using electrometric or photoelectric endpoint indicators which are more sensitive than the unaided eye.

Other salinity determinations

The titration method for measuring chlorinity has been replaced by measuring the electrical conductivity of seawater. Conductivity is directly related to salinity and can therefore be used for salinity determinations. Beginning about 40 years ago, **salinometers** (Fig 10.2) were devised in which the researcher could compare the conductivity of an unknown sample to standard seawater and thus determine the salinity. There are now many different types of salinometers, but they are all similar. They provide salinity determinations accurate to 0.01‰ with much greater reproducibility than titration. Salinometers also facilitate shipboard determinations, which are difficult by titration because of the ship's motion.

Another property of seawater that is a function of salinity is the refractive index: the ratio of the speed of light in a vacuum to, in this case, its speed in seawater (Fig. 10.3). A precision refractometer is used to compare the refractive index of the unknown seawater with a standard sample. Care should be taken to filter the water to ensure accuracy. If suspended mineral matter or organic debris is present, it affects the refractive index, even though it has no relation to salinity.

The relationship between salinity and density was mentioned in Chapter 4. This relationship also can be used to determine salinity, but temperature must be carefully determined and corrected for because of its marked effect on density. From Knudsen's table, the relationship between chlorinity and density is

$$\sigma_t = -0.069 + 1.4708 \ Cl - 0.001579 \ Cl^2 + 0.0000398 \ Cl^3.$$

A graphic illustration of this relationship is shown in Fig. 10.4.

Salinity of seawater can also be determined by the velocity of sound in water and the freezing point of water, although these methods are not very accurate. At constant temperature, the velocity of sound in seawater increases 1.3 meters/second per ‰ S. In the open ocean, therefore, a range of only a few meters/second would exist, making salinity determinations difficult. The freezing point of seawater is depressed as salinity increases. By precisely determining the temperature of freezing, it is possible to calculate salinity (Fig. 10.5). This is not a commonly used or accurate means of calculating salinity because of the precision required in measuring the freezing temperature.

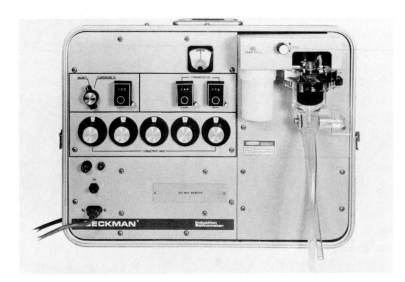

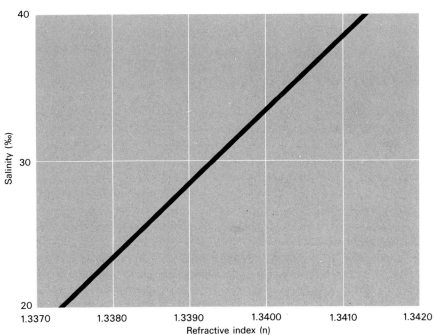

FIG. 10.3 *Relationship between refractive index and salinity of seawater. The actual position of the line depicting this relationship varies with water temperature but has the same slope.*

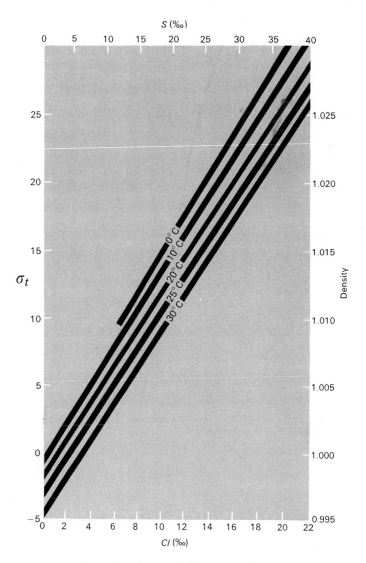

FIG. 10.4 *Relationship between density and salinity of seawater. (Modified from* H. W. Harvey, 1957, The Chemistry and Fertility of Sea Waters (2nd edition), London: Cambridge University Press, p. 128.)

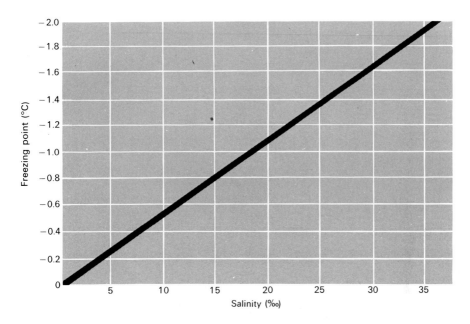

FIG. **10.5** *Relationship between salinity and freezing point of seawater. (After Harvey, 1957, p. 129.)*

Salinity distribution

Oceanic water has a generally uniform salinity because of nearly complete mixing by circulation. There are, however, some significant and predictable deviations from the normal salinity. These variations are both geographic and vertical in the ocean basins.

Most seawater has a salinity of between 33‰ and 37‰, with the mean about 34.5‰. Most areas of significant deviation from the above narrow range fall into two categories: (1) areas adjacent to land masses were there is dilution from runoff, and (2) areas of restricted circulation where evaporation causes increased concentration of dissolved constituents.

The term **brackish** is used to designate seawater of abnormally low salinity, generally less than 17‰, such as might be found near river mouths. Areas of abnormally high salinity (greater than 47‰) are called **hypersaline.** The Baltic Sea is an example of a large, partially restricted basin containing brackish water; less than 10‰ salinity at most locations. It receives a great deal of runoff and does not have open circulation with marine waters— thus the significant dilution. On the other hand, the Red Sea and the Persian Gulf are high-salinity basins, partly due to their restricted circulation with open

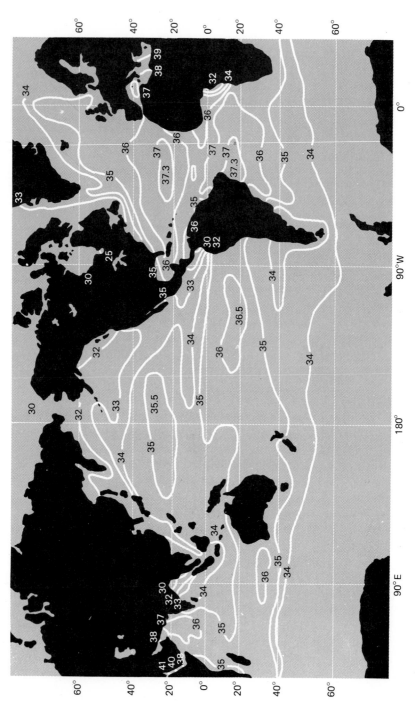

FIG. 10.6 Surface salinity of the world for the month of August. (After Pickard, 1964, p. 43.)

marine waters, but also due to lack of runoff. Evaporation exceeds precipitation, causing an increasing concentration of dissolved constituents.

Surface salinity. A study of world salinity distribution (Fig. 10.6) permits us to make some generalizations about surface salinities. The highest salinity in most ocean basins is near the center, because of a general lack of surface circulation near the center. For instance, the Sargasso Sea in the central North Atlantic is an area of relatively calm surface waters with a major circulation gyre moving around it. Similar areas are present in the South Atlantic, Indian, North Pacific, and South Pacific basins. The high-salinity area in the South Atlantic abuts against the eastern part of South America.

The ocean with highest salinity is the North Atlantic (35.5‰), and the North Pacific is the lowest (34.2‰). This is just the reverse of what one might expect, because there is a considerable amount of runoff concentrated in the relatively small North Atlantic, and yet the vast and relatively low-runoff North Pacific is lower in salinity. As yet, this is somewhat of a mystery. If oceanic waters of the entire Northern Hemisphere are considered, however, they are less saline than those of the Southern Hemisphere. The reason is that there is much more land providing runoff in the Northern Hemisphere, and the melting ice from the Arctic Sea and the Greenland ice cap overbalances the runoff and the melting of the Antarctic ice cap in the south.

There is a fairly regular relationship between surface salinity and latitude. Highest salinity values occur in subtropical regions (Fig. 10.7), with a general decrease near the equator and in higher latitudes. This pattern is

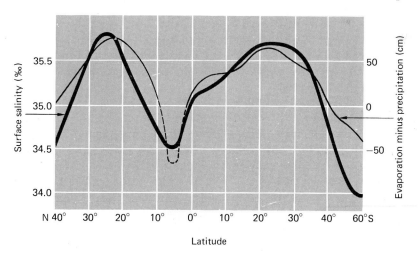

FIG. 10.7 *Distribution of surface salinity as compared with evaporation minus precipitation. Note the striking parallelism of the curves. (After Sverdrup, et al., 1942, p. 124. Copyright renewed 1970.)*

largely a result of more atmospheric precipitation in low and high latitudes, and less precipitation in the subtropics. There is also a minor decrease in salinity near polar regions, because of dilution by melt water from ice (Fig. 6.20).

Salinity changes with depth. Below a zone of a few hundred meters, salinity is quite uniform in the open ocean, within a range of less than 1‰. The upper water layers have a salinity range from about 33‰ to 37‰ in the open ocean, but below the upper layers, salinity falls between 34‰ and 35‰. The latitude and its climate are dominant in determining salinity of the surface layer. Because of the great influence of temperature on density, it is possible for cold, low-salinity water to be carried to great depths. This is the case adjacent to the Antarctic ice sheet (Fig. 6.20).

Effects of ice formation on salinity

Seawater freezes at negative Celsius temperatures; normal seawater freezes at $-1.9°C$. When water converts from a liquid to a solid at this temperature, the dissolved materials are not incorporated in the ice, which is simply H_2O. Such exclusion of impurities in formation of a mineral, ice being an example, is typical.

When ice forms, it causes a local and ephemeral increase in salinity because the dissolved constituents are concentrated in the remaining water. During formation of sea ice, some of this higher-salinity seawater is physically trapped in cavities between ice crystals.

ACIDITY AND BASICITY OF OCEAN WATERS

The terms acid and base are introduced as fundamental in beginning chemistry, but rarely is their importance in natural waters stressed. In the ocean these conditions may have considerable effect on organic growth and activity, and on chemical reactions.

A simple yet thorough definition of each is as follows. An **acid** is a proton donor and a **base** is a proton acceptor. Our common experience with an acid indicates a sour taste and the ability to dissolve many other substances. All acids contain hydrogen, and when combined with water they tend to dissociate. Whether an acid is termed strong or weak is determined by the degree of dissociation of the hydrogen ion. Strong acids like hydrochloric (HCl), nitric (HNO_3), and sulfuric (H_2SO_4) highly dissociate, while weak acids like carbonic (H_2CO_3) and acetic ($HC_2H_3O_2$) do not. When hydrochloric acid dissociates, it does so in the following manner:

$$HCl + H_2O \rightarrow H_3O^+ + Cl^-.$$

A base is a substance that has a soapy feel, tastes bitter, and can neutralize an acid. Chemically, all bases contain the hydroxide ion and yield

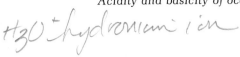

this on dissociation. Some common strong bases are ammonia (NH₄OH), potassium hydroxide (KOH), and sodium hydroxide (NaOH); but calcium hydroxide (Ca(OH)₂) is weak. A base dissociates similarly to an acid:

$$Ca(OH)_2 \rightarrow Ca^{2+} + 2OH^-.$$

The term **alkali** is commonly used synonymously with "base" for a solution containing abundant OH⁻.

The above discussion implies that H⁺ means acidic and OH⁻ means basic. In order to standardize terminology regarding these ions, a shorthand notation, pH, is used to express the acidity-basicity relationship. It is equal to $\log^1/[H^+]$, or $-\log$ of the hydrogen ion activity. The negative log of the hydrogen ion activity is very close to the hydrogen ion concentration and is used to express this relationship. The pH range is from 0 to 14; a pH of 7.0 (0.0000001 mole/liter) is neutral. Values less than 7.0 are acidic and those more than 7.0 are basic. An example will illustrate this relationship:

$$Na^+ + OH^- + H^+ + Cl^- = H_2O + Na^+ + Cl^-.$$

In the above reaction, a strong base (NaOH) is in solution with a strong acid (HCl). The ions of concern, OH⁻ and H⁺, combine to form water, since neither ion is in excess. Such a reaction is an example of neutralization of an acid with a base. Here the OH⁻ and H⁺ ions balance and a neutral pH of 7.0 is attained. The combination of OH⁻ and H⁺ continues until they attain equal concentrations of 10⁻⁷M, which is the point where neutrality is reached.

Distribution of pH in marine water
Although most natural environments have a pH between about 4 and 9, extreme conditions may exceed these values. Aquatic environments range from about 2.5 in some lakes in volcanic terrain to nearly 10.5 in evaporite lakes in arid regions. Seawater usually falls between 7.5 and 8.4, with a mean value of 7.8 for all oceans.

Minor deviations from the mean are commonly found near the water surface, where there is equilibrium between CO_2 in the water and that in the atmosphere. However, dissolved carbon dioxide combines with H_2O to yield carbonic acid (H_2CO_3). Therefore, surface areas have pH values of 8.1 to 8.3, and locally even higher values occur where photosynthesis is taking place at a great rate. This process reduces the CO_2 concentration and causes pH to increase (this process will be discussed in Chapter 12). Below the zone of photosynthesis, the amount of organic activity affects pH in that animals are taking up oxygen and expelling CO_2. This activity causes a depression of pH to nearly 7.5 (Fig. 10.8). This decrease occurs just below the zone of photosynthesis where animals are highly concentrated. There is also a gradual decrease in pH toward the ocean floor. In local areas on the ocean floor, pH may drop below 7.0 where oxygen is absent and H_2S is produced.

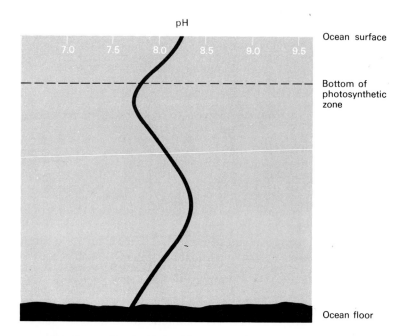

pH

7.0   7.5   8.0   8.5   9.0   9.5

Ocean surface

Bottom of
photosynthetic
zone

Ocean floor

FIG. 10.8 *Generalized pH distribution in the water column of the open-ocean environment.*

Hydrogen ion concentration is also affected by temperature and pressure changes, although they are not of much significance. Increase in temperature causes the hydrogen ion concentration to increase (pH decreases) at a rate of about 0.0003 pH unit per 10°C. Practically, therefore, this relationship can be disregarded, because the total temperature range in the ocean would provide so little pH change that it would approximate the limits of testing reliability.

Great pressure increase with depth causes an increase in the rate of carbonic acid dissociation. As a result, there is an increase in hydrogen ion concentration (pH decreases). This change is more significant than that caused by temperature and should be accounted for in detailed pH distribution studies. The average rate of decrease is about 0.02 pH unit per 1000 meters depth.

Determination of pH

A great many techniques exist for determining the pH of seawater. They range from simple acid-base indicators to delicate instruments which yield accurate measurements of hydrogen ion concentration. Simple colorimetric indicators are widely used because of the ease with which the tests are conducted and the simple apparatus involved.

Litmus paper is the most common acid-base indicator. It contains a vegetable dye that changes color above or below the neutral hydrogen ion concentration. Such a test is only of qualitative value. There are many indicator dyes that will change color within quite a narrow range of pH. If optimum conditions prevail, these can provide data almost as accurate as electrometric techniques. Most such indicators have a narrow range in which the color changes, usually less than two pH units. For example, phenolphthalein is a commonly used indicator for basic conditions. A red precipitate forms at a pH of 8.3. In order to obtain precise pH values with this or other indicators, one must compare the color with that of several solutions of known pH or with commercially available color-slide comparators. With such methods, an accuracy of 0.05 to 0.10 pH unit can be achieved.

Most pH measurements are by electrometric devices (Fig. 10.9) which are available in a variety of models, from small, portable field models to more precise laboratory equipment. In all types the same basic elements are used, a glass electrode in combination with a reference electrode. The glass electrode consists of an acid solution inside a special glass bulb. When this electrode is immersed in a solution (seawater), an emf (electromotive force) is produced between the inner and outer solutions that is proportional to the logarithm of the hydrogen ions in the external solution. The reference electrode has a constant emf, and the glass electrode varies with pH so that the voltage can be calibrated to determine the hydrogen ion level.

FIG. 10.9 *Laboratory-type portable pH meter that can be used on a-c or battery power.*

SELECTED REFERENCES

Martin, D. F., 1968, *Marine Chemistry*, New York: Marcel Dekker, Vol. 1, Chapters 3, 5, and 6. Laboratory techniques relating to analysis of various chemical characteristics of the ocean.

Riley, J. P., and R. Chester, 1971, *Introduction to Marine Chemistry*, New York: Academic Press, Chapter 2.

Williams, Jerome, 1973, *Oceanographic Instrumentation*, Annapolis: Naval Institute Press, Chapters 5 and 9. Brief description of modern techniques and instruments for determination of salinity and pH.

PRECIPITATION IN THE SEA 11

The solubility of various elements and ions in seawater covers a wide range. Some of these elements may reach a state of saturation or supersaturation and be precipitated from solution. Also, organisms use great quantities of dissolved constituents in building skeletons and other living tissues. Although inorganic precipitates are local and comprise a small portion of ocean sediments, a great quantity of sediment is contributed by organisms.

One of the first attempts to determine the age of the oceans was by salinity. It was theorized that if the annual rate of contribution by runoff was compared to the total salinity in the oceans, it would be possible to determine the age of the oceans. In 1898 John Joly determined that about 90 million years would account for the accumulation of the sodium then in seawater. Present data indicate that the rate of runoff that he calculated was too high, and a more accurate figure would be 250 million years. This is obviously not the age of the oceans, but a close approximation of the residency time (Table 9.2) of sodium. Joly's error stems from the fact that he did not take into consideration the precipitation of elements from seawater.

The above discussion implies that the oceans may constitute a large equilibrium system in which further addition of dissolved constituents via runoff (Fig. 11.1) will cause an imbalance, resulting in precipitation and subsequent return to the equilibrium state. This assumption is true in only the loosest sense. The observed gradual increase in total salinity with time indicates that a perfect balance does not exist. Also, complete uniformity, which must be present for a true equilibrium state, does not exist in the oceans. Variation in temperature, pressure, runoff, and circulation attest to this fact. Another important factor is the activity of living organisms, which can directly or indirectly cause an imbalance in the equilibrium.

A great variety of compounds may be precipitated from seawater, far too many to be treated in this chapter. In addition, some elements are commercially extracted directly from seawater. Three primary types of precipitation take place in the sea: (1) direct physicochemical precipitation, (2) biogenic precipitation, directly from organisms, and (3) precipitation caused indi-

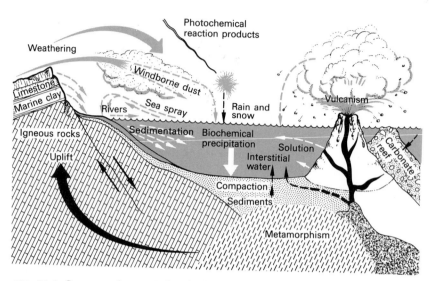

FIG. 11.1 *Sources of materials which are dissolved or suspended in seawater or which accumulate on the ocean floor. (After Anikouchine and Sternberg, 1973, p. 60.)*

rectly by activity of organisms which alters the water chemistry (Fig. 11.1). This chapter will treat only three of the major groups of compounds formed in the seas: evaporites, carbonates, and nodules. Evaporites are of economic significance, nodules have great potential for economic exploitation, and carbonates are important in geologic history and as skeletal material for organisms.

The following pages present a generalized view of precipitation in the marine environment. More complete treatment of this subject may be found in the references at the end of the chapter.

EVAPORITES

The process of evaporation is relatively simple; it is the conversion of a liquid into a vapor. By taking a volume of seawater and completely evaporating it, we find that there are some solids remaining. Close examination reveals that most of this residue is common salt (halite, NaCl). As the water is driven from the solution, the dissolved constituents become more and more concentrated. Precipitation occurs when the solubility limits of a particular compound are exceeded. This is an apparently simple chemical process: water vapor is removed from seawater, causing concentration of dissolved constituents until precipitation takes place. Of course, the process is far more complicated than this, because of interactions of the cations, anions, water temperature, and vapor pressure of the system.

Evaporite minerals occur in rather specialized environments in the oceans. In quantities they may be of considerable economic value, and in ancient rocks they give significant evidence about past environments. About 30 evaporite minerals are known but only four or five are abundant. These abundant evaporites are composed of a fairly small number of the major dissolved constituents of seawater: chlorides and sulfates of Na, Mg, Ca, and K.

Conditions for evaporite deposition

An ideal evaporite basin is one having restricted circulation with the open ocean, little or no runoff from adjacent land masses, shallow depth, and subtropical location. Warm temperatures enhance evaporation but are not as important to the process as drying winds. The ideal geomorphic location for evaporation would be on the leeward side of a mountain chain where moisture is removed from the atmosphere as the winds pass over the mountains. Some modern areas of marine evaporite deposition are the Gulf of California, the Persian Gulf, the Red Sea, and the Gulf of Karabugas. (The last named is not strictly a marine environment, as it is part of the Caspian Sea, but is an area of great evaporite deposits.) None of these environments is forming evaporite deposits of the magnitude that we find in the rock record.

Some of the factors influencing evaporation are subject to change with time, particularly with the seasons. As a result, there may be cyclic precipitation caused by cyclic changes in temperature, air circulation, and rainfall. These obviously affect the dilution or concentration of seawater. There are other processes which could cause changes in the relative composition of marine water, such as changes in the rate of chemical weathering on the land. For instance, the sodium supply via runoff might vary significantly in direct relation to temperature and rainfall as these are two important factors in chemical weathering. The point is that even though these changes may appear slight, it is possible for them to bring about significant differences in the rate of evaporation from the sea or in the mineral species being precipitated.

Evaporite sequence

When seawater evaporates, compounds precipitate in a definite sequence as salinity increases. The first precipitate is calcium carbonate (calcite), followed by dolomite. Both of these carbonate minerals are relatively unimportant as evaporites. Dolomite, in particular, does not presently seem to be forming in nature as a direct precipitate from seawater, but if it is present in an evaporite sequence it is probably the result of rapid replacement of calcite by the dolomite. Virtually all the carbonate crystallizes before formation of the important evaporite minerals—sulfates and chlorides.

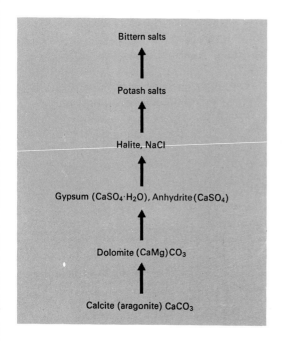

FIG. 11.2 *Standard succession of minerals from evaporation of seawater. The first mineral to precipitate is at the bottom.*

Continued evaporation yields calcium sulfate, which occurs in the hydrated form (gypsum, $CaSO_4 \cdot 2H_2O$) and also as an anhydrite ($CaSO_4$). Precipitation of gypsum begins when salinity increases to 3.35 times that of normal seawater (about 115‰). When the solution reaches nearly five times its original salinity, anhydrite replaces gypsum as the sulfate being precipitated. The above precipitates occur in this sequence at standard temperatures (25°C). If the temperature reaches 35°C no gypsum is formed; only the anhydrous variety precipitates. The anhydrite, however, is quickly hydrated to gypsum in the marine environment. If gypsum is buried and thus removed from the aqueous environment, the water may be lost and anhydrite formed again.

Halite (NaCl), the most abundant evaporite mineral in modern environments, begins to precipitate at salinities over 300‰. Anhydrite co-precipitates with halite until the solution reaches about five percent of its original volume. At this point, anhydrite stops precipitating and polyhalite, $K_2Ca_2Mg(SO_4)_4 \cdot 2H_2O$, is formed along with halite. The next evaporites to co-precipitate with halite are, in order, the **potash salts** and the **bittern salts.**

The sequence of evaporite minerals mentioned above was first determined by an Italian chemist named Usiglio in the mid-nineteenth century. He evaporated Mediterranean Sea water in the laboratory and established the foundation for a predictable evaporite sequence (Fig. 11.2). Beyond the precipitation of halite, Usiglio's data were not accurate; however, he obtained a sequence which compared with natural evaporite succession in rocks of Europe.

If a layer of seawater 1000 meters thick was evaporated at 25°C, the residue would be 0.047 meter of carbonates, 0.40 meter of gypsum, and 11.8 meters of halite.

PRECIPITATION OF CARBONATES

Although more than 50 different carbonate minerals have been recognized and described, only three are of significance in the modern marine environment. These are aragonite ($CaCO_3$), calcite ($CaCO_3$), and dolomite ($CaMg(CO_3)_2$). Aragonite and calcite make up a polymorphous series; that is, the minerals have the same chemical composition but different crystallographic structure. Aragonite is orthorhombic and calcite is hexagonal. Aragonite is an unstable form and in time inverts to calcite.

For many years it was thought that all calcium carbonate grains in modern environments and ancient rocks were derived from the skeletal material of organisms. Great interest in carbonate sediments and rocks has been generated in the past two decades by both the oil industry and academic researchers because of the great qualities of petroleum found in carbonates. As a result, we now know a great deal about the formation of carbonate minerals in the marine environment.

We are in an unusual period of geologic time for carbonate deposition. It is a time of little carbonate precipitation, at least since the beginning of the Cambrian Period. Areas where significant formation of carbonates is taking place now include the Bahama Banks—Florida Bay area, the Caribbean, the Great Barrier Reef, the Red Sea, and the Persian Gulf. Many small areas of carbonate deposition also exist, particularly the reefs of the low latitudes. Of all the areas of carbonate formation, the Bahama Banks is the largest and has received the most concentrated study. Virtually all aspects of the marine environment have been examined in order to obtain the details of carbonate precipitation. Such investigations are of particular importance because limestones ($CaCO_3$) are a common constituent of the rock record and an understanding of modern carbonate deposits will provide help in interpreting the origin and depositional environments of ancient carbonates.

Carbonates can be precipitated in three ways: (1) directly by organisms as skeletal material, (2) by direct physicochemical precipitation and, (3) directly, by organisms causing water conditions that facilitate carbonate formation. Despite accelerated research efforts, there is little agreement as to the relative importance of these three processes. Carbonate sediments are composed of skeletal fragments in a wide range of sizes, tiny needle-shaped crystals about 1 to 2 microns in length, and oolites, which are small spheres (less than 1 millimeter in diameter) composed of a nucleus and thin concentric rings of carbonate material.

Carbonate chemistry

Seawater contains a great deal of calcium; in fact it is essentially saturated and locally may be supersaturated. It is therefore not a shortage of Ca^{2+} which is restricting carbonate deposition. Carbonates occur in various forms with a mean accumulation of about 0.001 meter/year at 25°C. The species of carbonates include carbonic acid (H_2CO_3), bicarbonate ions (HCO_3^-), and carbonate ions (CO_3^{2-}). Their abundance is a function of pH and at normal marine conditions of pH (8.1) the bicarbonate ion is most abundant.

To understand how and why calcium carbonate is formed, we must look at a few simple reactions.

The chemical reactions of carbonate formation are relatively simple and usually proceed rapidly. For instance, if an acid is added to calcium carbonate,

$$CaCO_3 + 2H^+ \rightarrow Ca^{2+} + H_2O + CO_2$$

or

$$CaCO_3 + H^+ \rightarrow Ca^{2+} + HCO_3^-.$$

By adding a base, the reactions would be reversed:

$$Ca^{2+} + HCO_3^- + OH^- \rightarrow CaCO_3 + H_2O,$$

indicating that the hydrogen-ion concentration is important in the stability of an aqueous solution of calcium carbonate.

From the above equations it is obvious that carbon dioxide is also quite important in carbonate chemistry. Each reaction involves either free CO_2 or CO_2 as part of bicarbonate or carbonate ions. All these carbonate species are present in seawater, as is a small amount of undissociated carbonic acid (Fig. 11.3). Carbon dioxide from the atmosphere combines with water to produce carbonic acid:

$$CO_2 + H_2O \leftrightharpoons H_2CO_3.$$

This dissociates in seawater to produce bicarbonate ions which may further dissociate to carbonate ions:

$$H_2CO_3 \leftrightharpoons HCO_3^- + H^+,$$

$$HCO_3^- \leftrightharpoons CO_3^{2-} + H^+.$$

At normal pH conditions, seawater contains 93 milligrams/liter total carbonate. Figure 11.3 shows the distribution of the carbonate species with respect to pH. Note that free dissolved carbon dioxide occurs only below normal hydrogen ion concentrations for seawater. At low pH, the concentration of CO_3^{2-} becomes lower as it combines with H^+ to form HCO_3^-. This indicates that CO_2 is lowering the pH by forming carbonic acid which then

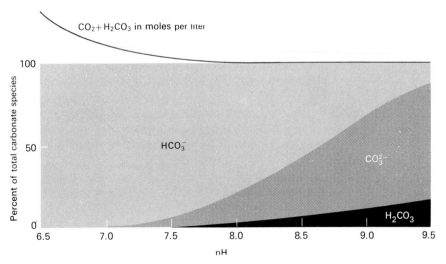

FIG. 11.3 *The effect of pH on the composition of carbon species in seawater at 16°C, 36‰ salinity. (Modified from Harvey, 1957, p. 153.)*

dissociates to form the bicarbonate ion. A mutually exclusive relationship exists between carbon dioxide and CO_3^{2-}. Therefore, in order to produce CO_3^{2-} and make it available for combination with Ca^{2+} it is necessary to remove CO_2 from the system.

Factors affecting carbonate precipitation
Most salts become more soluble as temperature rises; however, calcium carbonate acts in the opposite manner. As the temperature increases, the gases contained in the water become less soluble. In the preceding section, we saw that removal of carbon dioxide causes precipitation of $CaCO_3$. Thus a rise in temperature has two effects on precipitation of $CaCO_3$: (1) it decreases the solubility of Ca^{2+}, and (2) it causes a loss of carbon dioxide. The latter effect is more significant. The Bahama Banks area illustrates both effects. Water in the deep Florida Straits may upwell and flow over the shallow banks. At depth, this water is cold and contains a moderate amount of carbon dioxide. As it rises, it becomes warmer and thus provides the necessary conditions for carbonate precipitation.

These factors and others have led to the false assumption that *all* calcium carbonate is precipitated in warm shallow water between the thirtieth parallels. There are calcium carbonate sediments forming as far north as the fjords of Scandinavia and Alaska. Temperature, therefore, is but one of the factors which contribute to carbonate precipitation, and its effects may be overcome by other factors.

Pressure acts much as temperature does, in that it affects solubility of CO_2 and therefore that of $CaCO_3$. The result is similar to that produced by temperature; lower pressure lowers $CaCO_3$ solubility and enhances precipitation by decreasing the amount of CO_2 also. Surface changes in barometric pressure change the solubility of CO_2. Catastrophic events like volcanic eruptions can also cause changes in the partial pressure of CO_2. However, these occurrences are only occasional and local, so that atmospheric circulation negates most of their effects on CO_2 distribution.

Perhaps the most underrated cause of carbonate precipitation in the sea is that of **photosynthesis**. In the photosynthetic reaction,

$$6H_2O + 6CO_2 + \text{sunlight} + \text{minerals} \leftrightharpoons C_6H_{12}O_6 + 6O_2,$$

there is a considerable uptake of CO_2 by green plants. This causes the same reaction as inorganic loss of CO_2 to the atmosphere, namely $CaCO_3$ precipitation. Most areas of high carbonate precipitation, such as Florida Bay, the Bahama Banks, and reefs in general, are also areas of high photosynthetic production.

The reverse of the above equation, **respiration**, will cause CO_2 to evolve, thereby depressing the precipitation of calcium carbonate. Such a reaction will take place where oxygenated water is in contact with nonliving organic material or in an area where animals are respiring CO_2 in amounts which exceed its uptake by plants. This process will be discussed in detail in Chapter 13.

As evaporation takes place, it causes an increase in ionic concentration which also may enhance precipitation. It has also been demonstrated that agitation of water can cause precipitation, by driving off carbon dioxide. Production of ammonia by bacterial activity may cause carbonate precipitation, but this is a minor factor. Ammonia causes the pH to rise, thereby increasing CO_3^{2-} and facilitating formation of $CaCO_3$.

Precipitation by organisms

Both marine plants and animals utilize calcium carbonate as skeletal material. This may be in the form of calcite, aragonite, or **high-Mg calcite**, and commonly the mineral composition is uniform for a particular group of organisms. However, environmental factors cause a variation in the mineral composition of the materials in some groups.

Calcium carbonate skeletal material is found in organisms ranging from less than a micron to more than a meter in diameter. The most obvious carbonate skeletons are found in the coral framework of reef in the shallow areas of the low latitudes. Shells of many other invertebrates are associated with the reefs and shallow carbonate banks. A large number of carbonate-producing organisms occupy the noncarbonate portion of the crust, or float at shallow depths in the open ocean.

Carbonate skeletal materials have been abundant since Cambrian times; however, a distinct change in type occurred during the Cretaceous Period, when floating calcareous organisms become abundant and widespread. Much of the present deep-sea floor is covered with tests of these plants and animals. There are explanations for this sudden abundance of $CaCO_3$ in the open sea, but none seems completely satisfactory. One theory is that prior to Cretaceous times, the $CaCO_3$ was on continental shelves and shallow epicontinental seas, but since Cretaceous times, it has circulated throughout the oceans. This seems logical, but no proposed mechanism has satisfactorily explained its implementation.

Distribution and occurrence of carbonate minerals

As previously noted, there are three important carbonate minerals in the marine environment: calcite and aragonite ($CaCO_3$), and dolomite ($CaMg(CO_3)_2$). Dolomite, the least common of these, is crystallographically similar to calcium carbonate. Magnesium can substitute easily for calcium in the structure and it is common to have carbonate minerals with a wide range of Ca:Mg composition. However, the pure magnesium carbonate mineral (magnesite) is not known to occur in the modern marine environment. The

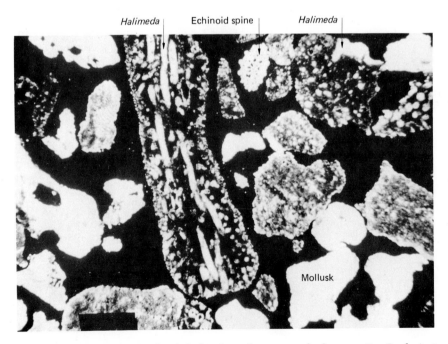

FIG. 11.4 *Photomicrograph of skeletal sand composed of aragonite. Scale is 1 mm. (Photo courtesy of E. A. Shinn.)*

term "high-magnesium calcite" is used for concentrations of greater than 5 mole-percent magnesium.

Apparently, a definite pattern of distribution exists, with calcite being the dominant skeletal mineral in deep-sea environments, and aragonite and high-Mg calcite in shallow marine environments. Deep-sea carbonates are generally considered to be completely organic in origin: tests of floating organisms. Shallow areas contain both organic carbonate debris and physicochemically precipitated carbonate. Most of this debris (Fig. 11.4) is

FIG. 11.5 *Examples of various calcareous green algae:* Halimeda *(left) and* Penicillus *(right). Approximately to scale.*

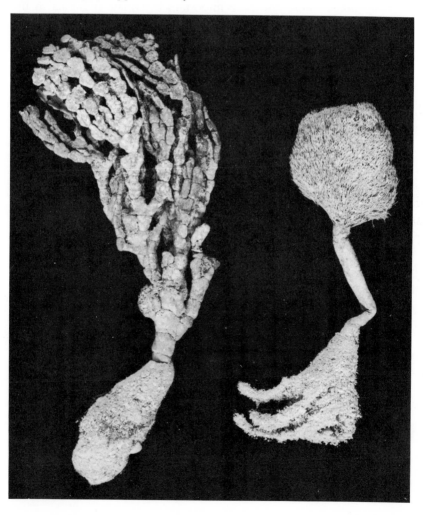

aragonite, with the exception of Foraminifera tests, which are calcite. Some mollusks have layers of calcite alternating with aragonite. The fine sediments and most organic debris are aragonite. High-Mg calcite is present in some mollusks but is most abundant in arthropods and echinoderms. Magnesium content of invertebrate skeletons seems to be directly proportional to the phylogenetic level of the group.

Two groups of algae contribute a great amount of shallow marine carbonates. Both the green (*Chlorophyta*) and red (*Rhodophyta*) algae contain calcareous varieties (Figs. 11.5 and 18.4). The coralline algae are a group of red algae which make aragonic material that has an easily recognized structure. Green algae occur in a wide variety of calcareous forms, with some of the debris being easily recognized as such. *Penicillus* (paintbrush algae) is one of the most abundant of the calcareous algae. It produces small, needle-shaped aragonite crystals that are virtually impossible to distinguish from the aragonite needles formed by physicochemical precipitation (Fig. 11.6).

(a)   (b)

FIG. 11.6 *Aragonite needles from (a) Penicillus and (b) inorganic precipitation show considerable similarity. (After P. E. Cloud, 1962, "Environment of Calcium Carbonate Deposition West of Andros Island, Bahamas," U.S. Geological Survey Prof. Paper 350, plates 4B and 5D.)*

PHOSPHATE AND MANGANESE NODULES

Exploitation of mineral resources in the sea will undoubtedly become more significant than it is at present. It may even become a fairly sizeable portion of our economic mineral production. Iodine and bromine are presently extracted directly from seawater, and plans as well as pilot studies have been made for determining the feasibility of extracting other elements. The sea floor may also provide minerals, but it is relatively inaccessible and the cost of recovery is prohibitively high at present. Two types of chemical marine deposits may be mined in the near future. These are phosphorite concretions on the continental margin areas and manganese nodules on the ocean floor.

Phosphorite

Phosphatic concretions on the ocean floor were first discovered by the *Challenger* Expedition in the early 1870s. They were described from Agulhas Bank south of the Cape of Good Hope in Africa. Similar deposits are presently known off the coasts of Japan, Argentina, eastern United States, and western North and South America. Undoubtedly their distribution is much more extensive, but lack of dredging in many areas prohibits our knowing the real extent of **phosphorite nodules**.

Phosphorus is economically important primarily as a fertilizer and for production of other chemicals. Various forms of the mineral apatite, $Ca_5(F \cdot Cl,OH)(PO_4)_3$ are found in commercial phosphorite deposits. The compound of commercial interest is P_2O_5, which may comprise up to 40 percent of the total. Sea-floor deposits are commonly about 20 to 30 percent P_2O_5. Bulk phosphorite is not expensive at the place of recovery; however, transportation is a major cost. Because of its present rather restricted production from land, the price of phosphorite is extremely high in many parts of the world. If recovery from the ocean were economical, most regions of the world could obtain phosphate at modest cost. The oceans are estimated to contain enough phosphate for several hundred years at the present rate of utilization.

Form, composition, and distribution of oceanic phosphorite. The best-known phosphorite deposits in the sea occur off the California coast. They exist as irregular nodular masses whose overall shape suggests accretionary origin. In color, the nodules are various shades of brown, and they range widely in size. The largest recovered nodule is nearly a meter in diameter; however, photographs of the shelf floor reveal much larger nodules. The size ranges down to less than a millimeter.

The internal structure of individual nodules also shows a wide variety of forms. Some are apparently homogeneous, but most show layering, or a conglomeratic texture. The layering is irregular in thickness and distribution. An oolitic structure with a few concentric layers surrounding a nucleus is present in some small nodules.

Nodules are found on the sea floor surface in many locales, but they are commonly restricted to areas with low sedimentation rate. In areas with a substantial influx of sediment, sufficient time is not available for accretion of phosphate minerals, and as a result, nodules are small or absent. Nodules are most abundant in areas where currents are relatively strong, such as the sides of escarpments, on the crests of local banks and ridges, on the walls of submarine canyons, and near the outer edge of the continental shelf. Phosphorite has been dredged from depths of as much as 3500 meters, at the base of the continental slope off South Africa.

Chemically, the phosphorite in the present ocean is comparable to currently exploited deposits in Idaho, Florida, and Tennessee. Nearly half is calcium oxide, about 30 percent is P_2O_5, and the remainder is composed of small amounts of oxides and carbonates (Table 11.1). Insolubles listed in Table 11.1 are residues after HCl treatment; they include various silicate minerals.

TABLE 11.1 *Chemical composition of phosphorite nodules.*
(From J. L. Mero, 1965, The Mineral Resources of the Sea,
New York: American Elsevier, p. 66.)

	Weight percentages	
Constituent	Forty Mile Bank off California	Agulhas Bank off South Africa
CaO	47.4	37.3
R_2O_3	0.43	9.4
P_2O_5	29.6	22.7
CO_2	3.9	7.1
F	3.3	—
Organic	0.1	—
Totals	84.7	76.5

Origin of phosphorite.   Distribution of phosphorite nodules permits some speculation of their origin. The *Challenger Report* indicates that such phenomena as upwellings, current anomalies, salinity changes on continental margin areas, and a large kill of organisms may be factors. When dead organisms settle to the bottom and begin decomposition, they provide a concentration of phosphorus in solution. Circulation out of this lower zone of decomposition on the continental margin makes it possible for phosphorus to precipitate, commonly on colloidal particles. These may be attracted to existing nodules or to other particles on the ocean floor, thereby producing the layering that is fairly common.

Phosphorite would be expected to form most commonly in the low pH oxygen-deficient conditions. The abundant decay of organic matter which yields phosphorus would provide such an environment. Most areas where phosphorite nodules are found have well-circulated and oxygenated water.

The origin of phosphorite formation is still largely a puzzle, but these and other observations provide some clues about the processes involved.

Manganese nodules

Like phosphorite, manganese nodules were also discovered by the *Challenger* Expedition. They were first located in the southeastern Pacific Ocean and have since also been found throughout the world ocean. Many marine geologists feel that manganese nodules are potentially one of the most important minerals in the sea. Studies are under way to determine the feasibility of recovering them on a commercial basis from the Pacific, where they are highly abundant.

The nodules occur in many sizes, from less than a millimeter to nearly two meters in diameter. Most are a few centimeters in diameter. In addition to the common nodular form, manganese oxide also occurs in slabs and crusts on rocks and shell debris. Nodules may be spherical, aggregates or clusters of nodules, or irregular masses (Fig. 11.7). Most are black in color, but they may be brown or tan. Their density averages about 2.50 grams/cubic centimeter.

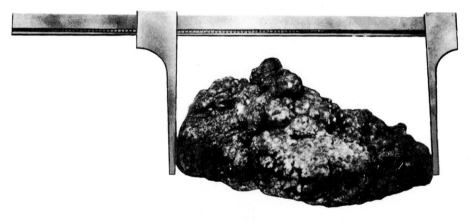

FIG. 11.7 *A large manganese nodule more than a meter across its longest dimension. (Photo courtesy of J. L. Mero.)*

Distribution of manganese nodules. Dredge hauls have yielded thousands of nodules from many locations, particularly in the central Pacific. However, these hauls indicate only that nodules are present; they do not tell

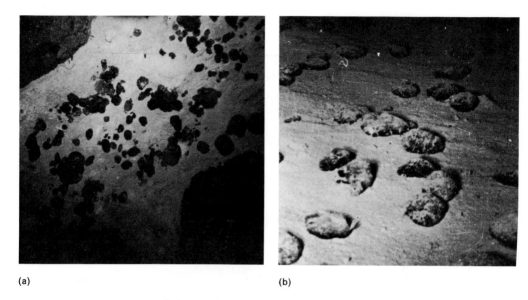

(a)   (b)

FIG. 11.8 *(a) Manganese nodules on calcareous ooze at a depth of 4560 meters in the southwestern Pacific Basin (field is 2 square meters). (Official photograph, U.S. Navy.) (b) Manganese nodules on the Bermuda Rise at a depth of 5000 meters (nodules are about 2.5 centimeters). (Photo courtesy of J. D. Hollister.)*

us anything about the abundance of nodules on the ocean floor or in sediment layers. Bottom photographs (Fig. 11.8) provide a fairly good means for estimating quantities on the surface. Nodules have been retrieved in some areas from dredges and cores, but bottom photographs of these areas reveal no nodules. Several bottom photos show ripples or other evidence of sediment movement; in such places, absence of nodules on the surface could be due to rapid burial. There are two large areas of surface concentrations in the Pacific (Fig. 11.9): a low-latitude belt trending east-west in the northern low latitudes and a large area of the southern Pacific. The Blake Plateau of the western Atlantic also contains a modest quantity of nodules.

Most manganese nodules come from the ocean floor at depths of about 4000 meters. They have been collected from about 100 meters off Japan to the great depths of Pacific trenches. Most nodules are found in association with deep-sea clays, calcareous and siliceous oozes, and bedrock. All these sources are indicative of sediment accumulating at a slow rate. Areas of highest abundance of nodules tend to have low rates of sedimentation. Estimates of the total amount of manganese nodules available on and near

FIG. 11.9 Distribution of manganese nodules in the Pacific Ocean. (Modified from Menard, 1964, p. 177.)

the surface of the ocean floor are in the tens of billions of tons. This exceeds present commercial reserves of manganese by hundreds of times.

Origin of manganese nodules.   Manganese, which is the major cation of concern, is supplied to the ocean via volcanic eruption, mineral decomposition on the sea floor, and runoff from land masses. Differences of opinion have existed as to which of these sources, if any, is the major contributor. Early ideas supported runoff from land as the most important; however, modern quantitative studies indicate that this source is inadequate for the present abundance and rate of accumulation of deep-sea manganese deposits. In recent years, vulcanism has been given increasing attention.

Manganese is circulated to remote areas of the ocean by oceanic mixing and currents, and by organisms which preferentially concentrate this element. Solution of tests of organisms or decay of living tissues on the ocean floor creates a supply of manganese. Actually, ocean water is saturated or nearly so with manganese and iron.

A common explanation for formation of manganese and iron oxides is that colloidal particles of these compounds exist in seawater, where cation concentration causes precipitation. As the colloidal particles are circulated near the sea floor they are attracted, because of their electrical charge, to hard objects, particularly metallic masses on the ocean floor. It has also been suggested that bacteria may play a role in manganese accumulation, but this has not been adequately demonstrated.

The rate of accumulation of these manganese compounds is quite slow, slower than most deep-sea sediments. Estimates range from a few hundredths of a millimeter to a millimeter per thousand years. The upper limit approaches the rate of other deep-sea accumulations. There are, however, examples of rapid accretion of manganese on various nuclei, some accumulating at a rate of centimeters every 100 years.

Composition of manganese nodules.   Although their name implies composition, these nodules are high in iron in addition to manganese and are sometimes called ferromanganese nodules. There are several minerals commonly present in the nodules, such as goethite, rutile, opal, barite, and apatite in addition to the manganese-oxide minerals. The manganese minerals have different structures and apparently represent different oxidation states. The opaque nature of these nodules is such that optical methods are of no value and x-ray techniques are necessary for identification.

Many nodules have been analyzed chemically, indicating a wide range of compositions (Table 11.2). Note that the average composition is high in iron and silicon as well as manganese. Also in the case of phosphorite nodules, the insoluble residues make up a significant part of the total.

TABLE 11.2 *Chemical composition of manganese nodules.*
(After Mero, 1965, p. 179.)

Constituent	Weight percentages		
	Maximum	Minimum	Average
MnO_2	63.2	11.4	31.7
Fe_2O_3	42.0	6.5	24.3
SiO_2	29.1	6.0	19.2
Al_2O_3	14.2	0.6	3.8
$CaCO_3$	7.0	2.2	4.1
$CaSO_4$	1.3	0.3	0.8
$Ca_3(PO_4)_2$	1.4	tr	0.3
$MgCO_3$	5.1	0.1	2.7
H_2O	24.8	8.7	13.0
Insoluble in HCl	38.9	16.1	26.8

The manganese nodules from the central Pacific are quite important also because of their copper and nickel content. Although present in concentrations of only about one percent, these elements are becoming increasingly scarce in our land-mining resources, giving these nodules much greater commercial value.

SELECTED REFERENCES

Borchert, Hermann, and R. O. Muir, 1964, *Salt Deposits—The Origin, Metamorphosis and Deformation of Evaporites,* London: Van Nostrand, Chapters 2, 4, and 5.

Broecker, W. S., and Taro Takahashi, 1966, "Calcium Carbonate Precipitation on the Bahama Banks," *Journal of Geophysical Research* **71**, 1575–1602. Detailed chemical study of Bahama waters to gain information on carbonate precipitation. Good chemistry background is helpful to thorough understanding of the article.

Cloud, P. E., 1962, "Behavior of Calcium Carbonate in Sea Water," *Geochimica et Cosmochimica Acta* **20**, 867–884. Thorough discussion of carbonate precipitation by one of the foremost advocates of physicochemical precipitation as a source for $CaCO_3$.

Cloud, P. E., 1965, "Carbonate Precipitation and Dissolution in the Marine Environment," *Chemical Oceanography,* J. P. Riley and G. Skirrow (eds.), New York: Academic Press, Vol. 2, pp. 121–158. One of the most thorough studies of the famous carbonate sediments of the Bahamas. Excellent illustrations but chemistry background is necessary to understand the "meat" of the paper.

Degens, E. T., 1965, *Geochemistry of Sediments—A Brief Survey,* Englewood Cliffs: Prentice-Hall, Chapter 3. Good general discussion of low-temperature chemical precipitation. Tremendous reference lists at the end of each section.

Horn, D. R., B. M. Horn, and M. N. Deloch, 1973, *Ocean Manganese Nodules: Metal Values and Mining Sites,* Washington, D.C.: National Science Foundation, Tech. Rpt. No. 4, NSFGX 33616. Provides updated information on known distribution of nodules.

Krauskopf, K. B., 1967, *Introduction of Geochemistry,* New York: McGraw-Hill, Chapters 3 and 12. Well-written general discussion of carbonate and evaporite precipitation.

Menard, H. W., 1964, *Marine Geology of the Pacific,* New York: McGraw-Hill, Chapter 8. Good treatment of occurrences of nodules on the Pacific Ocean floor.

Mero, J. L., 1961, *Sea Floor Phosphorite,* State of California Mineral Information Service, **14** (11), 1–12. General newsletter type article on phosphatic nodules in the Pacific. Excellent for the general reader.

Mero, J. L., 1965, *The Mineral Resources of the Sea,* New York: American Elsevier, Chapters 4 and 6. A one-of-a-kind book with strong emphasis on the phosphatic and manganese deposits on the sea floor. Many excellent photographs.

Stewart, F. H., 1963, *Data on Geochemistry,* Chapter Y, "Marine Evaporites," U.S. Geological Professional Paper 440-Y. Mostly phase diagrams and detailed compositions of various marine evaporites. Text material not generally appealing to beginning students.

12 NUTRIENT CYCLES IN THE OCEANS

The common precipitates in the sea are derived largely from the major seawater constituents. Some of the minor constituents of seawater are at least equally significant but for other reasons. A few of these are the so-called nutrient elements, which are essential to life in the seas. The most important of these are phosphorus and nitrogen, which are incorporated in living tissues, and also silicon, which is necessary for the formation of the tests (skeletons) of diatoms and radiolarians.

The nutrient elements are present in various states in seawater but in low concentrations. Because of their utilization by organisms, there is considerable seasonal and geographic variation that can be related to organic activity.

It has long been known that regardless of the concentration of their species in the sea, a constant relationship exists between phosphorus and nitrogen. This indicates that the uptake of nutrients by organisms is in direct proportion to the ratio in which they are present in the ocean, and that they are returned to the sea via decomposition at the same relative rates. The average ratio of nitrogen to phosphorus in the sea is about 7:1 by weight and slightly more than twice that (15:1) in number of atoms.

Before proceeding with the nutrient cycles, it will be necessary to elaborate briefly on the organic cycle in the sea (Fig. 12.1). Available nutrients such as phosphorus and nitrogen are utilized in the photosynthetic process and incorporated in plant tissues. The fact that this process requires sunlight dictates that it occur close to the water surface in the **euphotic zone**. Consequently, nutrient concentration in this zone undergoes maximum variation because of seasonal cycles of utilization by plants. Primary consumers, which are also concentrated in the euphotic zone, feed on plants and may in turn be fed upon by other consumers. Some nutrients are released from these organisms by way of excrement, but most of the nutrients are not released until decomposition of the organism after death. Much of the decay process is concentrated at or near the sea floor, which necessitates considerable transport of nutrients to make them available for photosynthesis in the near-surface euphotic zone.

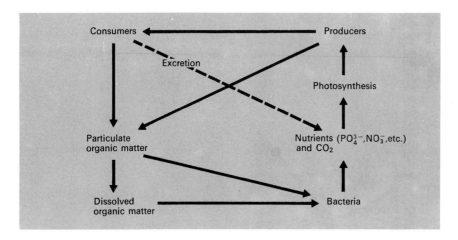

FIG. 12.1 *Organic cycle in the oceans.*

CIRCULATION OF NUTRIENTS

Movement of nutrient materials from their site of concentration near the ocean floor to the zone of utilization near the surface can be accomplished in a variety of ways. The obvious controlling factor is circulation in the oceans, particularly in a vertical direction.

Upwelling

Probably the most significant process that brings deep-water nutrients near the surface is upwelling (Chapter 6). Since upwelling depends on winds carrying surface waters away from or parallel to coasts, this process is of major significance on west coasts in the mid-latitudes. A similar effect is produced in the areas of the equatorial countercurrents. Both situations represent a divergence of surface waters, which causes water at depth to move upward, taking the place of the parting surface waters. The reverse situation gives the expected opposite results. Where surface waters converge and descend, they tend to carry nutrients down with them. As a rule, areas of diverging currents are high in nutrient content and therefore high in production of organic tissues, and converging currents yield opposite conditions. Upwelling may be seasonal, as along the California coast and the availability of nutrients also corresponds to such seasonal changes.

Turbulence and local currents

Coastal areas are high in overall organic activity because of the nutrient supply from adjacent land areas. Currents provide the major mechanism by which nutrients are transported along coasts and continental margins. Tidal currents are locally significant. They transport nutrients supplied by runoff and also cause disturbance of surface sediments, which are likely to contain high nutrient concentrations from decay of organic tissues. In a like manner, longshore currents (Chapter 6) may also be effective.

Wave action, particularly after major storms, is a significant means whereby nutrients are circulated and bottom sediment stirred. Such action promotes decay and increases nutrient availability. Downslope transport of bottom sediment along continental slopes and submarine canyons serves the same function. All these phenomena which tend to make nutrients available are located around the margins of ocean basins and are in large part responsible for the high organic production of these areas.

Density currents
The vertical movement of water is due largely to changes in its density in response to water temperature changes. This is particularly important in midlatitude areas, where significant seasonal changes in climate are common. Here surface waters are cooled during the winter months, with a corresponding increase in their density. The increased density causes surface water to sink, and deep water, which may be high in nutrients, moves toward the surface.

A different type of temperature-controlled circulation may also cause nutrients to be brought toward the surface. During winter months, the water on the continental shelf may be cooled. In shallow areas, this cooling produces a mass of heavier water which moves down the gradient of the shelf and displaces deep, nutrient-laden water toward the surface.

PHOSPHORUS CYCLE
Phosphorus may exist in the oceans in a variety of states (Fig. 12.2): dissolved phosphate and organic compounds, and insoluble and organic suspended phosphorus compounds. The average content of phosphorus in seawater is about 70 micrograms/liter, but there are significant deviations from this value, particularly near the surface.

Dissolved phosphate exists primarily as $H_2PO_4^-$ or HPO_4^{2-}; the simple phosphate ion (PO_4^{3-}) is comparatively rare. Organic compounds in the sea can be in the form of proteins, lipids, or sugars. Calcium and iron phosphates may be present as suspended colloidal particles. These colloids may become absorbed on sediment particles (Chapter 11) and be removed from the nutrient cycle.

Of all phosphorus in the sea, inorganic phosphorus comprises the majority of all species. These species may, however, undergo seasonal variations in their relative abundance. For instance, during the winter when organic production is low, about 90 percent of all phosphorus is inorganic

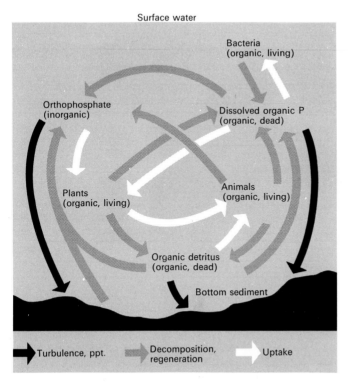

FIG. 12.2 *Phosphorus cycle in the marine environment. (Af-ter D. W. Hood, 1964, "Chemistry of the Oceans," Chem. and Eng. News, June 1964, p. 30A.)*

phosphate, whereas during the summer it may comprise less than half the total, with organic phosphorus compounds making up the majority.

Phosphate distribution

The vertical distribution of phosphate exhibits a predictable pattern in the open ocean. The distribution is quite similar for all three oceans, except that values for the Atlantic Ocean fall below those for the Pacific and Indian Oceans, for which the values are remarkably similar (Fig. 12.3). There are four layers of characteristic phosphate content. A thin surface layer is gener-ally low in phosphate except during times of low utilization by organisms. The second layer extends down to depths of a few hundred meters and shows the greatest increase in phosphate concentration. A thick layer ex-tends to a depth of about 2000 meters and exhibits maximum phosphate

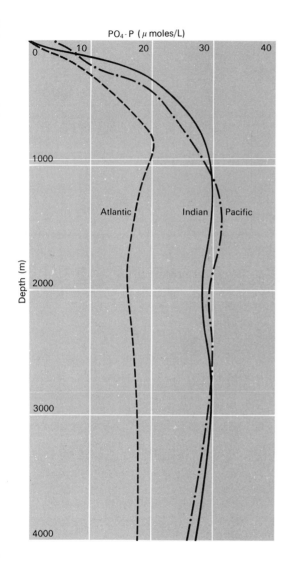

FIG. 12.3 *Vertical distribution of phosphate in the major oceans. (After Sverdrup, et al., p. 241. Copyright renewed 1970.)*

concentrations. Immediately below this is the bottom layer, also thick, and uniform in phosphate concentration. Values in the bottom layer are only slightly less than those in the zone of maximum concentration.

The layer of maximum concentration lies immediately below the euphotic zone. Decay takes place in this zone as dead plants and animals begin their descent to the depths of the sea. A corresponding low oxygen concentration is present in this layer, reflecting the utilization of oxygen by decaying organic matter.

Variations in distribution
Some changes in phosphate concentration are caused by seasonal and diurnal changes in organic activity. Most studies of these changes have been conducted in shallow coastal water, but similar results would be expected in the upper waters of the open ocean. Seasonal changes are largely determined by latitudinal climatic belts. The most pronounced changes occur in temperate latitudes because the greatest climatic changes occur in these areas. Cold temperatures and short day lengths during winter months yields high phosphate values, whereas increasing day length and light intensity in spring cause decrease in phosphate as plant growth accelerates. This condition continues until a thermocline is developed in the summer which inhibits mixing and plant growth is retarded as available phosphate is utilized.

Although some studies of phosphate concentration in the tropics reveal a seasonal variation, many others do not. Even where changes have been observed, they are moderate. One significant area of change is in the Indian Ocean, where the monsoon season causes distinct changes in phosphate content of the ocean. During the monsoon season (summer), winds cause a great deal of mixing of seawater, and phosphate content is high. After the monsoons end, peak concentrations are reached, and values then decrease to a minimum just prior to the next monsoon.

Diurnal changes in phosphate content are minimal except in shallow bays and coastal areas. Here the variations are directly related to sunlight and plant activity. Concentrations are lowest at times of peak plant activity and highest early in the morning before sunrise.

NITROGEN CYCLE
Like phosphorus, nitrogen is essential for life. It occurs in a wide variety of forms in the oceans, with molecular nitrogen (N_2) being the most abundant, accounting for several times the total of combined forms. Nitrite (NO_2^-) and nitrate (NO_3^-) are the most abundant combined forms; they represent the high oxidation states of nitrogen in the sea. Both plants and bacteria assimilate inorganic nitrogen during organic production (Fig. 12.4). Circulation and fixation of nitrogen provide means for replenishing the supply as growing organisms deplete it.

Two contrasting theories have been suggested for the origin of our atmosphere and of nitrogen in particular (Chapter 9). One postulates a primitive atmosphere composed in part of ammonia. As time proceeded, this compound broke down into nitrogen, which remained in the atmosphere, and hydrogen, which combined with oxygen to form the hydrosphere. A

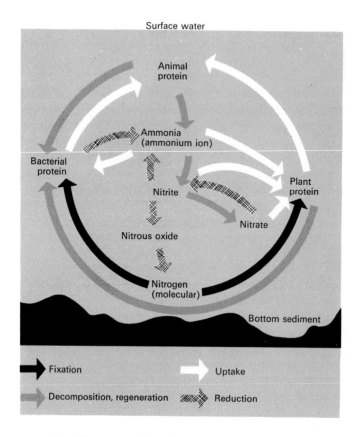

FIG. 12.4 *Nitrogen cycle in the marine environment. (After Hood, 1964, p. 31A.)*

second theory ascribes the origin of our atmosphere to volcanic activity, which yields gases composed of about 8 percent nitrogen. Continued volcanic activity throughout geologic time is thought to be the source of most inorganic nitrogen.

Nitrogen distribution

Vertical distribution curves for nitrogen (Fig. 12.5) conform quite closely to those of phosphorus. The curves have similar shapes and the same four zones are present, although relative concentrations are somewhat different. The maximum concentration peak in the Atlantic is quite pronounced, and Pacific Ocean concentrations are even higher than those of the Atlantic, but uniformly below those of the Indian Ocean.

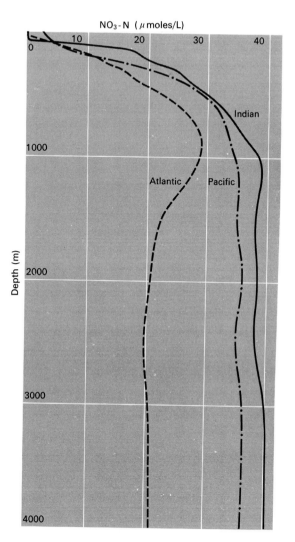

FIG. 12.5 *Vertical distribution of nitrate in the major oceans. (After Sverdrup, et al., 1942, p. 242. Copyright renewed 1970.)*

Differences occur in distribution of the various nitrogen species. Nitrate distribution is probably the most investigated of the species and shows a maximum at several hundred to a thousand meters (Fig. 12.5). Nitrite is fairly restricted in its distribution; it is concentrated in a thin zone near the thermocline. Ammonia (NH_3) is apparently rather uniformly distributed throughout the water column. Nitrate and nitrite-nitrogen increase greatly in bottom water as they proceed from the equator toward the Antarctic. Although phosphorus also increases in this manner, the rate of increase for nitrogen is greater. This may be attributed to the decomposition of organic nitrogen compounds during transport to the south.

Fluctations in distribution

Seasonal changes in nitrogen distribution are closely related to geography and organic activity. For the most part, nitrogen distribution in space and time tends to parallel that of phosphorus. Nitrates along continental shelves show some variation with the width of the shelf. The Atlantic shelf of the United States is fairly broad and here the thermocline restricts circulation during summer months. Off the California coast, where the shelf is narrow and upwelling is common, there is a fairly uniform distribution of nitrate. Away from continental margins, nitrate concentrations may decrease to the point that they limit organic production.

Ammonia and nitrite show seasonal variations which reflect both uptake by growing organisms and, particularly in the case of ammonia, activity of primary consumers. Several studies have shown a direct correlation between ammonia concentration and presence of primary consumers, as ammonia may be present from animal excretion (Fig. 12.4). Locally, ammonia from this source may provide the majority of the nitrogen requirements of a plant population.

Bacterial effects on nitrogen in the sea

The sea contains large numbers of bacteria in various marine environments: on the sea floor, attached to debris and organisms, and in a freely suspended state. Their abundance largely parallels that of other living organisms and bottom detritus. Bacteria tend to be more concentrated along continental margins than in the open sea, because of the concentration of nutrients and organic activity in these areas.

Changes caused by bacterial activity include decomposition of organic compounds, oxidation and reduction, and their own utilization of nutrients. Nitrifying bacteria oxidize ammonia to nitrites and nitrates, and are concentrated in the euphotic zone, and near or in the bottom sediment. Other types of bacteria are able to reduce nitrite and nitrate compounds and to liberate gaseous nitrogen. They are concentrated in bottom deposits of the sea.

Nitrogen-phosphorus ratio

The above discussion on nitrogen and phosphorus has emphasized the close similarity in the distribution of these elements. There are some geographic and seasonal variations in this relationship, although it is, in general, constant (Fig. 12.6). Given a ratio of 15:1 for the atoms of phosphorus and nitrogen, values from the three oceans exhibit a straight-line relationship. For

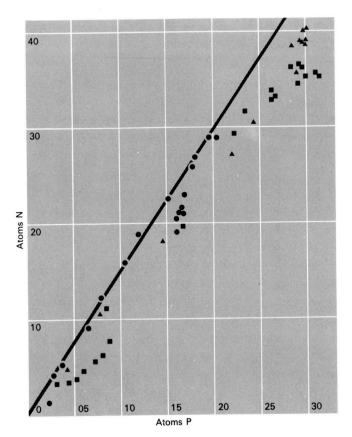

FIG. 12.6 *Relationship of nitrate to phosphate in the oceans. (After Sverdrup, et al., 1942, p. 243. Copyright renewed 1970.)*

this reason, it is possible to determine the concentration in one ocean when the others are known, even though nitrogen and phosphorus are minor constituents of seawater. Care should be exercised in making such calculations because exceptions to this constant ratio may occur where circulation is restricted or in surface waters. Continental shelf areas of the New England coast show ratios as low as 1:1 and in general lower than the open ocean, and winter ratios are generally lower than those for summer.

SILICON CYCLE

Silicon is not a nutrient element, but it is necessary for skeletal material in **diatoms**, one of the most abundant primary producers of food. It is also an important constituent of **radiolarian** tests. Silicon follows a cycle similar to the true nutrient elements because of its importance in diatoms particularly. As nutrient elements are taken up by diatoms in their development, so is the silicon to form the test. Silicon is derived from weathering of rocks of the earth's crust. In the ocean it occurs as various silicate ions, as suspended SiO_2, and in **clay minerals**.

The concentration of silicon in seawater has a wide range, from being undetectable to as much as 300 micromoles/liter. However, concentrations of silicon are considerably below saturation levels in the oceans, despite the relatively large source of silicon via runoff. It is therefore necessary to speculate on its low concentration with respect to its solubility. Such observable phenomena as formation of authigenic silicate minerals, removal by animals, and rapid availability to plants probably contribute to its low concentration.

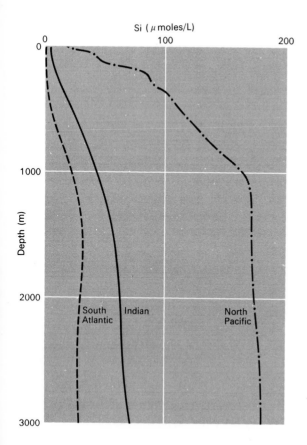

FIG. 12.7 *Vertical distribution of silicate in the major oceans. (After Sverdrup, et al., 1942, p. 245. Copyright renewed 1970.)*

Silicon distribution

Pacific Ocean values of silicon content are appreciably higher than those of the Indian or Atlantic Ocean (Fig. 12.7). The shape of the vertical distribution curves bears some resemblance to those of the nutrient elements in that low values are present near the surface and then the concentration increases to about 1000 meters. However, silicon does not show a distinct maximum at this level but continues at about the same concentration to near-bottom depths.

As might be expected, the nearshore areas are generally high in silicon due to runoff from adjacent land. Upwelling along the edge of the continental shelf serves to supplement these high values. In the ocean basins, there is a noticeable increase in silicon toward the Antarctic.

Seasonal changes in silicon content are closely aligned with those shown by other nutrient elements. Winter shows high values, but during the spring the rapid increase in organic production depletes the supply of silicon quite rapidly.

SELECTED REFERENCES

Armstrong, F. A. J., 1965, "Phosphorus," *Chemical Oceanography*, J. P. Riley and G. Skirrow (eds.), New York: Academic Press, Vol. 1, pp. 323–364. Extensive discussion of the behavior of phosphorus in the marine environment.

Armstrong, F. A. J., 1965, "Silicon," *Chemical Oceanography*, J. P. Riley and G. Skirrow (eds.), New York: Academic Press, Vol. 1, pp. 409–432. Similar to above but discussing silicon.

Barnes, H., 1957, "Nutrient Elements," *Treatise on Marine Ecology and Paleoecology*, J. W. Hedgpeth (ed.), Vol. 1, pp. 297–344.

Harvey, H. W., 1955, *The Chemistry and Fertility of the Sea*, Cambridge: Cambridge University Press, Chapter 3. General treatment of nutrient cycles and the effects of organisms on marine chemistry.

Martin, D. F., 1970, *Marine Chemistry*, New York: Marcel Dekker, Vol. 2, Chapters 5–7. Good and thorough treatment of nutrient cycles in the world ocean.

Raymont, J. E. G, 1963, *Plankton and Productivity in the Oceans*, New York: Macmillan, Chapter 7. Fairly extensive treatment of nutrient elements but most of the data cited are old.

Riley, J. P., and R. Chester, 1971, *Introduction to Marine Chemistry*, New York: Academic Press, Chapter 7.

Sverdrup, H. U., M. W. Johnson, and R. H. Fleming, 1942 (copyright renewed 1970), *The Oceans—Their Physics, Chemistry and General Biology*, Englewood Cliffs: Prentice-Hall, Chapter 7. Good chapter on nutrient cycles and distribution of nutrient elements. Data presented on charts and graphs are not up to date.

Vaccaro, R. F., 1965, "Inorganic Nitrogen in Sea Water," *Chemical Oceanography,* J. P. Riley and G. Skirrow (eds.), New York: Academic Press, Vol. 1, pp. 365–408. Detailed discussion of nitrogen in seawater with emphasis on the inorganic varieties.

PHOTOSYNTHESIS AND PRODUCTION IN THE OCEANS 13

The occurrence and distribution of certain elements necessary for life in the sea were discussed in the previous chapter. A logical sequel is the utilization of these and other materials in the manufacture of organic tissues. To understand interrelationships of various organisms in the sea, it is necessary to briefly describe the energy flow in the ocean system. It is therefore appropriate to consider here the First and Second Laws of Thermodynamics. The First Law states that energy can be neither created nor destroyed; it is transferred or transformed into different forms. The Second Law states that during energy transformation, energy changes from a concentrated to a dispersed condition.

Both of these laws are fundamental to a clear understanding of the production of living tissues and the succession of living organisms which are dependent on one another. In the formation of plant tissues, the energy from sunlight is utilized and combined with essential material from the sea. This process, known as **photosynthesis**, is one of the most important natural reactions:

$$CO_2 + H_2O + \text{minerals} + \text{sunlight} \rightarrow \text{organic matter} + O_2 + \text{heat}.$$

This reaction takes place in all photosynthetic organisms and it serves as the basis for all life in the sea except the **autotrophic bacteria**. The reverse of this reaction is respiration; in this process, organic matter and oxygen provide raw materials for the photosynthetic reaction (Fig. 13.1). These two fundamental processes make up the organic cycle.

Photosynthetic organisms can be visualized as forming the base of the ecosystem, with higher levels being occupied by animals which are the consumers (Fig. 13.1). Considerable differences exist in the **biomass** at each successive **trophic level**. For instance, the herbivores, primary consumers of the system, constitute much less biomass than the plants (producers). The same is true for carnivores, and so on. The reason for this is that in each successive level, a certain amount of energy is returned to the system in the

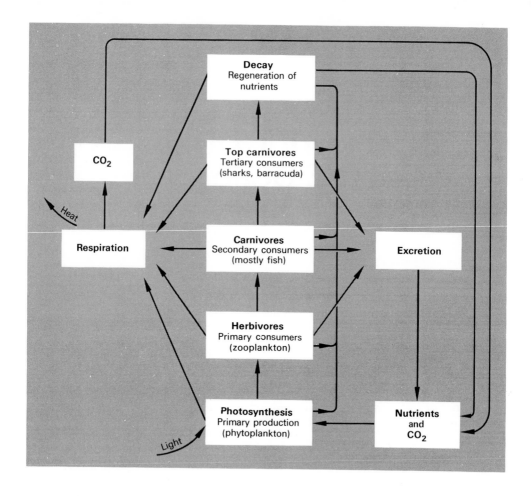

form of waste. This may be in the form of nutrient materials or carbon dioxide.

PRODUCTIVITY

The rate at which photosynthesis occurs is the real measure of productivity, and such a measure must include a time factor. The actual amount of green plants, the **standing crop**, is not a true measurement of productivity, but it is commonly related to it. A considerable biomass of photosynthetic organisms indicates high productivity, but the opposite is not always true. Productivity of an environment may be high, but the standing crop may not reflect this because of a high rate of consumption by herbivores.

Measuring productivity

It is often necessary for ecologists to know the rate of production for a given environment. This rate can be measured in a variety of ways; the method used is largely dictated by the environment under study. This is one of the few types of scientific investigation in which the marine scientist seems to have an advantage. In an aquatic environment there are many methods of

◀**FIG. 13.1** *Flow diagram showing principle components of the organic cycle in the world ocean.*

measuring plant productivity that are not applicable to the terrestrial environment. These involve chemical changes in the water due to variations in production. They cannot be used on land because of the more rapid diffusion and circulation of the atmosphere, and because of contamination by nonorganic terrestrial phenomena. In order to measure productivity, it is necessary to determine either the actual amount of organic matter formed during photosynthesis, or the utilization or formation of other constituents necessary to photosynthesis.

Oxygen content. The most commonly used index of productivity is the dissolved oxygen content of the water. Oxygen is produced along with organic matter during photosynthesis, and it is produced in amounts directly related to the rate of organic production. Determinations of the amount of oxygen being produced over a given period of time, therefore, provide close approximations of productivity.

Since the photosynthetic reaction is diurnally controlled by the position of the sun, as well as by its presence or absence, it is usually necessary to collect water samples for productivity determination over a 24-hour period, for example at 3-hour intervals (Fig. 13.2). This provides a good measure of the changes in production related to the sun's position.

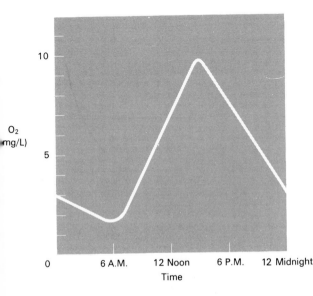

FIG. 13.2 *Typical diurnal distribution plot of the oxygen content of seawater.*

Harvest method.   A much less accurate method of determining productivity is to measure the acutal growth of a producing biomass. In this method a small area, for example a square meter, is surveyed and all plants are collected, weighed, and replaced in their original position. The procedure is repeated at selected time intervals and the growth of the producing biomass with time is the productivity. Such a method is merely an estimate and is restricted to shallow areas where collecting is possible. It only accounts for benthic production, and the removal and replacement of plants alters their rate of growth somewhat.

Carbon dioxide assimilation.   Equally important to the production of oxygen in the photosynthetic reaction is the assimilation of carbon dioxide. It is, therefore, possible to determine plant productivity by measuring changes of CO_2 content.

From the discussion in Chapter 11, it is apparent that pH changes reflect changes in various carbonate species. With modern pH meters which are sensitive to the nearest 0.002 pH unit, small changes can be detected. However, because of the decrease in pH with temperature increase, it is necessary to distinguish pH changes due to temperature from those caused by CO_2 uptake. Actually H_2CO_3, HCO_3^-, and CO_3^{2-} all may provide carbon for conversion to organic matter during photosynthesis. The buffering capacity of seawater may tend to mask changes in hydrogen ion concentration. For this reason, a fairly large uptake of CO_2 is necessary to produce enough pH change to give accurate measurements.

Another problem that exists is the precipitation of calcium carbonate. In an area where this occurs, it is not possible to use the CO_2 uptake as a production index unless inorganic carbonate changes can be distinguished.

The concentration of various species in the carbon dioxide-carbonate system shows significant diurnal changes which are directly related to photosynthesis (Fig. 13.3). Free carbon dioxide is present during respiration at night but disappears as photosynthesis takes place during the daylight. The total carbonate, particularly HCO_3^-, follows a predictable relationship to daylight during a 24-hour period.

Carbon-14 fixation.   A slight modification of the carbon dioxide technique is the use of the radioactive carbon-14 isotope (^{14}C). A quantity of radioactive carbonate is added to a known volume of seawater. The plants suspended in the water continue to photosynthesize for a given period of time, after which the water is filtered and the organisms retained. A Geiger counter is then used to measure the radioactivity in the organic tissues. In order to have a precise measure of the actual uptake of ^{14}C by the plants, a "dark" sample from the same water is also checked for ^{14}C content, and the difference

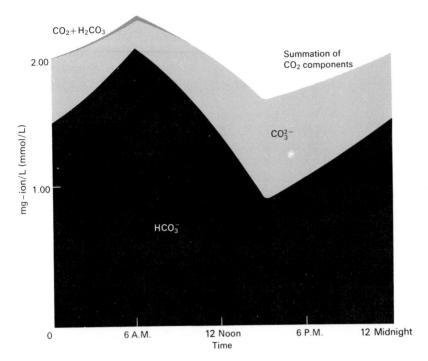

FIG. 13.3 *Changes in the carbon dioxide system through a 24-hour period. (K. Park, D. W. Hood, and H. T. Odum, 1958, "Diurnal pH Variation in Texas Bays and Its Application to Primary Production Estimation," Publs. Inst. Mar. Sci. Univ. Tex.* **5,** *60.)*

between the two samples represents the total production for that period of time. The ^{14}C method is the most sensitive technique for measuring net production, but it does not provide any data on respiration.

Nutrient uptake. The loss of nutrient elements such as phosphorus and nitrogen to organic matter via photosynthesis is also a measure of productivity. Water samples are collected periodically and analyzed for the nutrient elements. The accuracy of this method is not good because some phosphorus is lost to inorganic precipitates as well as to living organisms, and some is recycled via excretion or decay of dead tissue. During spring or other times of significant uptake by plants, such methods of analysis provide better estimates of productivity than at other times.

ENVIRONMENTAL FACTORS IN PRODUCTIVITY

Many time-related and geographic variations in the marine environment cause fluctuations in the rate of photosynthesis. These may include changes caused by diurnal and seasonal cycles, circulation of nutrients, or depth and clarity of water. Any or all of these may cause variation in light intensity and/or supply of raw materials.

Being a sunlight-induced process, photosynthesis shows a characteristic daily cycle (Fig. 13.2). During the daylight hours, organic production takes place and exceeds respiration, but after sunset, production stops and respiration continues. In order for a marine community to show continued growth and development, there must be a net production over respiration. As the seasons change and therefore the day length changes, there is significant corresponding change in the daily production at a given location. The incident angle of sunlight, which varies with the seasons, also causes changes in productivity.

Primary organic production is limited to the upper 100 meters or so of the ocean because of limitations imposed by light penetration. The 100-meter depth is a general average, and there is considerable range on either side of this boundary. The change in light penetration with depth can be determined accurately by photoelectric cells which measure the amount of light reaching a particular depth. Light intensity is measured when and where water samples are collected, and thus production can be directly correlated with light intensity. Maximum efficiency of light utilization in photosynthesis occurs about 20 meters below the water surface (Fig. 13.4).

There is a wide variety of phenomena in the oceans that cause local changes or geographic variations in light penetration. Generally, light penetration is less on the continental shelf than in the open ocean. This phenomenon is a simple function of turbidity in coastal areas where runoff, wave action, and circulation stir bottom sediments and inhibit light penetration. However, turbidity is compensated for by the high abundance of nutrient elements, so that production is actually quite high in coastal areas.

Wind is a factor in production in that it causes waves, currents, and therefore turbidity. It also produces choppy water which may cause more reflection of the sun's rays than calm water, and thus it may slow down photosynthesis. Wind also causes some problems in measuring production, particularly by the oxygen method. Agitation of water causes considerable diffusion of oxygen between the sea and the atmosphere.

Clouds may be a major factor in production locally and for short times. It is standard practice to record the percentage of cloud cover when taking water samples for oxygen analysis. Such information may be useful in explaining anomalous O_2 values on the diurnal curve. The same can be done for wind velocity.

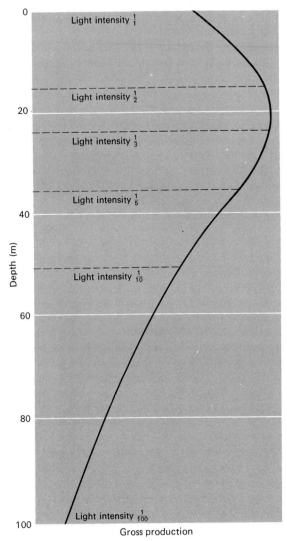

FIG. 13.4 *Vertical distribution of light intensity and gross production.*

Temperature of the water is an important indirect factor in primary production. It is considered indirect because temperature change does not directly cause increase or decrease in the rate at which photosynthesis proceeds. The primary effect of temperature lies in the formation and deterioration of the thermocline and its influence on nutrient availability (see Chapter 12). The famous spring plankton blooms are the result of nutrient upwelling and increased production because of water becoming isothermal. Once the thermocline is stabilized, nutrient upwelling is uncommon.

Temperature is also important in the solubility of gases involved in photosynthesis. Both CO_2 and O_2 are more soluble in cold water than warm. If all other factors are the same, photosynthesis sould be enhanced by cold water because of the relative abundance of CO_2.

PRODUCTION IN SELECTED AQUATIC ENVIRONMENTS

Some general characteristics are present in many aquatic environments. It is appropriate to briefly compare both marine and nonmarine water bodies to gain a broad insight into their relative productivity. Each environment mentioned below is considered with respect to its average characteristics; exceptions exist for nearly all the generalities described. *Streams* for the most part are high in production, but there is considerable variation seasonally and geographically. Being a direct source of runoff from the adjacent land, they have the first chance at utilizing available nutrients. Of course, nutrient content varies greatly from place to place in response to presence or absence of soil and fertilizers, and in response to type of soil. Highest production would be expected where chemical weathering proceeds rapidly and develops rich soils.

Circulation of streams is good, so that mixing provides fairly homogeneous distribution of nutrients. During the winter months when runoff is low or nonexistent, productivity goes down. Ice cover on the streams is also a factor in limiting production, because of its impedance of light transmission and also diffusion of carbon dioxide and oxygen.

Production in *lakes* is largely dependent on the same factors as in streams, but with somewhat more variability. The depth factor must be considered because the depth of many lakes exceeds the euphotic zone. The net effect is to decrease production because much greater volumes of water dilute the nutrients. Productivity in lakes is, in general, inversely proportional to the depth. Lakes Michigan and Superior, for example, are deep, well circulated, and low in nutrients. Such lakes are called **oligotrophic** and are characterized by low productivity. Shallow lakes commonly are the opposite, having abundant nutrients and high productivity. Such lakes are called **eutrophic.**

Pollution is becoming increasingly important as a deterrent to production. Many lakes and streams, particularly in industrial areas, are becoming supersaturated with organic debris and nutrients. The organic materials uses up oxygen in the decay process and may use so much that the water body can no longer support life.

Coastal bays and *estuaries* are among the most productive of all environments. These water bodies are shallow, receive large amounts of nutrients from runoff, and are generally well circulated. Nutrients may be restricted in arid regions where runoff is sporadic, and pollution may be a problem if circulation is restricted. Tidal currents are the primary source of circulation and are supplemented by wind-driven currents and waves.

Reefs have the greatest rate of production of any modern environment. They may occupy only a small area, but the production per unit area is great. It is enhanced by the concentrating effect of shallow water over the reef. Water circulation, particularly vertical movement adjacent to reef areas, provides a supply of nutrients, and water is well oxygenated. Reefs characteristically are covered by clear water, which is conducive to very efficient utilization of light. Within the living reef there is a general zonation of production, with the highest rates around the edge, particularly on the windward side, and lower rates near the central portion. The zonation is a result of the nutrients being utilized immediately as they move across the reef.

The open ocean shows a low overall production rate due primarily to its great depth. All production is concentrated in the thin euphotic zone, which is only about 2 or 3 percent of a vertical column of water to the ocean floor. Each ocean basin has its own characteristic production rate, with some local variation.

PRODUCTIVITY IN OCEAN REGIONS
Production rates in the open ocean shows a general decrease from the coastal margins to the central basin areas (Fig. 13.5). It is the expected pattern, as shown in previous discussions. In addition some productivity patterns are related to the major surface circulation cells (see Chapter 6). For example, the surface convergence of westward-moving waters at the equator causes narrow zones of highly productive water far out in the open ocean.

In general, tropical ocean waters have low production rates and show little variation with changing seasons of the year. A notable exception is the narrow zone of water adjacent to the equator, which is relatively high in production. Rates here are consistently two or three times those of other tropical waters. Near the subtropical areas, where there is some seasonal stratification, production proceeds at slightly lower rates than near the equator.

Relatively stagnant oceanic areas in the center of circulation gyres, such as the Sargasso Sea, also have low production rates. The primary reason is the general lack of nutrients caused by low to moderate circulation and the great distance from land masses. Another contributing factor is that in such a stagnant basin the thermocline is likely to be more stable, thereby preventing vertical movement of nutrients.

The most variable open ocean areas are in the temperate latitudes. Seasonal changes produce corresponding changes in production. However, these variations may not be regular and predictable from one year to the next,

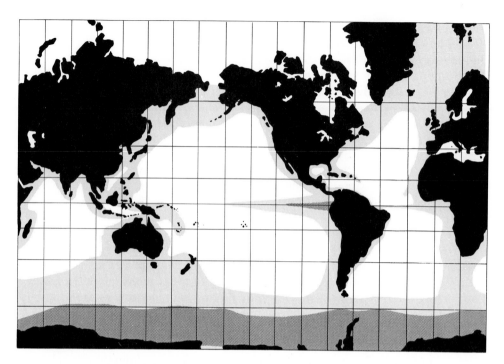

FIG. 13.5 *General distribution of relative productivity. Heavy shading indicates areas of hightest production; light shading indicates moderate production.*

due to climatic variations over a few years. It is necessary, therefore, to have data from several successive years in order to come to any general conclusions. Some regional changes in productivity are present in such areas as the North Sea, Labrador Sea, and other partially isolated portions of major basins. In these areas circulation with the major basin is not complete and productivity may be higher or lower than the major basin due to varying nutrient supply and seasonal mixing.

The polar areas show striking contrast in productivity, with the Arctic being low and the Antarctic extremely high. There is distinct seasonal change in the Arctic, with fairly high rates during times of open water, but the ice cover limits photosynthesis to such an extent that the overall rate is low.

The high productivity of the Antarctic is due to the unique circulation of the area. Deep, nutrient-laden water moves to the south from the low latitude areas (Chapter 12). This water rises to the euphotic zone in response to the northerly movement of intermediate and near-surface water. As a result, the supply of phosphorus and nitrogen is so great that it cannot be depleted by photosynthesis, despite the tremendous productivity of the area.

SELECTED REFERENCES

Coker, R. E., 1954, *This Great and Wide Sea*, Chapel Hill: University of North Carolina Press, Chapters 14 and 15. Short introduction to productivity and metabolism in the oceans.

Nielsen, E. Steemann, 1963, "Productivity, Definition and Measurement," *The Sea*, M. N. Hill (ed.), New York: Wiley, Vol. 2, Chapter 7. Excellent description of various techniques for determining productivity in the sea.

Odum, E. P., 1959, *Fundamentals of Ecology*, (2nd edition), Philadelphia: W. B. Saunders, Chapter 2. Brief and general discussion of some biochemical cycles in the oceans.

Odum, H. T., and C. M. Hoskin, 1958, "Comparative Studies on the Metabolism of Marine Waters," *Publ. of the Institute of Marine Science*, **5**, 16–46. Rather specific article, but one which includes a great deal of information on methods of productivity measurement.

Raymont, J. E. G., 1963, *Plankton and Productivity in the Oceans*, New York: Macmillan, Chapters 7–10. Lengthy and comprehensive treatment of productivity in the marine environment.

Riley, J. P., and R. Chester, 1971, *Introduction to Marine Chemistry*, New York: Academic Press, Chapter 9. Excellent and detailed discussion on productivity.

Ryther, J. H., 1963, "Geographic Variations in Productivity," *The Sea*, M. N. Hill (ed.), New York: Wiley, Vol. 2, Chapter 17. Mostly devoted to latitudinal changes in productivity.

Strickland, J. D. H., 1965, "Production of Organic Matter in the Primary Stages of the Marine Food Chain," *Chemical Oceanography*, J. P. Riley and G. Skirrow (eds.), New York: Academic Press, Vol. 1, Chapter 12. One of the most extensive discussions of primary production currently available. Thorough understanding requires some background in the subject.

BIOLOGICAL OCEANOGRAPHY

PART **IV**

BIOZONES 14

The sea has been described as a uniform environment, but this is true only relative to the terrestrial environment. Actually there is a wide variety of ecological niches in seawater and on the sea floor. Such variables as temperature, depth, pressure, currents, nutrients, and light penetration may limit distribution of certain organisms. In addition, the different sediment properties are quite important in the distribution of organisms which live on the ocean floor.

The first attempt to zone the distribution of organisms in the oceans was by Edward Forbes during the summer of 1834 in the North Sea (see Introduction). Since that time there have been many different classifications of marine environments.

Terminology in these classifications was not standardized at first, and there was much confusion in the literature as a result. These inconsistencies were largely eliminated through the efforts of the National Research Council Committee on Marine Ecology and Paleoecology. Formed in 1940, its efforts led to a monumental two-volume treatise in 1957 (see Barnes reference at end of Chapter 12).

The oceanic realm consists of two broad environments: the **benthic** or bottom environment, and the **pelagic** or water environment. Within each of these there is a variety of environments, each with a fairly narrow range of characteristics.

PELAGIC ENVIRONMENTS

Organisms that live in the ocean but not on the bottom occupy the pelagic environment. Commonly referred to as the "water environment," it extends from low-tide level to the deepest part of the oceans. Within the pelagic environment there are two major subdivisions; the **neritic** province, which covers the continental shelves, and the **oceanic** province, beyond the shelves (Fig. 14.1). Although a depth of 200 meters has been used as the boundary between these, it is only an approximation, as the depth at the shelf edge has a wide range (see Chapter 2). It is really impossible to draw a line that

239

separates these environments, because chemically and biologically there is a gradual transition from one to the other. However, the true neritic environment is distinct from the true oceanic environment. Chemically, there are more nutrients in the neritic environment and it shows local and temporal variation, whereas the oceanic environment is more homogeneous and is generally low in nutrients. There are also distinct faunal and floral differences which will be discussed in succeeding chapters.

Waters of the neritic province and their organisms are better known than oceanic waters because of their shallow depth and proximity to land. Organisms that occupy these shallow waters must have at least moderate tolerance to variation in the environment. Storms cause turbulence, runoff causes extreme local chemical changes, and currents distribute nutrients. These and other phenomena contribute to highly productive waters. In the neritic environment, the shallow depth would not have high pressures, virtually all the water is within the photic zone, and thermoclines are less stable than in the open ocean. These variables may cause distinct differences from the fauna and flora in the oceanic environment.

Beyond the continental shelves, the oceanic environment extends to depths in excess of 10,000 meters. Within this environment, there are vertical subdivisions, but geographically the environment is relatively unchanged. The four subdivisions based on depth ranges are the **epipelagic**, which corresponds in depth to the neritic or about 200 meters below sea level, the **mesopelagic**, which extends to 1000 meters, the **bathypelagic**, extending to the average depth of the ocean floor (about 3800 meters), and the **abyssopelagic** or extremely deep zone (Fig. 14.1). The boundaries between them are indefinite and somewhat variable. Each zone has its own distinctive characteristics and associated organisms.

The upper epipelagic zone closely corresponds to the photic zone, but extends slightly below it. In the open sea, this is the zone of highest productivity, and although it represents only a small percentage of the total volume of the pelagic environment, it contains most of the life. Temperatures are generally high and subject to seasonal and latitudinal variation.

The moderate depth range of the mesopelagic zone is characterized by maximum change in temperature. Because it is below the zone of light penetration, no plants are produced here, and the primary consumers rely on falling detritus for food. Many of the small floating animals migrate diurnally from this zone to the epipelagic zone to feed on plants.

Actually the bathypelagic zone is the uppermost truly uniform zone of the ocean. It is essentially constant in its temperature, absence of light, and in most other ways. The only appreciable changes are caused by density gradients which induce deep-water currents. The abyssopelagic zone displays the same general features. Both the bathy- and abyssopelagic zones are sufficiently deep so that great pressures cause enough viscosity change to

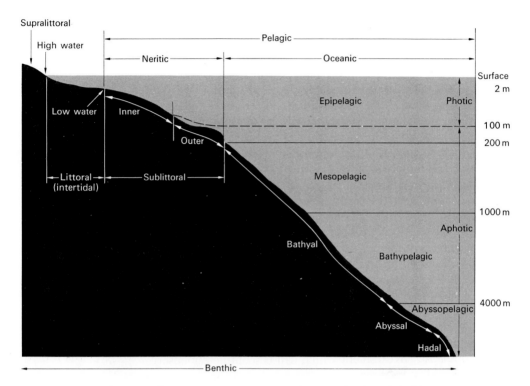

FIG. 14.1 *Classification of marine environments. (After J. W. Hedgpeth (ed.), 1957, Treatise on Marine Ecology and Paleoecology,* Geological Society of America Memoir 67, vol. 1, p. 18.)

partially impede movement of some organisms. Animals at these depths are adapted to these great pressures and would not survive if raised to shallow depths where pressure is low.

BENTHIC ENVIRONMENTS

The benthic or bottom environments of the ocean are less homogeneous than their pelagic counterparts because they incorporate bottom characteristics with features of the overlying water. In addition, the bottom environment between high and low tide is unique in that it does not have a pelagic counterpart.

The intertidal zone, generally called the **littoral** zone (Fig. 14.1) by marine biologists, is perhaps the most rigorous of all marine environments. Some may argue that because this environment is not always covered by seawater it is not properly a marine environment. However, the organisms which occupy this zone must tolerate marine salinities, and it is therefore considered a transitional environment to the marine realm.

There is actually a relatively narrow zone above normal high tide which has specific characteristics and is occupied by a distinct assemblage of organisms. This supratidal or **supralittoral zone** is subjected to spray of waves or inundation during storms. Most of the time, it is exposed to the atmosphere. A few algae and other plants have adapted to the unique conditions of this environment; it is also a present site of dolomite formation (Chapter 11).

The littoral zone (Fig. 14.1) also requires tolerant organisms, since it is alternately wet and dry as tides move up and down. However, the cyclic nature of this alternating condition is regular, and many organisms are specially adapted to survive in it. Mobile organisms can move with the tides and some clams do so, always found buried in the substrate at the strand line. Some animals bury themselves in the moist substrate during low tide to prevent desiccation, and there they wait for the tide to come in, bringing food and relief from drying conditions. Certain algae which live on the substrate can retain enough moisture in their tissues to survive the period of exposure during the tidal cycle. The littoral zone is actually the best known of all marine environments because of its complete accessibility. It is the only marine environment which can be completely observed.

The **sublittoral** is better known as the continental shelf, and it is also relatively well known compared to deeper benthic environments. This zone contains the densest population of bottom-dwelling organisms. It is commonly subdivided into inner shelf and outer shelf, with the boundary being the base of the zone of light penetration (Fig. 14.1). As mentioned before, the photic zone does not extend as deep over the shelf as in the open ocean because of bottom turbulence on the shelf. However, the shelf environment may be entirely within the zone of light penetration in some areas, and there this subdivision would not apply. Where both shelf conditions are present, there is a distinct difference in the organisms in each, reflecting the change from a zone of production (inner shelf) to one without living plants.

Beyond the continental shelf, the benthic environment is generally stable and homogeneous. The exception would be on the continental slope and rise, where there may be moderate movement of sediment and the substrate is fairly unstable. The **bathyl** environment includes the slope, rise, and some of the ocean floor. Originally, this was termed the azoic (without life) zone by Forbes, but now we have collected and photographed living organisms from all depths. The substrate at these depths is fairly homogeneous in both texture and composition. There is little sediment movement and lower rate of accumulation with respect to shallower depths. These factors contribute to a relatively uniform bottom and a stable environment.

The **abyssal** and **hadal** zones (Fig. 14.1) include the abyssal plains and oceanic trenches, respectively. The fauna are restricted by extreme conditions of pressure, cold, and darkness, and general lack of abundant food supply. Organisms living in these environments are carnivores and scavengers who feed on detritus that falls from above. The substrate is uniform and quite soft in most places. The details of these and other deep-sea sediments will be included in Chapter 23.

15 ECOLOGICAL FACTORS IN THE MARINE ENVIRONMENT

Ecology is most commonly defined as the study of organisms in their relationships to each other and to their environment. Marine organisms live in a relatively uniform and stable environment compared to those that occupy a terrestrial environment. Even so, there is considerable variation in conditions, particularly in the littoral and sublittoral environment. A wide range of conditions varies with time, whereas others are not time dependent. Ecological factors that have an effect on the growth and distribution of marine organisms can be conveniently categorized into physical, chemical, and biological types.

Regardless of the niche or mode of life of a marine organism, there are numerous ecological factors which control its distribution and development. In a single chapter it is not possible to treat them in anything but a general and superficial manner. Although it is necessary to discuss these factors separately, the reader should keep in mind that each individual organism is constantly exposed to the entire system of factors and their complex interactions.

PHYSICAL FACTORS

Light
Directly or indirectly, light controls the distribution of the majority of marine organisms and it is the single most important ecological factor in the sea. In addition to its obvious importance in photosynthesis, sunlight is also directly important for vision and heat, and indirectly for chemical and physical changes. There is considerable range in the depth to which light rays penetrate seawater, but regardless of this there is selective absorption of certain waves in the spectrum.

For obvious reasons, vision is quite important to animals in searching for food and shelter and for escaping predators. Many advanced invertebrates, particularly arthropods, and fish have visual organs, whereas some of the lower invertebrates have light-sensitive organs of a less sophisticated nature.

The availability and intensity of light at various depths is subject to both periodic and nonperiodic changes. Such periodic changes are primarily diurnal and seasonal, with many organisms responding to them. Nonperiodic changes include cloud cover, roughness of the water surface, turbidity of water, and ice cover. The sea actually is subjected to greater nonperiodic changes in illumination than is the terrestrial environment. All of these changes are important because they exert control on the rate of photosynthesis and also affect the distribution of nonphotosynthetic organisms.

Temperature

Throughout most of the oceans' total volume, the temperature is nearly constant, with a range of less than 5°C. This volume includes nearly all of the subthermocline portion of the oceans. In that environment temperature changes have little effect on the organisms present. If they are living there, they can tolerate the cold, and the temperature does not fluctuate markedly under any conditions, with respect to time or space. Above the thermocline it is just the opposite; temperature varies with time (diurnally and seasonally) and with respect to latitude. Probably the most significant geographic effect of temperature is the high diversity of species in low latitudes. Distribution of certain taxa is controlled by temperature, and organisms can be classed as **stenothermic** or **eurythermic**. One of the best known temperature-restricted groups is that of reef-forming corals, which are generally restricted within the 20°C isotherm (Fig. 15.1).

Growth and reproduction of a population above the thermocline is likely to be temperature dependent. Under optimum temperatures there would be nothing to restrict functions of the organisms and they would develop and reproduce normally. If the optimum range was exceeded somewhat, the population would probably tolerate the conditions, but there might be a stunting of their growth or other detrimental effects. The temperature might even change to exceed the limits suitable for reproduction, so that, whereas individuals in the population might tolerate the change, the population would be doomed for want of propagation.

There seems to be a general relationship between temperature and size of an individual, and also the length of its life. One of the general axioms of marine ecology is that larger organisms tend to be in areas of low temperatures. There are many exceptions to this rule, however, with diatoms and benthic Foraminifera being prime examples.

Within a species, higher temperatures seem to reduce the length of life. Temperature changes may have significant effects on the respiration rate or metabolic rates of organisms. Extreme temperature conditions, particularly high temperatures, have caused mass mortality in certain species of the affected area. It is thought that death of the individuals was caused by lack of oxygen due to the high temperatures driving it off to the atmosphere. In-

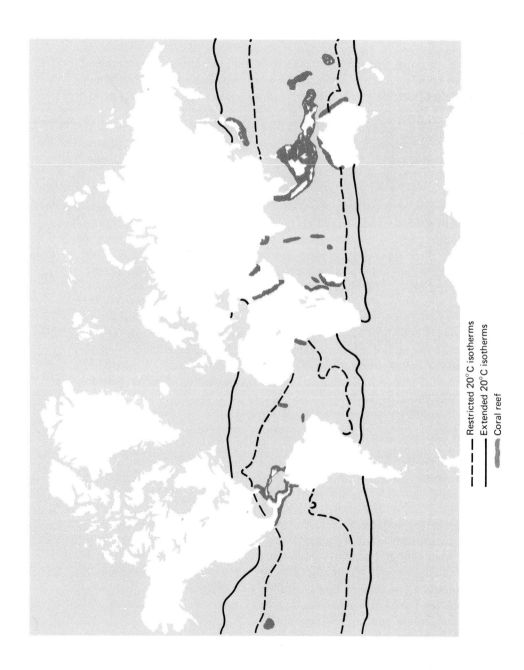

Restricted 20°C isotherms
Extended 20°C isotherms
Coral reef

◄**FIG. 15.1** *Restriction of coral reef development by temperature. Dashed lines represent restricted 20°C isotherms; solid lines represent extended 20°C isotherms. (After J. W. Wells, 1957, "Coral Reefs," Treatise on Marine Ecology and Paleoecology, J. W. Hedgpeth (ed.), Geological Society of America Memoir 67, vol. 1, plate 9.)*

direct effects of temperature changes include a reduction in salinity tolerance because of less than optimum temperatures.

Mobility of floating and swimming organisms is somewhat affected by variations in viscosity and density caused by temperature changes. Such minor changes do not present a real problem for fish, but the relatively helpless floaters may be displaced markedly from their normal niche.

Salinity

Like temperature, salinity is almost uniform in the open ocean. As a result most marine organisms are adapted to life in water of constant salinity and can tolerate only slight changes in its concentration. Such organisms are called **stenohaline**. In contrast are the animals that occupy marginal areas of the sea where runoff and evaporation may cause considerable and erratic changes in salinity. The organisms occupying these areas must be able to tolerate wide-ranging salinity conditions (**euryhaline**). There are also organisms that are stenohaline but may be adapted to low salinity (brackish) or high salinity (hypersaline) conditions.

Body fluids in most marine invertebrates have a concentration that is the same as seawater. On the other hand, bony fish, crabs, and some worms maintain salinity of body fluids that is higher or lower than seawater. When organisms of these types are placed in an environment of salinity different from that of their internal fluids, they must adjust by osmosis in order to bring their system into the necessary relationship with the environment. The invertebrates are permeable to water, so they can readily gain or lose it depending on the concentration of their environment. In the bony fishes, for example, excess salts are secreted through the gills and water is taken in by swallowing.

Salinity variation causes another significant obstacle for many marine organisms by substantially changing the specific gravity of seawater. Such changes are of most consequence to floating organisms (plankton) that require support of their tissues by the buoyancy of seawater.

Pressure

Perhaps the most predictable and uniform ecological factor in the oceans is that of pressure. Water itself compresses very little, so that density changes related to pressure are negligible. What is significant is the pressure exerted on organisms and on gases, both in seawater and in the organisms themselves. The swim bladders of fish are quite compressible and will decrease to about half their volume in ten meters of descent. This change in swim-bladder volume is not of any apparent harm unless the fish ascends to the point that gas in the bladder expands and ruptures the bladder. Some fish can expel gas to compensate for decreasing water pressure.

Most marine organisms tend to live within a rather narrow range of water depth and, therefore, pressure. However, some organisms may experience changes in depth of a few hundred meters during daily vertical migration; this means pressure changes of 20 or more atmospheres. Such vertical tolerances are characteristic of some fishes, cephalopods, and plankton. Certain organisms live in deeper water at low latitudes than they do at higher latitudes, and they do not migrate vertically.

There are also indirect effects of pressure on organisms, particularly its effect on gas solubility. Carbon dioxide is increasingly soluble with depth, due to pressure, and consequently is not available at great depths for organisms that build skeletons of calcium carbonate. Pressure also exerts an additional minor influence on organisms by increasing the viscosity of protoplasm and slightly changing the rates of certain body functions.

Substrate

In the ocean the type of substrate ranges widely among various unconsolidated sediments and solid surfaces such as rock, shell, or even metal and wood. Bottom dwellers and some swimming organisms may be limited by the type of substrate. Plants require a firm substrate of sediment or a solid surface. Some grow in sandy or firm mud sediments where the plant is held by roots or rootlike structures, whereas others, along with attached animals, adhere to rocks, pilings, or other solid surfaces. Certain forms of barnacles prefer rough surfaces. Boring organisms, such as certain clams, can penetrate a variety of rock or wood types and some worms are found only in wood.

The vast majority of oceanic substrate is unconsolidated sediment which ranges from coarse gravel to fine muds; it may be soft or firm. Organisms that move slowly over the bottom require a moderately firm surface to prevent them from sinking. The snail *Aporrhais pespelecani* has large fingerlike protuberances that provide considerable surface area, thereby preventing the shell from sinking in very soft mud; they act somewhat like snowshoes. Burrowing organisms are more sensitive to sediment texture and also to substrate stability. Shelled animals that burrow prefer sand or mud substrate and will not be found in gravel. Certain worms that build armor on their tubes require coarse particles of sand or broken shell material. Some varieties of Foraminifera build their shells (tests) of sand grains of various composition. In both of the above situations the substrate composition could limit the distribution of the organism.

The substrate acts as a camouflage for many bottom-dwelling fish and invertebrates. Certain fish are almost impossible to see if they are on the bottom type for which their coloration and markings are designed; however, if they ventured onto different bottom types, their bodies would probably be highly conspicuous. Actual physical shelter is provided by the substrate for many micro-organisms that live within the sediment. It also provides a collecting area of nourishment for all bottom-feeding organisms. Some pass the sediment through their digestive tract, utilizing organic debris and excreting the mineral matter. Others have the ability to separate nutrient materials before ingestion.

Currents

Circulation of water may be detrimental to some organisms and quite beneficial to others. Floating organisms and their larvae rely on water movement to transport them and thereby facilitate migration of the various taxa. In addition to transporting organisms, currents carry nourishment and sediment, and resupply gases or nutrient elements that are taken directly from the seawater by the organisms. Currents that cause sediment to go into suspension may be harmful in that they reduce light penetration and the sediment may clog the filtering apparatus of certain animals. Conversely, the stirring up of bottom sediment may cause organic debris to become suspended and therefore available as nourishment for organisms. Strong local currents may be disastrous to some of the bottom dwellers by transporting them to an uninhabitable environment.

By and large, currents provide many services to organisms. The most significant is that of supplying food for immobile creatures. Such organisms rely entirely on currents for their food. Circulation is necessary to carry away wastes and resupply the environment with essentials that are preferentially taken up by organisms. It also provides the homogenization of ocean waters that makes the environment essentially uniform and thus allows considerable movement by planktonic and nektonic organisms without ecological barriers. Seasonal changes in current patterns may cause seasonal changes in the faunal composition of a particular region.

Waves

The only parts of the marine environment in which waves are ecologically significant are the nearshore and intertidal zones. With the exception of the reef areas, this zone is restricted to the margins of land bodies. The considerable range in wave size causes a variation in the area affected by waves. Like currents, waves are capable of stirring up sediment. They also provide a hazard to some bottom dwellers, which must be well attached to their substrate to avoid displacement. Environments such as jetties, piers, or rocky cliffs where waves break with considerable force harbor organisms which are especially adapted to this continual pounding. Sea urchins, barnacles, and similar animals are found here along with some types of algae. The amount of wave force can restrict some of the more delicate forms. Some species of reef corals thrive on the high level of physical energy created by waves, and actually grow in that direction.

The spray of waves and the resulting swash back and forth across the intertidal zone is beneficial to many organisms that occupy this rigorous environment, particularly those organisms that have become adapted to the upper part of the littoral zone. Food is supplied to intertidal organisms by the sweep of the waters back and forth.

Organisms may be influenced by internal waves and their propagation. This has been found to be of particular significance in some commercial fish.

Indirectly, waves are significant to organisms in that they aid interchange of gases between the atmosphere and the sea. This is particularly important for maintaining proper oxygen and carbon-dioxide levels.

Tides

The cyclic rising and lowering of sea level that we call tides provides two distinct conditions which are factors in littoral ecology: significant currents and the alternating emergence and submergence of the littoral zone. Superimposed upon the daily cycle is the lunar cycle, so that there are monthly changes in the littoral environment as well.

Tidal currents are to a large extent controlled by the topography of the littoral zone and the local tidal range. In an area of small range and a straight, gently sloping bottom, tidal currents are insignificant in terms of the ecology of the environment. A contrasting situation is found in the Bay of Fundy, where there is considerable tidal range and a very wide littoral zone. Here, there are many permanent or semipermanent tidal channels which are ecologically quite distinct from the tidal flat environment. A similar situation is found along the western coast of Andros Island in the Bahamas or on a major delta. These tidal channels are areas of higher physical energy than the tidal flats and also contain a somewhat different community of organisms.

The organisms found in tidal channels are submerged almost all the time, so that there are digitate areas of the littoral zone that contain fauna and flora like those in the lower part of the littoral zone but that extend in some cases miles inland from the edge of the sublittoral zone. Tidal flat areas extend over most of the littoral zone, in places where environmental distinctions can be made within the zone. Tidal flats are exposed to the atmosphere for much longer periods of time than tidal channels and contain a more rigorous and highly adapted community. Generally, the diversity of taxa is limited in such rigorous environments.

Tidal currents are important for most of the same reasons as the more permanent marine currents. They provide circulation of food, nutrient elements, dissolved gases, and waste materials.

Tidal pools, which are small enclosed bodies of water that are regularly affected by tides, also provide a unique environment. Water level fluctuates according to tidal movement, providing a flushing of the temporarily stagnating pool. Salinities increase while tides are low, so that euryhaline organisms inhabit such areas. They are continuously submerged, which is unique in the littoral zone.

The alternating emergence and submergence of the littoral zone requires special adaptations of the organisms occupying this zone. Desiccation is the most critical problem, and organisms living there have capabilities for storing moisture. To a great extent, this involves evaporation-prevention features such as gelatinous coatings on plants or shells on animals.

Activities of many animals take place in regular patterns that follow tidal cycles. These include feeding, reproduction, and migration and are not confined to the littoral zone or to any single mode of marine existence.

Acoustics

Although there is still much research to be done, it appears that the transmittal of sound through seawater is an ecological factor worthy of some consideration. First of all, several varieties of marine organisms such as mammals, fishes, and crustaceans produce sounds. Fluctuations in the density of seawater can cause changes in the absorption of these sounds and therefore the distance over which they travel. Second, certain commercial fish appear to react to sounds emitted by echo-sounders and to boat noises in a conditioned-reflex fashion.

CHEMICAL FACTORS

Hydrogen ion concentration (pH)

Like many of the previously discussed physical factors, pH is remarkably uniform throughout most of the marine environment. The changes brought about by the slight fluctuations in concentration that do occur are not especially significant. Hydrogen ion concentration is rarely, if ever, a limiting factor in the marine environment. It is known, however, that physiological changes can be caused by pH changes.

Marine plants can tolerate or exist in environments with pH ranging between 5.0 and 10.0. In situations such as small, completely restricted pools or embayments, a luxuriant algal growth may cause the pH to rise considerably as the result of photosynthesis. Many animals can tolerate the same wide range.

Carbon dioxide
Most carbon dioxide in seawater is obtained via respiration of plants and animals. It is, of course, a necessary ingredient in the photosynthetic process; however, there is no evidence that concentrations ever become low enough to limit plant production. One of the most important functions of carbon dioxide is its role in the precipitation of calcium carbonate: both biochemically and physicochemically. The abundance of carbon dioxide is sufficient in all marine environments, although phytoplankton blooms cause a sudden and marked decrease. Intense pressure and low temperatures at depth increase the solubility of calcium carbonate, thus restricting the formation of carbonate skeletal materials in these environments. As a result of the lack of photosynthesis, there is anomalously high carbon dioxide concentration.

Oxygen
Almost all forms of life require oxygen. Some types of bacteria can do without it; they live in anaerobic conditions. Locally, significant variation in oxygen abundance may be limiting under certain concentrations; however, areas where oxygen is absent are quite rare. In some deep fjords or parts of the Black Sea, there is no oxygen, and as a result, no animal life. Commonly, the boundary between oxygenated and anaerobic conditions corresponds appoximately to the sediment-water interface.

Photosynthesis, atmospheric contributions, and circulation provide an adequate oxygen supply. A number of animals have the ability to overcome temporary oxygen deficiencies by using glycogen as a source of oxygen. Others can sustain themselves in anaerobic water for a few weeks, some by going into a type of short-term hibernation in which they are inactive and do not feed. When high temperatures are combined with an oxygen deficiency, respiration rates increase, thereby compounding the problem.

Burrowing organisms, such as clams or some worms, may occupy anaerobic sediment while utilizing oxygenated water from above the substrate.

Nutrient elements
Phosphorus and nitrogen are rarely present in low enough concentrations to be limiting factors in the marine environment. These two elements are not only in constant ratio in water (Chapter 12) but also are taken in by organisms in the same constant ratio except in certain species. For instance, some algae concentrate phosphorus. There is a direct relationship between the amount of phosphorus available, the rate of photosynthesis, and temperature. Uptake of phosphorus by primary producers makes it available to the consumer organisms that feed on photosynthetic organisms.

Many bacteria concentrate nitrogen. Various forms of nitrogen are present in seawater and are utilized directly by certain organisms, particularly the green plants. Other elements are necessary for certain organisms. These elements include potassium, iodine, copper, and strontium. Silicon is required by diatoms and radiolarians for their tests but there does not seem to be a shortage of this element in seawater.

BIOLOGICAL FACTORS

Dispersal and migration
The ability of a certain species to disperse its kind to all possible habitats that are available is a significant factor in determining its ability to survive in the sea. Attached bottom-dwelling forms, both plant and animal, have to rely entirely on the success of planktonic larvae, eggs, or spores to be carried by currents to areas favorable to their growth and development. Because of this rather inefficient method of distribution, tremendous numbers of the embryonic forms are necessary in order for some of them to settle in satisfactory environments. It is not uncommon for an adult oyster to produce several million eggs at a time. This is in contrast to some types of fish which lay a few tens of eggs that hatch into fairly active fry that can operate effectively on their own almost immediately upon hatching.

Swimming or vagrant bottom-dwelling animals may also have planktonic larvae or eggs, for example, the mollusks, echinoderms, and most fish. Bottom-dwelling fish generally lay their eggs on the shallow shelf area where they hatch, and the fry become floaters until they can swim with enough power to overcome prevailing currents.

By far the greatest amount of dispersal within a population takes place in the mobile larval and other immature forms, but adult individuals also contribute to dispersal of the species. With the exception of attached forms, they wander about in search of food and/or shelter. In doing so, they can and frequently do extend their area of distribution when favorable conditions are encountered. There are numerous physical and chemical barriers to this type of migration and dispersal; most of them are factors considered in the previous two sections of this chapter. Many of them are, of course, barriers to floating larvae also.

Food
Next to oxygen, food is probably the most vital necessity of marine organisms. Virtually all of their time and most of their energy is spent in search of it. (Plants are not considered here because their food, carbon dioxide and nutrient materials, were covered in the previous section.)

There are a number of ways in which marine animals feed, and the sum total of them utilizes virtually all available sources of food. Many single-celled animals engulf their food, which consists of single-celled, photosynthetic organisms, by surrounding it with their ameboid protoplasm.

Seawater contains a large amount of particulate organic detritus in suspension in addition to many forms of tiny plankton. Many different kinds of marine animals feed on these food sources by filtering out the solid materials. Filter feeders include many small floating animals, shellfish, worms, sea lilies, and others. This type of feeding is not very selective and the rate cannot be adjusted to correspond to food concentration in the water. Also suspended sediment may be taken in with food particles. When suspended organic debris is plentiful, the animal cannot utilize it as fast as it enters the digestive tract, and consequently much potential food is passed through unaffected by the organism.

Many bottom dwellers and some swimming animals are scavengers that feed selectively on available plant and animal debris. In addition, there are predacious carnivores that feed on specific types of animals. Many varieties of fish and certain bottom-dwelling animals belong to this group. Of particular interest are some of the predatory snails that feed on other snails, clams, or barnacles by penetrating the shell and withdrawing the flesh. This is apparently accomplished by a combination of rasping and secreting an acidic substance on the shell.

Predators
Almost all organisms in the sea are the prey of some other organisms except perhaps for the top carnivores, the sharks, and even they may be attacked by each other under certain conditions.

Virtually all animals have adaptations that enable them to survive to propagate their kind. In the case of many of the microscopic animals, sheer numbers are about their only protection. Larger or more advanced organisms may have extreme mobility, protective armor, camouflage, or poisonous organs. Most fish are quick and agile, some invertebrates have armored exoskeletons, and some have stinging nematocysts. Many bottom-dwelling fish such as flounder and stonefish blend in quite well with the substrate. Many other types of camouflage and also mimicry are common among some of the less mobile and less armored animals.

Space
There is little problem among the floaters and swimmers as far as crowding in the sea is concerned. The only significant limitation on these organisms in this respect is overdemand on the food supply; there is no problem of lack of space to physically support the population. Such is not always the case with

bottom-dwelling organisms; however, it is not always possible to distinguish between crowding for space and crowding for the food supply. Actually there may be competition for other things as well: oxygen, sunlight, or a desirable substrate type.

Competition for these necessary elements may take place between individuals of the same or different species. Crowding that might be detrimental in one environment would not be so in another due to differences in availability of light, food, and other necessary elements of the environment. Plants must compete for light and may shade one another under crowded conditions. Burrowing filter feeders may be so closely spaced that there is not enough food to allow maximum development of each individual. Similar crowding of other organisms can cause stunted individuals.

SELECTED REFERENCES

Friedrich, Hermann, 1969, *Marine Biology* (translated from German), Seattle: University of Washington Press, Chapter 3. Good treatment of ecological factors, especially physical ones.

Hedgpeth, J. W. (ed.), 1957, *Treatise on Marine Ecology and Paleoecology,* Geological Society of America Memoir 67, Vol. 1, Chapters 6–11. Several chapters on individual ecological factors in the marine environment, each written by a specialist.

Kinne, Otto (ed.), 1975, *Marine Ecology, A Comprehensive, Integrated Treatise on Life in the Oceans and Coastal Waters,* New York: Wiley. An excellent multi-volume treatise on the subject; designed for the specialist in marine biology.

Moore, H. B., 1958, *Marine Ecology,* New York: Wiley, Chapters 2–4. A unique treatment of ecological factors in the marine environment from which much of the present chapter is condensed. Moore's treatment suffers somewhat from lack of general discussion on each topic.

Tait, R. V., 1968, *Elements of Marine Ecology,* London: Butterworths. Fairly recent general text which is totally lacking in photographs.

Vernberg, W. B., and F. J. Vernberg, 1972, *Environmental Physiology of Marine Animals,* New York: Springer-Verlag. Good treatment of marine ecology with emphasis on physiological aspects.

16 PLANKTON

Most organic life in the ocean is planktonic, some throughout the life cycle and some only temporarily. **Plankton** are organisms that float, drift, or have feeble swimming abilities (insufficient for substantial horizontal migration against oceanic or tidal currents).

Although most plankton are microscopic or submicroscopic, some are large. Both plants (**phytoplankton**) and animals (**zooplankton**) are included in this category of organisms, with plants being restricted to the photic zone. It is within this zone in the open sea that photosynthesis by the phytoplankton occurs.

PHYTOPLANKTON

Most phytoplankton in marine water belong to relatively few phyla; however, there is considerable diversity within these phyla. Phylaplankton are mostly unicellular and microscopic. The vast majority are in the phylum Chrysophyta, sometimes called the yellow-green algae, which include diatoms and coccolithophores. Diatoms have an external skeleton or **test** which is highly siliceous and, after the death of the organism, makes up the deep-sea sediment called diatomaceous ooze.

Diatoms range in diameter from a few microns to nearly a millimeter. Because of their small size, the electron microscope is used to observe their delicate structures and features. In addition to wide variation in size between different species, there may be a large size range within a given species. Shape also ranges widely from one species to another (Fig. 16.1), with somewhat spherical and symmetrical forms the most common. Many diatoms secrete a type of oil, less dense than water, in order to regulate their vertical position in the sea.

It has been suggested that diatoms are actually sinking throughout their life cycle, and that after passing beyond the depth of the photic zone they die. Although some reach the bottom to form diatomaceous ooze, most are consumed in upper waters by herbivores.

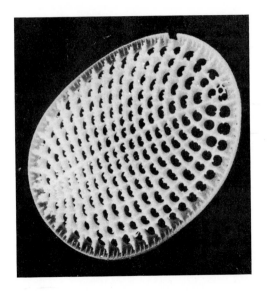

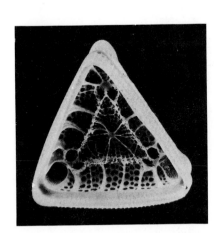
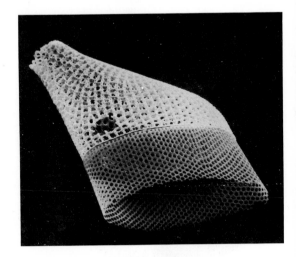

FIG. 16.1 *Selected diatoms showing a few of the many shapes assumed by these organisms. (From G. Dallas Hanna, "Nature's Opaline Gems," and G. Dallas Hanna and A. L. Brigger, "Stereoscan Microscopy of Diatoms," Pacific Discovery, October 1968. Photos courtesy of Engis Equipment Company, Morton Grove, Illinois.)*

Coccolithophores are the size of the smaller diatoms, a few microns in diameter. They have a calcareous skeleton. The complete individual commonly shows a radial type of symmetry (Fig. 16.2a). After the organism dies, the skeleton disaggregates into several tiny and similar plates (Fig. 16.2b). These comprise a significant amount of the calcareous oozes on certain parts of the ocean floor.

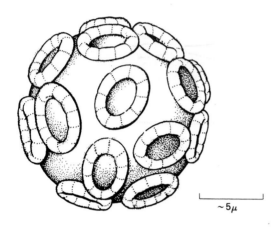

(a)

(b)

FIG. **16.2** (a) Complete coccolithophore individual, and (b) single coccolith plates. (Photos courtesy of W. W. Hay.)

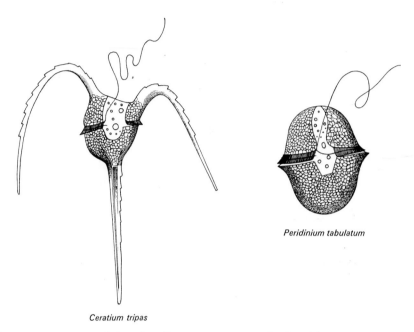

Peridinium tabulatum

Ceratium tripas

FIG. 16.3 *Typical dinoflagellate species. Note the hairlike flagellum on each individual. (After G. E. MacGinitie and N. MacGinitie, 1968, Natural History of Marine Animals (2nd edition), New York: McGraw-Hill, p. 102.)*

The third major group in the phytoplankton is the dinoflagellates (Fig. 16.3), which are in the phylum Pyrrophyta. These are one-celled organisms, and not all of them are photosynthetic. The dinoflagellates are even more diverse than diatoms and are unique among phytoplankton in that they have feeble powers of mobility. Such mobility is produced by hairlike flagella, present in all dinoflagellates. These organisms do not have skeletons of mineral matter and hence do not contribute to oceanic sediments.

In addition to the tremendous amount of microscopic phytoplankton, there is also one common pelagic plant which is multicellular and large. This is Sargassum, a type of brown algae (Phaeophyta) which has long branching stems and small stalked bladders. The bladders assist the plant in floating and also provided the origin of its name. They resemble small clusters of grapes, and "sargassum" comes from the Portuguese word for grape. Sargassum begins its life in attached form in low latitudes, but it is broken loose by wave activity and drifts at the surface with the ocean currents. Great quantities are present in the dead water of the North Atlantic, and the name Sargasso Sea has long been applied to this area.

Phytoplankton distribution

All parts of the world's oceans contain phytoplankton, but there are both geographic and vertical variations in distribution of the three primary types and also within each of these types. A common way to categorize the distribution patterns is by latitude, that is, by the arctic, temperate, and tropical zones. Diatoms, for instance, flourish in cold water and sometimes are so abundant that the water is discolored. (Such phenomena, termed plankton blooms, are not restricted to the diatoms.) On the other hand, dinoflagellates and coccolithophores prefer warm water.

There are certain taxa of phytoplankton which are characteristic of coastal areas, whereas others prefer the open sea. In most cases, however, there is general tolerance for either coastal or pelagic conditions. Salinity may be a factor in phytoplankton distribution in that some are **stenohaline**; that is, they have a narrow range of salinity tolerance.

Seawater conditions that affect its density are of extreme importance to phytoplankton distribution for obvious reasons. Temperature and salinity are of prime concern because they cause significant density changes and as a result affect vertical distribution of plankton. There are adaptations for the floating mode of life in both the diatoms and the dinoflagellates, such as thin cell walls and many delicate protuberances to increase the surface area. Seasonal changes in form counteract density changes in water caused by temperature fluctuations.

ZOOPLANKTON

Planktonic animals include a much more diverse assemblage than the phytoplankton, although they are fewer in numbers and constitute less total mass. The zooplankton comprise most of the primary consumers in the oceans. A much wider range of size and complexity is present in this group than in the phytoplankton, although most zooplankton are also microscopic to submicroscopic. Many types of invertebrates have larval stages that are planktonic, and numerous other animals float and drift throughout their life cycle.

Virtually all marine animals are part of the zooplankton at some time in their life cycle (Table 16.1). In terms of numbers and diversity the phyla Protozoa and Arthropoda are the most abundant of the zooplankton. The Coelenterata, Mollusca, and Echinodermata contribute highly to the meroplankton in the form of planktonic larvae.

The phylum Protozoa includes the Foraminifera and Radiolaria, which have tests and do contribute to deep-sea organic oozes. Foraminifera, particularly the genus *Globigerina*, make up most of the protozoans' contribution to the zooplankton. These organisms have delicate porous calcareous tests

TABLE 16.1 *Important zooplankton phyla.*

Scientific name	Common name
Protozoa	Single-celled animals
Colenterata	Jellyfish, corals
Ctenophora	Comb jellies
Annelida	Segmented worms
Chaetognatha	Arrowworms
Mollusca	Clams, oysters, snails (mostly larval forms)
Arthropoda	Insects, crabs, shrimp (larval and adult forms)
Echinodermata	Starfish, sand dollars, sea urchins (mostly larval forms)
Chordata	Fish (juveniles)

which have spines and many protruding pseudopodia (Fig. 16.4). The many spines and the bulbous shape provide a large surface area and assist the organism in floating. Planktonic foraminifers are barely visible to the naked eye.

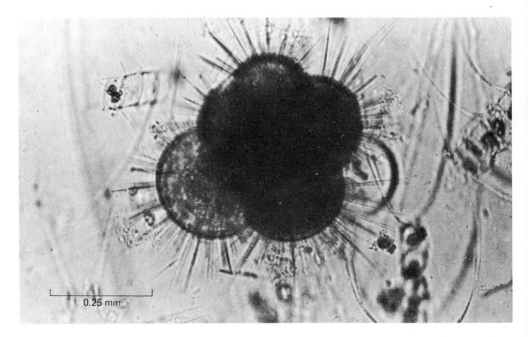

0.25 mm

FIG. 16.4 *Example of a living foraminifer. The many projections are mostly delicate spines. (Photo courtesy of Ralph Buchsbaum.)*

5 cm

(b)

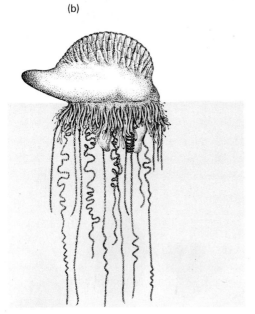

(a)

FIG. 16.5 *Portuguese man-of-war (Physalia) has stinging tentacles (a) for capturing its prey and a large float which is blown by the wind. The float is usually a few to several centimeters in length (b).*

The radiolarians are siliceous protozoans considerably smaller than foraminifers, but the largest type can be seen with the naked eye. The skeletal structure is fairly complicated and has numerous pseudopodia and delicate spines similar to globose Foraminifera.

One of the diverse and significant phyla in the zooplankton is the Coelenterata, which includes the largest of the planktonic organisms. Corals, anemones, and jellyfish, some of which are always planktonic and some of which are not, are in this phylum. In general, the anatomy of coelenterates is simple, with each individual being saclike in form. Coelenterates are characterized by an alternation of generations between a medusa (free-moving stage) and polyp (attached stage). In considering plankton, the medusa stage is the only one of concern.

Hydrozoans as a group are quite small and most of us are not aware of their presence. Some of the largest zooplankton are also hydrozoans; these are the common marine representatives of this group. Such large animals include the poisonous Portuguese man-of-war *(Physalia)*. This creature has a special adaptation to the pelagic environment in that it has a large gas-filled sac (medusa) that acts as a float (Fig. 16.5). The tentacles of *Physalia* may be several meters long and contain toxic nematocysts which inflict severe pain on contact. These are used to capture prey for food.

Probably the best known planktonic coelenterates are the jellyfish (Scyphozoa). The medusa stage dominates and is the one seen floating at or near the surface of marine waters. This bell-shaped structure may be almost 50 centimeters in diameter and has long tentacles containing stinging nematocysts (Fig. 16.6). Jellyfish have feeble locomotive powers which they use to move vertically in the water. The ability to move comes from a muscular system which can contract and relax the bell-shaped part of the medusa, causing a pulsation which moves the animal.

The Ctenophora is a group of animals having a somewhat more advanced structure than the coelenterates; they are wholly marine and are soft bodied. All ctenophores are **luminescent** and thus can be easily seen at night in surface waters. The common name "comb jellies" is applied in part because of their obvious jellyfishlike bodies. The combs are eight ciliated bands (Fig. 16.7) which enable the ctenophores to keep themselves in an upright position and to move to various depths.

There are various worm phyla which have some species in the planktonic realm. The phylum of segmented worms (Annelida) is a significant contributor to the plankton because its larvae are prominent in the coastal and neritic zooplankton. One class of these, the Polychaeta, includes many common marine forms. Only a few species are planktonic as adults, however, because most are bottom dwellers.

One of the most important phyla in the plankton, as well as in the nekton (true swimming organisms) and benthos, is the Mollusca, which

264

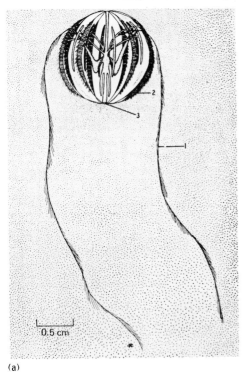

(a)

(b)

FIG. **16.7** *(a) Comb jelly (Ctenophora), with tentacles (1), combs (2), and mouth (3). (Modified from MacGinitie and MacGinitie, 1968, p. 145.) (b) Ctenophores, which are about 1 centimeter in diameter. (Courtesy of the Carolina Biological Supply Company.)*

◀**FIG. 16.6** *Common bell-shaped jellyfish floats about near the surface and catches food with its tentacles. They are commonly several centimeters in diameter. (Courtesy of the Carolina Biological Supply Company.)*

contains the common shell animals. The form that is of consequence in the plankton is the larvae of these organisms which resembles the larvae of the annelids. As the larval form matures, a small shelled organism develops which drops to the bottom, where the animal spends most of its life cycle. In addition to these forms, there are some mollusks that are planktonic as adults.

The Pteropoda is the most important group of planktonic mollusks and they are snails (Gastropoda). They are primarily soft bodied, but they have small shells. Pteropods (Fig. 16.8) are quite abundant and their small shells (tests) constitute "pteropod ooze" in low-latitude oceanic areas.

Echinoderms, which include starfish, sand dollars, sea urchins, and sea lilies, are not important contributors to the plankton, although historically they are of interest. Larval stages of this phylum are common in neritic plankton, and the search for these forms led to the first use of a plankton net and the study of plankton in general.

By far the most numerous and diverse phylum in the plankton is the Arthropoda (jointed appendages) which comprise more than three-fourths of all known animal species in the sea and on land. Although the phylum is diverse and contains several classes, only one, the Crustacea, is common as plankton.

The Ostracoda are an important neritic plankton group in the Crustacea. They are generally small bivalves (less than five millimeters in diameter) that

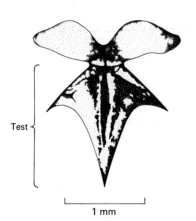

Test

1 mm

FIG. 16.8 *Pteropod individual. These tiny planktonic animals contribute their calcareous tests to deep-sea sediments called pteropod ooze.*

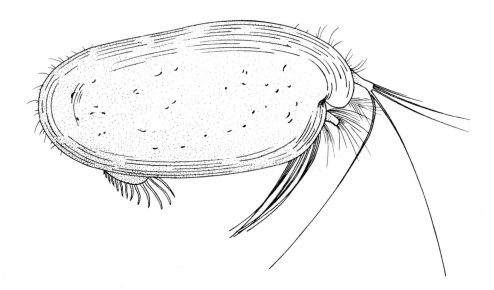

FIG. 16.9 *A member of the ostracods, which comprise a significant portion of the shallow-water plankton. Actual size is 1–3 millimeters in length.*

have feeble locomotive powers (Fig. 16.9). One large species, *Gigantocypris mulleci,* is nearly three centimeters long and occurs at great depths in the open sea.

Copepods are the most abundant of the crustaceans and are present throughout the oceans at all seasons. They are an important part of the marine food chain because they serve as a primary food source for many small carnivores. The bodies of copepods are elongate and segmented, with paired appendages and antennae (Fig. 16.10). Most are only two to four millimeters long, but one species, *Calanus hyperboreus,* reaches nearly a centimeter in length. Copepods feed primarily on diatoms and particulate organic matter.

By far the largest of the crustaceans are the Malacostraca, which includes crabs, lobsters, and shrimp. This subclass is represented by eight orders in the plankton, including mysids, isopods, amphipods, euphausids, and decapods. Mysids are holoplanktonic (spend their whole life cycle as plankton) and generally occur in the neritic plankton, although a few species are present to 4000 meters. They look like small shrimps and most are a few centimeters in length. Mysids are filter feeders and are fed upon by flounder, shad, and other shallow-water fishes.

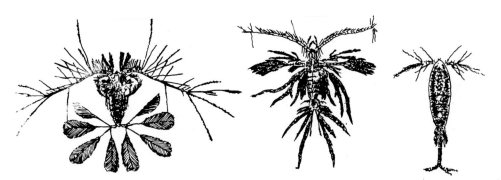

FIG. **16.10** *Various copepod forms. The many long and delicate appendages are for feeble locomotion and also aid in floating. (After R. E. Coker, 1954, This Great and Wide Sea, Chapel Hill: University of North Carolina Press, p. 227.)*

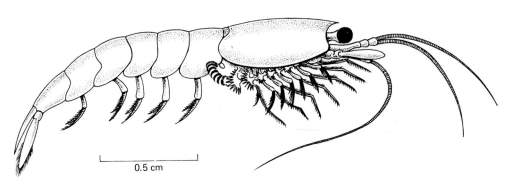

0.5 cm

FIG. **16.11** *Euphausid-type arthropod. These organisms comprise much of the "deep-scattering layer." (After R. D. Barnes, 1968, Invertebrate Zoology (2nd edition), Philadelphia: W. B. Saunders, p. 491.)*

Euphausids along with copepods are the most important of all zooplankton. They are entirely marine, two or three centimeters long, and look much like tiny shrimps (Fig. 16.11). Some feed on tiny zooplankton and others are filter feeders. The euphausids serve as the primary food source for many whales and are commonly referred to as "krill."

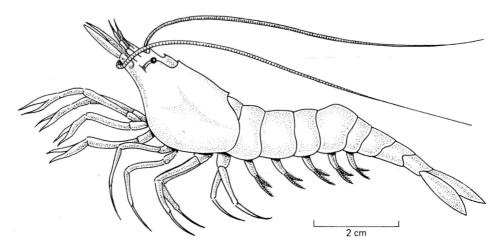

FIG. 16.12 Shrimp, a representative decapod crustacean.

Many species of the decapods such as the crabs and lobsters, leave the plankton as adults. The various types of shrimps (Fig. 16.12) make up most of the planktonic portion of this order. They represent the largest of the crustaceans, with most being several centimeters long. Adult shrimp are actually borderline members of the plankton, in that they have fairly good swimming abilities. The crabs and lobsters contribute to the plankton in the form of their zoea larvae.

The phylum Chordata is also abundantly represented in the plankton. Hatchlings of fish belong in this category and comprise large portions of the plankton locally in shallow waters. This is true only of the bony fishes; the cartilaginous types (rays and sharks) do not occur in the plankton at all.

Zooplankton distribution

Planktonic animals are probably more widely distributed throughout the oceans than any other organisms. Basically, two factors permit such widespread occupation: (1) their planktonic habit enables the animals to be carried wherever the water transports them, and (2) their independence of light permits them to occupy the entire vertical range of the ocean. Theoretically, there should be no area of the oceans without some zooplankton, and in fact this seems to be true.

Even though migration is rapid and the open ocean environment is relatively uniform, there are many physical, chemical, and biological factors which limit the distribution of most zooplankton. Only a few species can be considered truly worldwide in their distribution. Following the general pattern of taxonomic diversity, there are relatively few species in cold waters of the high latitudes, whereas tropical waters contain a wide variety of zooplankton. It appears that the primary limiting factor in their distribution is temperature. Commonly, these animals are carried some distance beyond their normal area of distribution and optimum temperature range. The organisms may be able to stay alive in the colder waters, but cannot reproduce.

Most zooplankton occupy shallow and near-surface depths because their primary food sources, phytoplankton and other smaller zooplankton, are in this zone. There are numerous zooplankton at all depths, however, and they seem to be largely temperature-controlled, with certain species apparently being associated with particular water masses. However, distribution at depth is still somewhat speculative because of insufficient good data.

Coastal and neritic zooplankton differ in general compositon from those of the open sea. One fundamental distinction is that the neritic fauna include significant forms of various larval stages of bottom-dwelling organisms. A few types of zooplankton, of which ostracods are probably the most significant, are abundant in the brackish water of coastal bays and estuaries, and are associated with river deltas.

The ability of zooplankton to migrate vertically is one of their most important characteristics. Such migration is primarily a diurnal phenomenon, with the animals coming near the surface at night and descending during the day. This ability is present in all major groups which comprise the zooplankton. The *Challenger* Expedition showed that there was considerable difference between the samples collected at night and those taken during daylight hours. Since the time of that expedition, there has been considerable time and effort devoted to diurnal migration studies of zooplankton. Most marine biologists agree that these animals are acting in response to changes in light intensity.

The deep scattering layer, which is common in many oceanic areas, is of interest here because it is a largely planktonic assemblage which migrates diurnally. For many years, beginning in 1942, an unusual sound-reflecting layer has been picked up on shipboard echo sounders. It is a band, a hundred meters or so thick, of relatively diffused reflecting objects and it moves toward the surface at night. Organisms which are dominant are the small lantern fish and the euphausid crustaceans. The purpose of their upward migration is for feeding in the productive photic zone.

PLANKTON COMMUNITIES

In the upper layers of the pelagic realm, there are distinct planktonic communities in the low latitudes (tropical) and in the high latitudes, the latter being divided into boreal (Arctic) and Antarctic. There is considerable transition between zones with mixed populations occurring. The primary distinction between these is the presence or absence of seasonal changes. The cold-water communities depend upon seasonal production by phytoplankton, whereas in the tropical zone this can take place year-round. The zones below light penetration and therefore production show little or no seasonal variation. The surface or upper pelagic planktonic communities are independent due to their dominance over primary producers, whereas the deeper communities are all dependent in terms of their energy sources.

Vertical movements characterize many types of zooplankton. There are essentially three types of regular vertical migration; that which is tied to various stages in the life cycle, seasonal migration, and diurnal or daily migration. All of these vertical migrations are designed to take advantage of food resources.

As a consequence of the above general characteristics of plankton distribution, it is common to consider three types of plankton communities based on vertical zonation. These are (1) the epipelagic or surface types, (2) the interzonal types, and (3) the bathypelagic or deep types. Phytoplankton and many varieties of herbivorous zooplankton comprise the surface communities. In addition to the dominant microplankton types, some of the larger plankton such as the various types of jellyfish and related coelenterates are in this group.

The interzonal groups spends a portion of their life cycle in the surface zone and another portion in the deep-sea environment. This group is divisible into two types of species, one which feeds at the surface and spends less time at depth, and the other which reproduces and develops in the deep sea but migrates upward for short periods, usually at night. The first group consists of only a few species; however, in high latitudes they comprise 50 to 90 percent of the total planktonic biomass. The lower group contains many species but with less biomass than its upper counterpart.

Many of the truly deep-sea organisms may be considered as interzonal. Some species live at about 5000 meters depth but may rise to the upper layers on occasion to feed on phytoplankton. Larval stages of deep-water organisms may occupy the surface waters.

COLLECTING PLANKTON

There is no fundamental difference between techniques used for collecting phytoplankton and those for collecting zooplankton, aside from the inherent differences in size and distribution. Planktonic organisms are dispersed throughout the entire volume of the sea; so for collecting purposes it is necessary to concentrate them in some way. Most of the wide variety of equipment that is used strains the seawater through a net of some type which collects and concentrates the organisms.

There are three fundamental sizes of plankton, and this necessitates specific types of collecting devices. The smallest of the plankton are **nannoplankton**, sometimes called centrifuge plankton. They range in size from about 5 to 60 microns and include the smaller dinoflagellates, diatoms, and all coccolithophores. Most planktonic organisms fall into the microplankton or net plankton category, which includes organisms from the upper limits of nannoplankton to about two millimeters, although this upper limit has no sharp boundary. Macroplankton are those which can be seen with the unaided eye; they range up to several meters in maximum dimension.

There are several problems that must be overcome or at least kept to a minimum when collecting planktonic organisms. Many plankton are soft bodied, so that collecting techniques must be as gentle as possible in order to prevent crushing the individuals and thereby rendering them difficult or impossible to identify. The wide vertical and geographic range of plankton coupled with the straining type of collecting techniques make precise location of the sample difficult to determine.

Nannoplankton

Standard plankton nets have pores which are too large to prevent nannoplankton from passing through. Special fine mesh nets are now available that will collect larger diatoms and dinoflagellates. Most methods for collecting nannoplankton, however, entail obtaining a volume of water and extracting the tiny organisms in the laboratory rather than *in situ*, as is necessary with most nets.

Those nannoplankton which are soft bodied, such as the dinoflagellates, must be handled carefully to prevent damage to the individuals. In the laboratory, a common way of concentrating the organisms is by centrifuging a quantity of the water sample. Filters are now available with extremely fine pores and small variability in pore size.

Nannoplankton with comparatively resistant skeletons, such as diatoms and coccolithophores, can be handled differently. The organisms in a sample can be killed and allowed to settle to the bottom of the container. Then the bottom water and residue can be filtered or centrifuged to further concentrate the organisms. Water may be forced through a fine filter to yield a residue of organisms.

Microplankton

Plankton nets of various sizes, shapes, and construction are the most common means of collecting plankton. All types of nets collect the organisms by filtering water through a silk cloth which retains microplankton. This silk is

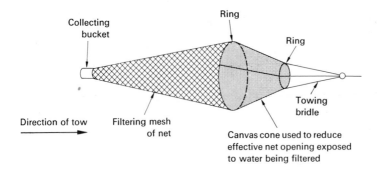

FIG. 16.13 *The Hensen plankton net, which is one of the more easily used and uncomplicated nets. (After A. C. Duxbury, 1971, The Earth and Its Oceans, Reading, Mass.: Addison-Wesley, p. 359.)*

of a special weave so that the mesh ranges from an individual opening of 0.054 millimeter to 1.364 millimeters.

Most nets have a similar gross structure: a conical net with a ring to keep the net open and a small container to catch the organisms which are funneled toward the cone's apex (Fig. 16.13). Such a net may be hauled vertically while the ship is at rest, resulting in a composite sample of the entire water column. However, this method provides no clue as to the depth zonation of plankton. A similar net can be towed horizontally while the ship is under way. By accurately controlling the ship's speed and knowing the amount of line payed out, it is possible to determine the depth of the net fairly well. Some nets have vane devices that keep the apparatus at a uniform depth. In order to quantify the sample, a small current-monitoring wheel is placed at the mouth of the net. It is possible to determine the volume of water passing through the net and from this calculate the density of the various plankton populations.

One of the big problems with the simple plankton net is placing the net at a given depth without contaminating the potential plankton sample as the net is lowered to that depth. Several types of opening and closing nets have been developed to alleviate the problem. The Clarke-Bumpus sampler (Fig. 16.14) is one of the most widely used quantitative nets. It has a butterfly valve which can be opened by a messenger that trips a release spring. The valve and a flow meter are housed in a metal tube at the end of the net, which is supported by a rigid frame. A series of Clarke-Bumpus samplers can be used simultaneously and opened at the desired depth, much like a series of Nansen bottles. The samplers must be towed slowly to avoid damage to the equipment, and as a result large crustaceans or other animals in the zooplankton may be able to avoid the net.

High-speed samplers of various types are used to avoid this possibility; they can collect plankton samples while the ship is under full power, perhaps moving from one station to another. Most of these samplers are

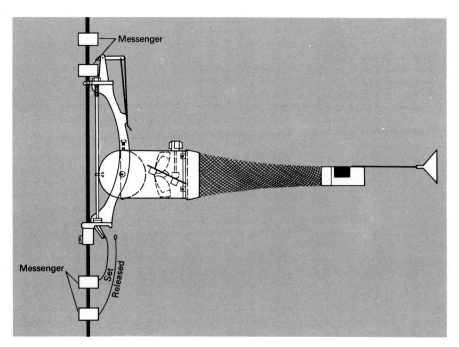

FIG. **16.14** *Clarke-Bumpus plankton sampler, which is capable of being opened and closed at depth by messengers in much the same manner as a Nansen bottle.*

elongate, metal, torpedo-shaped tubes which have a net or filter inside (Fig. 16.15).

A special high-speed sampler, the Hardy continuous plankton recorder, is designed to be towed behind commercial ships as they cross the ocean. By

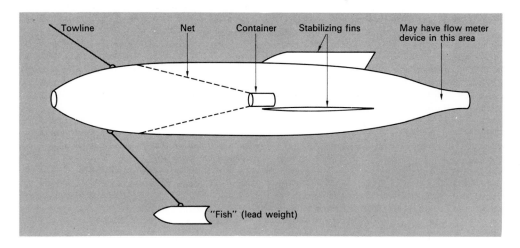

FIG. **16.15** *Torpedo-shaped, high-speed plankton sampler.*

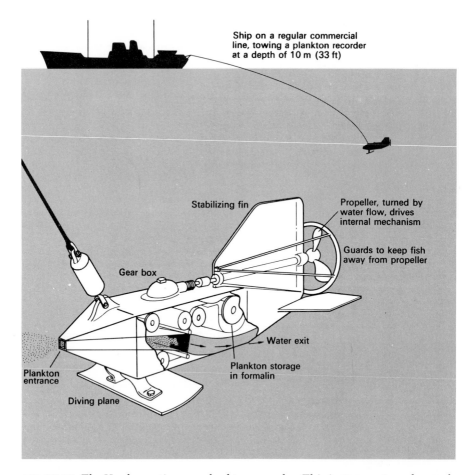

Ship on a regular commercial
line, towing a plankton recorder
at a depth of 10 m (33 ft)

Stabilizing fin

Propeller, turned by
water flow, drives
internal mechanism

Guards to keep fish
away from propeller

Gear box

Water exit

Plankton
entrance

Plankton storage
in formalin

Diving plane

FIG. 16.16 *The Hardy continuous plankton sampler. This instrument can be used while the ship is at full speed. A quantitative sample is collected in a preservative, but the individuals are somewhat crushed by the rollers. (After R. S. Wimpenny, 1966,* The Plankton of the Open Sea, *New York: American Elsevier, p. 271.)*

doing this regularly, it is possible to determine any seasonal or long-range changes in plankton composition. The sampler is shaped somewhat like an airplane with a small square opening at the nose (Fig. 16.16). Water passes through and the organisms are caught on gauze which is moving on a conveyor-belt type of apparatus. The gauze with the collected organisms is wound on a spool in a tank of preservative, commonly formalin. The roll of gauze is then easily removed and unrolled for examination in the laboratory. The distance traveled by the sampler can be calculated by using the speed of the ship, which determines the rate that the spools turn.

Another way of obtaining a plankton sample from a particular depth is to pump the water up to the ship and there pour it through the net. This is particularly advantageous for sampling plankton near the bottom where it is impossible to control a towed net. Such a method provides accuracy in both the depth and the amount of water sampled.

Macroplankton

Generally the same samplers and nets used for microplankton are also used for larger organisms. In fact, when collecting planktonic organisms, it is difficult to separate the two except, of course, by using a very coarse mesh and eliminating the smaller individuals. The opposite procedure is impossible unless the individual is too large to fit in the open end of the net.

The quite large zooplankton may be conveniently collected at the exclusion of small types by an otter trawl or other similar net. These nets, somewhat cone-shaped with coarse netting, are kept open by two vanes, one at each side (Fig. 16.17). Shrimp or small planktonic fingerlings of fish are collected this way, as are the true swimming organisms.

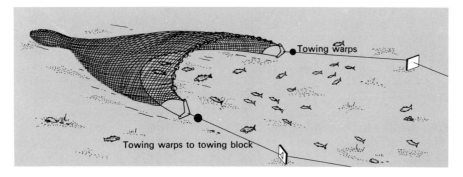

FIG. 16.17 *Modern otter trawl, which can be quite large. Paddles keep the net open, thus eliminating the need for a rigid frame at the open end as in the beam trawl (see Fig. 17.17).*

SELECTED REFERENCES

Barnes, H., 1959, *Oceanography and Marine Biology—A Book of Techniques,* New York: Macmillan, Chapter 1. A well-illustrated book on a wide variety of sampling and observational techniques and apparatus.

Barnes, R. D., 1968, *Invertebrate Zoology* (2nd edition), Philadelphia: W. B. Saunders. Excellent treatment of invertebrates from standpoint of morphology and physiology. Little information on living habits.

Briggs, J. C., 1974, *Marine Zoogeography,* New York: McGraw-Hill. Rather comprehensive treatment of the subject but suffers from heavy emphasis on fish and insufficient illustrations.

Buchsbaum, Ralph, 1948, *Animals Without Backbones* (revised edition), Chicago: University of Chicago Press. Old but classic general book on invertebrates with excellent photographs of many zooplankton.

Ekman, Sven, 1953, *Zoogeography of the Sea,* London: Sidgwick and Jackson. Classic volume on marine zoogeography but it suffers from being out of date. Tremendous bibliography of literature available to that time.

MacGinitie, G. E., and Nettie MacGinitie, 1968, *Natural History of Marine Animals,* (2nd edition), New York: McGraw-Hill. Useful treatment of marine animals in general, strongly emphasizing the invertebrates. Illustrations generally poor.

McConnaughey, B. H., 1974, *Introduction to Marine Biology,* St. Louis: C. V. Mosby, Chapters 4, 7, and 8. Readable and well-illustrated textbook for undergraduate students.

Raymont, J. E. G., 1963, *Plankton and Productivity in the Oceans,* New York: Macmillan, Chapters 5, 11–15. Good detailed account of plankton, with emphasis on function and distribution of specific taxa. Assumes a general knowledge of invertebrates.

Wimpenny, R. S., 1966, *The Plankton of the Open Sea,* New York: American Elsevier, Chapters 2, 3 and 11. Thorough and readable volume devoted exclusively to plankton. A major drawback is the lack of any subheadings within a chapter, which makes the search for specific data laborious.

NEKTON *17*

The sea has long been an important economic resource of many countries. At least up until this date that economy is based largely on fish, which along with a few other animal groups make up the **nekton.** There is really no distinct separation between nekton, or true swimming organisms, and many of the plankton. In many groups it is simply a matter of size and maturity within a species that enables it to move from the feeble swimmers (plankton) to the truly nektonic way of life.

Individual control by an organism over its movement is the single most important adaptation that nekton have. It enables them to move where they choose under their own power, at least within the limits of their tolerance of the various environmental characteristics. A swimming organism can search for its food instead of relying on currents, it can better avoid predators, and it can evacuate an area if environmental conditions change.

Taxonomically the nekton are not diverse with respect to other modes of marine life. There are no plants that are true swimmers, and the animals all fall in a few phyla, with the vast majority being in the Chordata. This phylum includes the sharks and rays (class Chondrichthyes), true fish (class Osteichthyes), sea turtles (class Reptilia), and the marine mammals (class Mammalia). In addition, some of the Mollusca (adult squid) and large Crustacea (some shrimp) also have swimming capabilities.

FISH

Among the fish there are two essentially different habits; those that live near the bottom are called **demersal** fish, and those that have no connection with the bottom are called pelagic. Demersal fish may look much like pelagic fish or they might have specific physical adaptations to bottom living. In either case the type of bottom sediment, its texture and composition, may be important to demersal fish, who generally feed and lay their eggs on the bottom.

Primitive fish

The class Cyclostomata, which includes the lampreys and hagfishes, is the most primitive of the fishlike animals. The entire group is jawless and has sucking-type mouths. They are demersal organisms that feed along the substrate with somewhat of a vacuum-cleaner action. The cyclostomes differ from true fish in these respects and also in that they do not have paired fins. In the geologic past there were many jawless fish, and this group can provide modern scientists with clues to the evolution of fishes.

Lampreys have slender bodies about one meter long, with a round mouth at the most anterior part of the body (Fig. 17.1). Small teethlike protuberances line the mouth and cover the tongue. A lamprey feeds by attaching its mouth to a fish and rasping the flesh until bleeding occurs. The blood is its primary food.

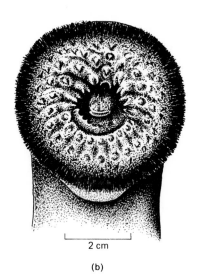

FIG. **17.1** *(a) Sea lamprey (Entosphenus) and (b) close-up of its sucker mouth with horny teeth. (Photo courtesy of the Carolina Biological Supply Company; sketch after Sir A. Hardy, 1959,* The Open Sea: Its Natural History, *Boston: Houghton Mifflin, p. 176.)*

2 cm

(a)   (b)

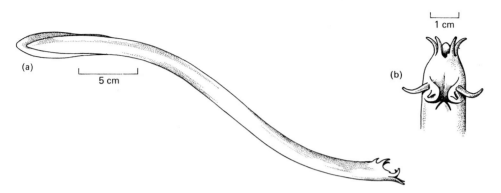

FIG. 17.2 *(a) Sketch of a hagfish, and (b) close-up of its mouth surrounded by tentacles. (After Hardy, 1959, p. 176.)*

The hagfish is similar in overall appearance to a lamprey (Fig. 17.2) but is about half the size. In contrast to the lamprey it is entirely marine. Morphology of the two groups is similar. Hagfishes also have a rasping tongue and commonly feed on wounded or dead fish by burrowing into the flesh; then they ingest all but the bones and skin. They are a problem to commercial line fisherman because they can destroy fish caught on the lines.

Modern cartilaginous fishes belong to the elasmobranch subclass of the Chondrichthyes. Sharks and rays belong to this group, which, like the Cyclostomata, is composed of living fossils. Elasmobranchs do not have an air bladder, as do the modern bony fishes. The absence of this important organ makes it necessary for elasmobranchs to expend a great deal of energy in moving vertically in the water. They must keep swimming to avoid settling on the bottom, whereas fish with air bladders can remain almost motionless without sinking. As a result of this internal difference, there are external modifications on elasmobranchs such as the sharks and dogfish. The tail is shaped in such a way that it propels the animal upward and forward. The front of the elasmobranch is lifted by the inclined pectoral and pelvic fins. (Fig. 17.3). The shark is thus able to move through the water with a lift

FIG. 17.3 *Outline of an elasmobranch, showing its adaptation for overcoming the lack of an air bladder while swimming. (Arrow points in the direction of thrust.)*

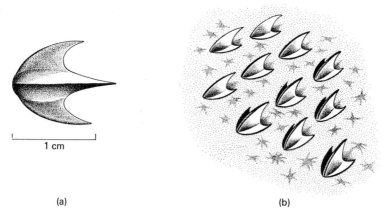

FIG. 17.4 *(a) Close-up of a shark denticle, and (b) sketch of how denticles are arranged on a shark's skin.*

similar to that of an airplane. Rays have fins that are drastically modified to provide great surface area and act like wings.

Elasmobranchs have mouths that are located ventrally and teeth that are imbedded in skin, not in bone; the teeth may be replaced several times. Elasmobranchs also have individual gill slits that are visible and not well protected. They must continually move to provide circulation of water and therefore oxygen, through these gill slits. Instead of scales the elasmobranch's skin is covered with small denticles (Fig. 17.4) which serve as a protective covering much like scales. They do not lay eggs but have internal fertilization and bear live young.

The most common elasmobranchs are the sharks, which are among the most feared of all animals. They are mostly flesh eaters and some are nearly the size of large whales. Certain species, such as the basking shark (*Cetorhinus maximus*) and the whale shark (*Rhincodon typus*), exceed 10 meters in length. Sharks in general are scavengers with a highly developed sense of smell. Blood has an exciting effect on them, and injured organisms are particularly likely to be attacked by certain species, such as the tiger, white, and hammerhead sharks (Fig. 17.5).

Skates and rays (Fig. 17.6) are specially adapted elasmobranchs whose perctoral fins are greatly enlarged for propulsion. They flap their pectoral fins like wings and glide smoothly through the water. This group is well constructed for its demersal mode of existence. They are usually well camouflaged on the dorsal side, and some types can actually change their coloration to match their environment.

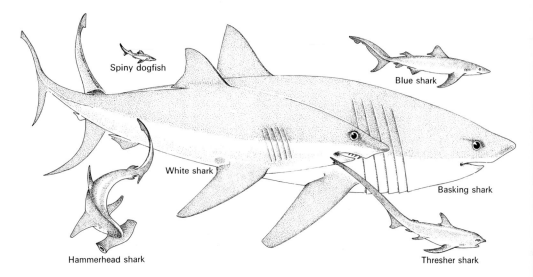

Spiny dogfish

Blue shark

White shark

Basking shark

Hammerhead shark

Thresher shark

FIG. 17.5 *Various types of sharks drawn on the same relative scale. The basking shark may reach a length of 12 meters.*

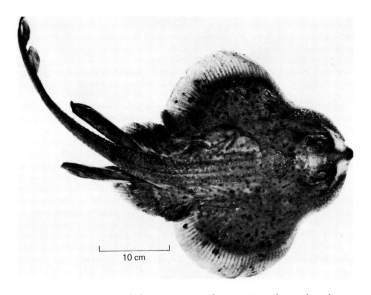

10 cm

FIG. 17.6 Raia, *one of the common skates. Note how the shape and coloration are suited to a demersal mode of life. (Photo courtesy of the Carolina Biological Supply Company.)*

Some rays have special features which are unique. The electric ray (*Torpedo nobiliana*) is capable of inflicting an electric shock to stun its prey. The stingray (*Trygon pastinaca*), found in shallow coastal waters, is a hazard to swimmers or waders. It has a long tail with a serrated spine which it drives into an enemy or intruder, producing a large and painful wound.

Bony fish

The bony or true fish comprise the class Osteichthyes, which includes about 100 families. It is not possible in a superficial treatment such as this to do justice to the wide variety of interesting types. The discussion presented here will be limited to some general features of three kinds of bony fish: (1) bottom dwellers, (2) pelagic fish, with emphasis on those of economic significance, and (3) abyssal types.

The presence of an air bladder in the bony fish is a great advantage, as it makes efficient propulsion possible: all their energy is utilized for forward motion. Paired fins are used primarily for stabilization. The arrangement of the mouth and nostrils is strikingly different from elasmobranchs, with the mouth in the extreme front of the body on most species. Gills are protected by a gill-cover, and instead of denticles the bony fishes have scales which are arranged like shingles on a roof and provide a fair amount of protection to the fish. There is also a striking contrast with elasmobranchs in the reproductive methods, as bony fish lay masses of eggs which are fertilized after laying, and which may yield up to thousands of **fry.**

The class Osteichthyes includes two orders: the Teleostei, which includes almost all bony fish, and the Chondrostei, which includes the sturgeon. The latter are extremely large and valuable fish that breed in rivers, and it is there that they are caught and their eggs (caviar) extracted. Actually, these are also living fossils and by far the most primitive of all bony fishes. Their bones change to cartilage with time and they have many armored plates on their back and sides. The sturgeon feed on bottom invertebrates.

Bottom dwellers. The bony fishes contain a large number of demersal species, some of which have the typical torpedo-shaped body and feed only on the bottom, while others have flat bodies specially adapted for bottom dwelling.

Two of the most important types of demersal fish are the cods and the flatfish. Included in the first group are cod, haddock, ling, and whiting. The cod (Fig. 17.7) is probably the most important bottom-dwelling food fish. Along with haddock and whiting, it belongs to the genus *Gadus*, which contains ten species. The cod is found in fairly shallow waters, usually less than 600 meters. They feed on crustacea when small, but change to a diet of small fish as adults. Haddock, although belonging to the same genus, feed

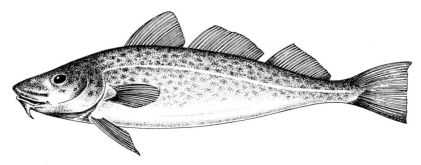

FIG. 17.7 *Codfish* (Cadus callaris), *one of the most important commercial fish.*

mostly on invertebrates, and consequently both may occupy the same waters. A large cod, which may be a meter long, lays as many as several million floating eggs.

The flatfish, which include halibut, flounder, and sole, have become extremely adapted to their existence on the ocean floor. The young of the group look much like typical fish with bilateral symmetry. During early development the body becomes flattened and one eye migrates across the head until both are on the same side. The mature flatfish is flat and nearly elliptical, with a grotesque arrangement of eyes and mouth (Fig. 17.8). The upper side, which may be the left or right depending on the type, is pig-

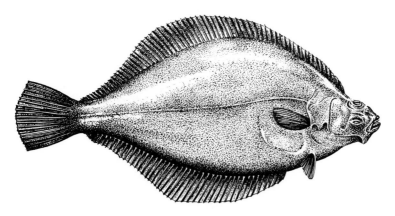

FIG. 17.8 *A typical bottom-dwelling flatfish* (Parophrys vetutus). *(After MacGinitie and MacGinitie, 1968, p. 416.)*

mented, whereas the underside is not. Some species of demersal fish can change their color and markings drastically to camouflage themselves. This along with their ability to partially bury themselves in sand makes them quite inconspicuous on the ocean floor.

Pelagic fish.   Pelagic fish receive the most concentrated efforts of commercial fisherman. Among these the common groups are (1) the herringlike fish, which include herring, sardines, shad, and menhaden, (2) the salmon, and (3) mackerel-like fish, which include tunas, bonitos, and the various types of mackerel.

The herring and their close relatives (Fig. 17.9) are usually less than half a meter in length; they are schooling-type fish and are extensively fished, particularly in the North Atlantic. They are used in a variety of ways for food, and also for fertilizers and oils. The herring and sardines are primarily consumer foods, whereas menhaden are used for fertilizers, oil, and catfood. Herring lay thousands of eggs, which settle to the bottom. Upon hatching, the yound feed on diatoms and dinoflagellates, but they change to a carnivorous diet as they mature.

FIG. 17.9 *The common mackerel* (Scomber scombrus), *one of the typical herringlike fish.*

Ocean salmon include few species and, like some of the previously discussed fish, they breed in fresh water. They may reach more than 25 kilograms in weight and exceed a meter in length. One of the unique features of the salmon is the change in morphology that takes place in the males as they move to fresh water for breeding (Fig. 17.10). The colors change and the jaw and dorsal area take on a grotesque shape. After spawning, most of the adults die; those that survive eventually make their way back to the sea.

The mackerel-like fishes are the second most important economic group next to the herring. Mackerel, bonito, and albacore are fairly small, usually less than half a meter in length, but the tuna are quite large and commonly

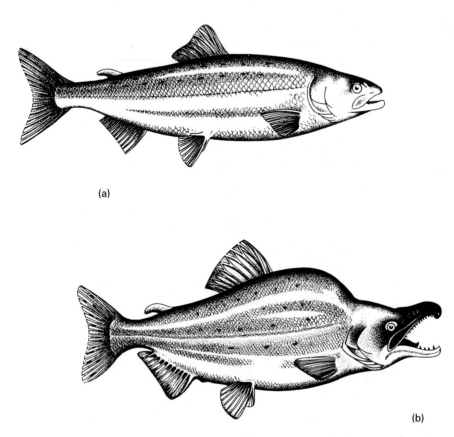

(a)

(b)

FIG. 17.10 *Striking difference in the male salmon between the time of marine occupation (a) and migration to fresh water for spawning (b). (After G. V. Nikolsky, 1963,* The Ecology of Fishes, *New York: Academic Press, p. 165.)*

weigh more than a hundred kilograms. Mackerel swim in large schools and are caught in fairly shallow waters along the coasts. They feed mostly on crustaceans, particularly a certain genus of copepod (*Calanus*). Mackerels lay up to 500,000 floating eggs over deep areas at the edge of the continental margin; then they return to the shallow shelf environment.

The bluefin tuna (*Thunnus thynnus*) is one of the most beautiful and spectacular fish in the sea. Also a schooling fish, they commonly travel at the surface with the dorsal fin extending above the water surface. The tuna is well constructed for rapid movement through the water. It feeds mostly on smaller fish such as herring and mackerel.

Abyssal types.   The rigors of living at great depths have produced a wide variety of strange and exotic abyssal fishes. There are about 40 families of entirely deep-sea fishes and a large number of deep-sea species of other groups. It will therefore be necessary to confine this discussion to general features of these animals and point out some unique adaptations that are present in certain species.

The swim bladder of many deep-sea fishes is considerably reduced in size or is disconnected from the intestine to prevent escape of gas under high pressures at depth. Deep-sea fishes are usually black or dark in color, without scales, and contain a variety of luminescent organs. The males of several varieties are only a small fraction of the size of their female counterparts. The absence of light at great depths has caused modification of visual organs. One genus (*Bathytroctes*) has telescopic eyes and some others have lost their optic organs. A few deep-sea fish have developed specialized pectoral fins which are sensitive to touch and replace visual organs. A deep-sea shark (*Spinax niger*) has a series of luminous organs along its sides; the lantern fish (*Myctophum*) also has many luminous photophores along its body. A few deep-sea fishes have a luminescence which is produced by symbiotic bacteria.

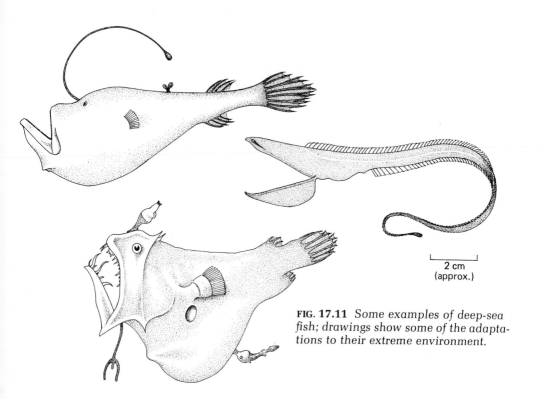

2 cm
(approx.)

FIG. 17.11 *Some examples of deep-sea fish; drawings show some of the adaptations to their extreme environment.*

Food is not plentiful in the dark abyss, and some species have developed unusual methods of obtaining their nourishment. The angler fish (Fig. 17.11a) has a long flexible stalk on its dorsal side. The stalk has a fleshy luminous lure on its end, enabling the fish to "catch" food by attracting other fish. This lure is dangled over the large mouth of the angler fish and the curious smaller fish are gulped down. One family of deep-sea fish has a tremendously large mouth (Fig. 17.11b) and an elastic stomach. These special features enable the fish to eat an animal much larger than itself. It is adapted to this environment of sparse food supply by being able to devour a large creature, since it may not be able to find another victim for some time.

OTHER NEKTON

Certain large Arthropoda, such as some shrimp, have the ability to swim; the only other significant invertebrates in the nekton are in the phylum Mollusca. There are two different types of cephalopods, the squid and the pearly nautilus, which belong in this category. Both move by a type of jet propulsion.

FIG. 17.12 *The common squid (Loligo), one of the few nektonic members of the mollusk phylum. (Photo courtesy of the Carolina Biological Supply Company.)*

Squid are soft-bodied animals with an internal skeleton called a pen. Their body is torpedo shaped with ten tentacles surrounding the mouth and two eyes near the mouth (Fig. 17.12). Like their relative the octopus, the squid can emit a dark fluid which assist them in escaping their predators. They move by expanding and contracting the mantle cavity with strong mantle muscles. Most species of the common squid (Loligo) are a few to several centimeters long; however, the giant squid (Architeuthis) may exceed 15 meters and is the largest known invertebrate.

The nautilus (Nautilus) is the last surviving genus of the subclass Nautiloidea. Five species are known, all from the southwestern Pacific

Ocean. Its soft anatomy bears some resemblance to the squid, but it is encased in a planispirally coiled shell (Fig. 17.13). In addition to the protection it offers, the shell acts as a buoyant type of balance. The nautilus propels itself by ejecting a jet of water.

Amphibians are not present in the sea, but other higher vertebrates such as reptiles and mammals do occupy this environment. It is commonly agreed that terrestrial ancestors of these three classes of vertebrates gave rise to the marine counterparts. Today there are a fair number of air-breathing vertebrates which have become well adapted to aquatic marine conditions. Reptiles, however, were considerably more abundant in the geologic past than they are now.

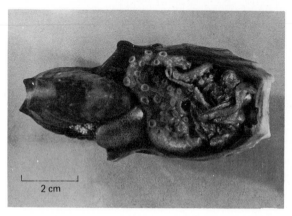

2 cm

FIG. 17.13 *The pearly argonaut, a modern nautiloid.*

Modern oceanic reptiles are restricted to sea turtles and a few snakes. The limbs of marine turtles are modified as paddles (Fig. 17.14) which enable them to swim with ease over tremendous distances. Even though they are primarily marine, they must lay their eggs in the sand along a beach. Such turtles are quite large, with one species reaching two meters in diameter and weighing about 500 kilograms. Those that are vegetarian stay near the coast where food is available, whereas the fish and crustacean eaters may be found hundreds of kilometers from the shore.

Sea snakes have developed the ability to be completely independent of land. They bear live young and thereby avoid the nesting problem of turtles. In order to facilitate mobility, sea snakes have developed a flattened tail which is used like a fish's tail.

Although mammals as a whole are best suited for a terrestrial existence, some have become well adapted to the sea. The sea cows and dugongs live in marginal marine areas. They have no true hind limbs and have flipperlike forelimbs for locomotion. One group of mammals which includes seals and walruses is of significance in the marine community and is also of economic value. These marine mammals have modified appendages for swimming. These modifications, although of extreme value in water, make it difficult for the individuals to move about of land where they must go to breed. Within this group these are three families, easily separated on the basis of their ears. Sea lions and fur seals, in one family, have small external ears, but the true seals and walruses do not.

FIG. **17.14** *A marine turtle, one of many animals in danger of becoming extinct in the near future. (After Hardy, 1959, p. 259.)*

The most widely distributed and most abundant marine mammals include bottlenosed dolphins, porpoises, and whales. This group is the best adapted of all mammals to the marine environment. They are completely removed from all land connections, even bearing their young at sea. The most striking and obvious difference between the whales and fish is the tail, which is horizontal in whales and vertical in fish. Whales, unlike most land mammals, have no external ears; this feature improves their streamlined shape for swimming.

Whales (Fig 17.15) include the largest of all known animals, even exceeding the great size of dinosaurs. Specimens over 30 meters long and weighing more than 100 tons have been reported. A large part of the whale's mass is the thick coats of blubber which serve as insulation and as a place to store food.

There are two types of true whales, the right whales, sometimes called the whalebone whales, and the toothed whales, which include porpoises and

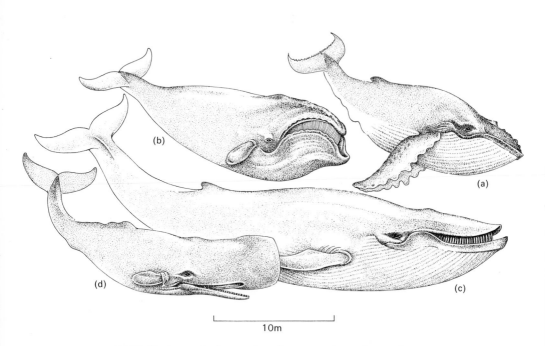

10m

FIG. 17.15 *Various whale species, drawn to show their relative sizes: (a) humpback whale, (b) right whale, (c) blue whale, and (d) sperm whale. (After Hardy, 1959, p. 277.)*

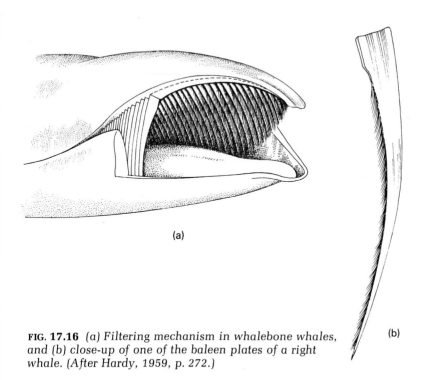

(a)

(b)

FIG. 17.16 *(a) Filtering mechanism in whalebone whales, and (b) close-up of one of the baleen plates of a right whale. (After Hardy, 1959, p. 272.)*

whales. The right whales have a complex straining apparatus of whalebone (baleen) which is fixed to their upper jaw (Fig. 17.16) and used to strain small zooplankton such as copepods and euphausids. The toothed whales, which may have teeth on one or both jaws, include the small varieties of whales as well as some large ones. The sperm whale (Fig. 17.15) has teeth on the lower jaw only and feeds primarily on squid. Porpoises and bottlenosed dolphins, which can be distinguished on the basis of their noses, have teeth on both jaws and feed on fish.

DISTRIBUTION OF NEKTON
The control that a swimming organism has over its distribution is of considerable importance because the chance of survival to maturity for reproduction is enhanced. The general assumption is that true nektonic organisms are the most advanced of the marine animals. However, even though swimming organisms have the ability to control their movement, they are responsive to most of the same variables that control distribution of the plankton. Such factors as temperature, salinity, oxygen supply, and nourishment are still

important. The one factor which is of major significance to plankton but not to nekton is seawater density. Differences in density are rather easily overcome by most fish and other nektonic animals.

Some fish, such as the sturgeon, salmon, and lampreys, regularly migrate to fresh water. Although they may be found in the open sea, they are more common to the coastal areas. A few species have remarkable tolerance for salinity changes. The genus *Fundulus* (killifish) can tolerate salinities ranging from 0‰ to almost 200‰.

Distribution of bottom-dwelling fish is controlled by many of the same factors that control pelagic fish, with the added consideration of the character of the bottom. The existence of demersal varieties is dependent on their ability to find sufficient nourishment, and their food supply is closely related to bottom composition. A scavenger feeder is not as dependent on a particular bottom type as is a detritus feeder or a fish that relies on specific organisms for its diet. Some bottom materials are suitable for burrowing animals, some for plants, and some for crawling organisms; as a result these characteristics are important in determining the distribution of the fish which feed on such creatures. Likewise, a bottom sediment in which detrital particles are rapidly oxidized will not support detritus feeders. Suitable nesting conditions are necessary for those fish that lay eggs on the bottom. These requirements compound the limiting factors that make demersal fish and other demersal creatures more restricted in distribution than are pelagic types.

NEKTON COMMUNITIES

Because of their dominance in the nekton, only fishes will be considered here. Only two of the major categories will be discussed, the coastal and the deep-sea varieties. Like the plankton, they can be treated conveniently by their depth zonation.

Coastal fishes

Coastal varieties live primarily within the continental shelves of the world and are commonly either at or near the surface or are bottom dwellers. Surface types include the herring and mackerel, whereas cod, flounder, and plaice are more demersal. Areas where the continental shelf is quite wide, such as the Grand Banks off of the east coast of Canada, contain large populations of these fishes.

One of the most important factors in the distribution of coastal fishes is water temperature. Typically, coastal-fish communities correspond closely to isotherms. The Arctic and Antarctic fauna is found at temperatures below 6°C, the temperate fauna is between 6° and 20°C, and the tropical fauna is restricted to areas above 20°C. Tropical communities contain the greatest diversity of species; these include the tunas, bonitas, albacores, swordfishes, and flying fishes. In addition, the coral-reef areas have their own beautiful and exotic varieties. In colder waters, such types as cod and herring dominate. They occur in much larger numbers although the variety of types is low. The truly cold-water communities show considerable contrast between hemispheres. In the Antarctic there is a rather diverse fauna, whereas in the Arctic there is a poor fauna in terms of both numbers and diversity.

Deep-sea communities

The deep-sea fishes are those living at depths below about 250 meters, out of the zone of significant light penetration. They are many times more diverse as those in surface waters. Estimates indicate that there are some 2000 species in deep water as compared to about 200 in surface water.

In the mesopelagic zone (Fig 14.1), there are many varieties of small fish that are related to the herring and salmon, as well as many types of lantern fish. In addition, there are nearly 100 species of angler fish and deep-sea eels. Most of these mesopelagic species are about 15 centimeters long as adults; however, one species reaches over a meter in length. Fishes in this twilight zone have larger, sensitive eyes.

Below about 1000 meters, the abundance of fishes declines markedly, although about 1000 species occupy this black zone of the ocean. About 75 percent of these bathypelagic fishes have some type of light organ; fishes in this zone have no eyes. The bulk of the fish at this depth are the angler fish.

COLLECTING NEKTON

Various collecting devices for nekton have been developed almost exclusively through the efforts of the fisheries industry. Nets of some type are employed in virtually all methods where great quantities of individuals are present; however, some varieties of hook-and-line fishing are also used.

Flounder, halibut, cod and other bottom-living fish were collected for many years by a beam trawl pulled by a trawler or other medium-sized vessel. The apparatus consisted of cone-shaped, coarse-mesh net attached to a sledlike frame (Fig. 17.17). In recent years commercial fishermen have replaced this trawl with the otter trawl (Fig 16.17). The main difference is that the otter trawl has an otterboard on each side of the net which keeps it open. An otter trawl can be much larger than a beam trawl, which is limited to a width of about eight to ten meters.

A purse seine may be employed to capture near-surface pelagic fishes such as herring, sardines, or menhaden. Small boats monitor the net, which is cylindrical in shape with floats at the top and weights at the bottom (Fig. 17.18). The net may be 100 meters or more in diameter. A purse line on the bottom is drawn tight to trap fish in the net, which along with its catch is then taken aboard a large ship. This is a somewhat cumbersome operation,

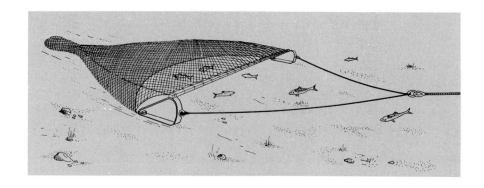

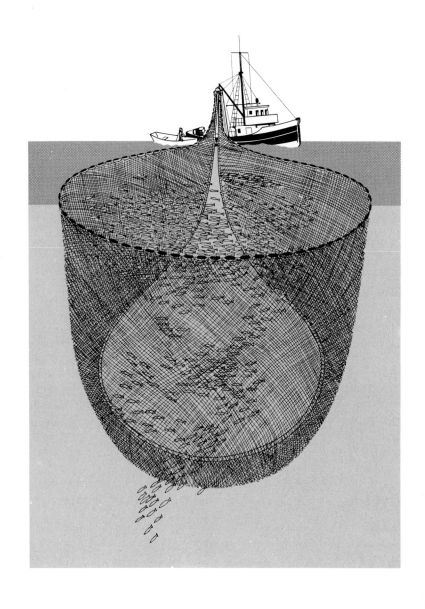

◀**FIG. 17.17** *Old-fashioned beam trawl, now rarely used in modern commercial fishing.*

◀**FIG. 17.18** *A purse seine operates on the same principle as a handbag with a drawstring.*

and nowadays a drifting gill net is often used instead to catch these pelagic fishes. This net also has floats to keep it near the surface but it is strung out a few hundred meters in a single direction and allowed to drift in an area of known fish concentration (Fig. 17.19). Fish swim into the net and are held fast in the mesh by their gill covers. The net is then hauled aboard and the fish extracted.

The hook-and-line method of fishing is used commercially under some conditions. Demersal fish such as flounder, halibut, and cod may be taken by

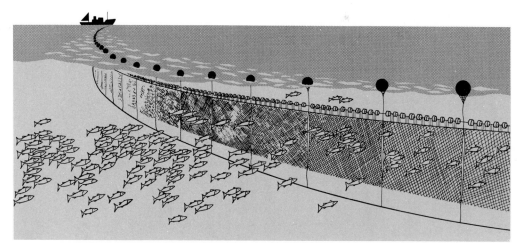

FIG. 17.19 *Drifting gill net in operation. The fish swim into the net and are caught in the mesh by their gills so that they cannot free themselves. The size of the mesh determines the size of the fish that are caught.*

laying a long line with many baited hooks on the bottom for a designated period of time. The ends of the line are marked by buoys and anchored to the bottom. A small vessel, called a long-liner, retrieves the line and its catch. This method has been larely replaced by trawling, but small boats and crews still operate with this technique. Some tuna fishing is also done by hooks and lines from shipboard. Men equipped with short lines and stout poles actually snag the tuna when they are in schools at the surface.

SELECTED REFERENCES

Briggs, J. C., 1974, *Marine Zoogeography*, New York: McGraw-Hill. Much emphasis on world distribution of fishes.

Cushing, D. H., 1975, *Marine Ecology and Fisheries*, London: Cambridge University Press. Quantitative approach to marine ecology with applications to fisheries.

Greenwood, P. H., 1975, *A History of Fishes* (3rd edition), New York: Wiley. Comprehensive treatment of fishes but lacking in photographs.

Hardy, Sir Alister, 1959, *The Open Sea: Its Natural History, Part II, Fish and Fisheries*, Boston: Houghton Mifflin. Excellent treatment of fish and other nekton, including their general ecology and distribution. Illustrations good. Examples are mostly from British Isles and North Sea areas.

Marshall, N. B., 1966, *The Life of Fishes*, New York: World Publishing Company. Excellent treatment of fishes and how they live.

Nicol, J. A. C., 1960, *The Biology of Marine Animals*, New York: Interscience. Somewhat advanced treatment with emphasis on physiology of marine organisms.

Nikolsky, G. V., 1963, *The Ecology of Fishes*, New York: Academic Press. Thorough treatment of fish ecology, particularly the physiological aspects. Examples and bibliographic citations are mostly Russian.

Perlmutter, Alfred, 1961, *Guide to Marine Fishes*, New York: New York University Press. Complete key for identification of marine fish, with good description of each species. Illustrations are undetailed line drawings.

Walford, L. A., 1958, *Living Resources of the Sea*, New York: Ronald Press, Chapters 13–16. First part of book is general treatment of conservation and management of marine resources. A few chapters are devoted to general description of major nektonic groups. Illustrations are nonexistent except for a few distribution maps.

BENTHOS 18

Organisms that live on the ocean floor are called **benthos**. They approach the plankton in diversity. This mode of marine existence includes a wide variety of both plants and animals, although the vertebrates are almost completely absent. Within the broad benthic category are two types of bottom occupation: those organisms, both plant and animal, that are attached to the bottom (**sessile** benthos) and those animals that can move about (**vagrant** benthos).

Benthic organisms are directly or indirectly dependent on the bottom for their food and, in many cases, their shelter. Composition and texture of the bottom are variables which may limit distribution of certain organisms, as is the chemical environment within the bottom sediment. Sessile organisms are attached by roots or holdfast structures. It is therefore necessary for the bottom to have the proper texture and firmness in order to provide a suitable place of attachment. Unstable conditions caused by strong currents are not suited to sessile habitation because of shifting substrate. Vagrant bottom dwellers are dependent on bottom characteristics in somewhat different ways. Many animals are scavengers and feed on organic material and other nutrients contained in the sediment. Burrowing and boring organisms are controlled largely by substrate texture and to some extent by its stability. A burrower has difficulty existing in an environment where sediment is constantly being carried into its burrow or where it is exhumed by bottom currents carrying away the sediment.

PLANTS
Plant diversity on the sea bottom, considerably less than in terrestrial environments, is restricted to the algae, which are the dominant form, and a few angiosperms. Gymnosperms are completely absent from the sea.

Algae
The vast majority of all benthic plants fall within four phyla (Table 18.1): the Cyanophyta (blue-green algae), the Chlorophyta (green algae), the

TABLE **18.1** *Benthic marine algae. (From Dawson, 1966, p. 10.)*

Phylum	Marine species	General size
Cyanophyta (blue-green)	150	Microscopic
Chlorophyta (green)	900	Microscopic to massive
Phaeophyta (brown)	1500	Microscopic to massive
Rhodophyta (red)	4000	Microscopic to massive

Phaeophyta (brown algae), and the Rhodophyta (red algae). All these groups are essentially worldwide in their distribution; however, there is considerable variation depending on climate, water depth, and other factors.

Cyanophyta. The blue-green algae are inconspicuous but are found almost everywhere in brackish and marine environments. They are very simple in form, with a filamentous character being prevalent. An individual filament is microscopic, but generally they occur in clusters of many individuals. Blue-greens appear to be the most primitive of all algae and are the oldest, having been found in Precambrian rocks. They have tremendous tolerance to variation in marine conditions and are able to live in areas of low light intensity. High or low salinity, prolonged exposure to the atmosphere, and extreme temperatures do not seem to prohibit growth of blue-green algae. A common modern occurrence of these filamentous algae is in leathery mats or mound-shaped structures called stromatolites (Fig. 18.1) which are dominantly in intertidal zones. Similar structures are known through much of geologic time. It is worth noting that in the phylum of blue-green algae there is a complex terminology which differs with each book on phycology.

Chlorophyta. The green algae are not abundant in marine waters, but they are the dominant algae in fresh water. Most marine forms are macroscopic; however, a few of these are fairly small and difficult to identify with the naked eye. There are three orders of tropical green algae and three which are common in temperate water. Most marine green algae can be recognized by their color, although some calcareous varieties have a chalky appearance.

The family Codiaceae contains several genera of benthic green algae that are calcareous; they are major contributors to bottom sediment in shallow tropical areas. Such genera as *Halimeda*, *Penicillus*, and *Udotea* (Fig. 11.4) are particularly abundant. They are rather large (several centimeters high) and have specialized holdfasts to anchor themselves.

Phaeophyta. Of all marine plants, the brown algae are the most totally marine, with only three fresh-water forms. Marine forms are commonly large and make up the bulk of coastal seaweeds in mid- and high-latitude waters.

FIG. 18.1 *Cross section of a core showing desiccated algal stromatolites near the top. These structures are constructed by filamentous blue-green algae. (Photo by M. A. H. Marsden.)*

FIG. **18.2** *Large kelp-type brown algae* (Nereocystis luetkeana). *(Photo courtesy of R. W. Pippen.)*

Their characteristic brownish color is due to a special pigment that masks the color of chlorophyll. Included in this group are the kelps, the largest and most complex of the algae. They are predominantly cold-water plants, common along the northern coasts of the Atlantic and Pacific Oceans. Within this order are *Macrocytis, Nereocystis* (Fig. 18.2), and *Pelagophycus*; they are the largest algae and form commercial kelp. A maximum length of 40 meters is reached by *Nereocystis,* which lives on rocky surfaces in shallow water. Many other brown algae are of commercial significance, particularly in the Pacific area. Some of the seaweeds are edible and are consumed by people in oriental countries. They are fed to domestic stock animals in the British Isles and Scandinavian countries. Undoubtedly the most valuable product of the brown algae is algin, a compound found in the cell walls and comprising as much as 2.5 percent of the plant by weight. Algin is an absorbent material which is used as an emulsifier or film-forming colloid in ice cream, cosmetics, paints, drugs, and latex.

Rhodophyta.   More species of red algae are present in the sea than all other algae combined. They are present in a variety of forms and are essentially worldwide in distribution, the highest concentrations being in tropical areas. Many are massive, calcareous forms. Red algae as a group exist at the greatest depths of all plants; they have been found alive at 200 meters. They can live there because of their ability to utilize the deep-penetrating light rays in the blue and violet end of the spectrum for photosynthesis. Red algae live in shallow water as well and are even common in the intertidal zone.

One of the largest families of the red algae is the Corallinaceae, or coralline algae (Fig. 18.3), which are calcareous. This group includes two general forms, an encrusting and nodular type and one which is jointed. Many species of these coralline forms are associated with reefs and other shallow tropical areas.

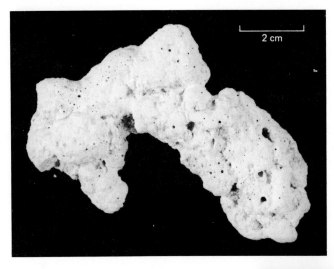

(a)

FIG. 18.3 *Two common corralline red algae:* (a) Lithothamnium *and* (b) Goniolithon.

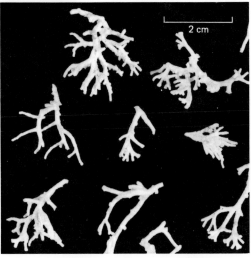

(b)

Flowering plants

Marine flowering plants are restricted to two families of angiosperms which include only eight genera. Most of these sea grasses are tropical except for the common *Zostrea* (Fig. 18.4), which ranges as far north as Alaska and Greenland. Eel grass, as this genus is better known, is common in shallow waters to

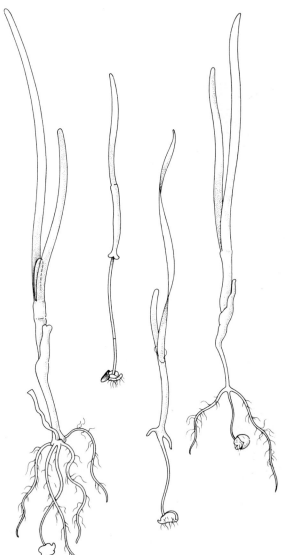

FIG. 18.4 *Seedlings of Zostrea, the common sea grass of Pacific and Atlantic coastal waters. (After E. Y. Dawson, 1966, Marine Botany—An Introduction, New York: Holt, Rinehart and Winston, p. 263.)*

depths of about 15 meters. This plant along with turtle grass *(Thalassia)* are the two most common marine angiosperms. Turtle grass (Fig. 18.5) is a more tropical type, found throughout the Gulf of Mexico. Both of these grasses are important as food for various shallow-water organisms and serve as a shelter for many others.

There are three other common flowering plants that live in the marine environment, but they are only partly submerged. These include high marsh grass *(Juncus)* and low marsh grass *(Spartina)*, which form coastal marshes, and mangroves. Mangroves occur throughout tropical and subtropical inter-tidal areas, where they are highly adapted to changing water levels. Their prop roots are a common sight about the bottom during low tide.

FIG. 18.5 Thalassia, *or turtle grass, which thrives in the shallow waters of the Gulf of Mexico.*

ANIMALS

All truly benthic animals are invertebrates. They are tremendously diverse in both form and function (Table 18.2) and are found throughout all ocean-bottom environments. By far the greatest concentration is on the continental-shelf areas where food is plentiful; however, in recent years deep-sea collecting and photographs have indicated that there is a much greater diversity and amount of deep-water benthic life than scientists had realized.

TABLE 18.2 *Common marine benthic groups.*

Scientific name	Common name
Sarcodina	Single-celled ameboid animals
Porifera	Sponges
Coelenterata	Corals and anemones
Annelida	Round worms
Mollusca	
Pelecypoda	Clams, oysters, scallops
Gastropoda	Snails
Arthropoda (Crustacea)	Barnacles, lobsters, crabs
Echinodermata	Sand dollars, sea urchins, starfish, sea lilies

There are two distinct subdivisions of benthic animals: the **epifauna**, which live on the bottom surface, and the **infauna**, which are burrowers and borers. Some animals may occupy both modes of life, but even these are dominantly of one type or the other. A great many of the major groups that were discussed in Chapter 16 on plankton (see Table 16.1) are also significant as benthos.

Sarcodina.   The Foraminifera of the phylum Sarcodina occur in much greater variety in the benthos than in the plankton. Instead of delicate globose forms, the benthic forams (foraminifers) are more heavily shelled, have no delicate protuberances, and occur in a great variety of shapes (Fig. 18.6). Most types are microscopic, although some of the larger tropical species may be a centimeter in diameter. The tests (shells) of most foraminifers are calcium carbonate, although there are some species that have agglutinated tests which are composed of small detrital sediment grains held together by organic material. Foraminifera in general are sensitive to changes in environmental conditions, and as a result most species are quite restricted in their distribution. This is in direct contrast to the ubiquitous nature of many planktonic species.

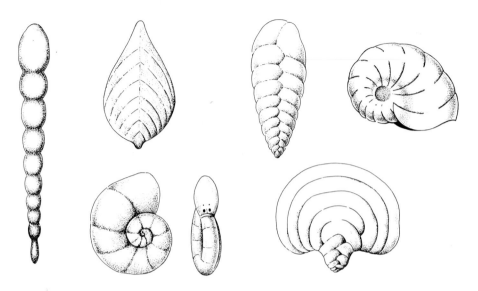

FIG. 18.6 *Various types of benthic Foraminifera.*
Actual size is about 0.5–1.0 mm.

Porifera.   Marine representatives of the phylum Porifera are of some economic importance. The sponges which constitute this phylum are simple, multicellular organisms (Fig. 18.7) that occupy sessile positions on the sea floor. Ciliated cells cause circulation of water through the organism, and organic debris is extracted for nourishment. Dependence on water circulation and a somewhat firm bottom may restrict the distribution of sponges.

Coelenterata.   The phylum Coelenterata includes the benthic classes of Hydrozoa and Anthozoa. Hydrozoans are small and inconspicuous members of the marine benthos that grow attached to rocky substrates. They have many tiny, delicate branches that are used to collect organic particles from seawater and give the animal a fernlike appearance (Fig. 18.8).

Anthozoans resemble flowers and include the anemones, corals, and sea fans. They may be solitary or colonial. Anemones (Fig. 18.9) are soft, fleshy, solitary anthozoans whose bright colors can be seen through shallow tropical waters. The mouth is surrounded by many flexible tentacles with stinging cells used to stun invertebrates or small fish, which are then ingested.

Corals closely resemble anemones in their soft anatomy, but they may be either solitary or colonial. Colonial varieties are dominant and are an impor-

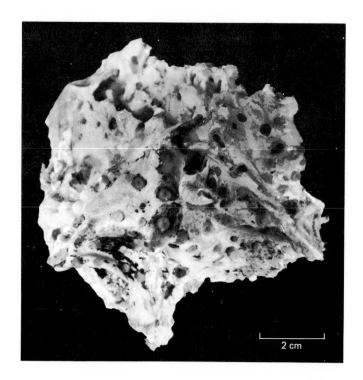

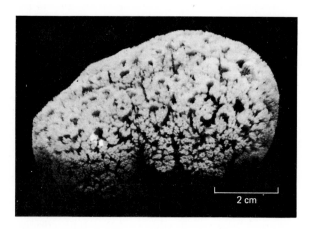

FIG. **18.7** Examples of modern sponges.

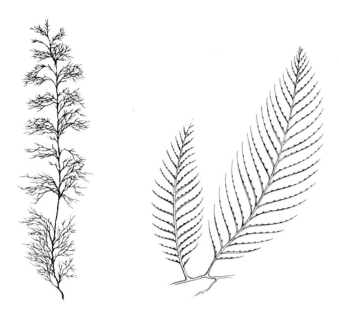

FIG. **18.8** *Typical hydrozoans (sea ferns). (After Hardy, 1959, p. 96.)*

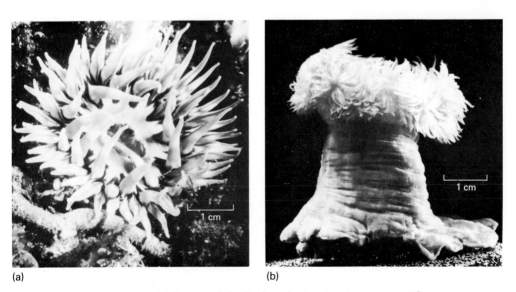

(a)　　　　　　　　　　　　　　　　　　　　(b)

FIG. **18.9** *Sea anemones, which are soft-bodied coelenterates, trap prey with their many tentacles. (Photos courtesy of (a) Charles Birkland and (b) the Carolina Biological Supply Company.)*

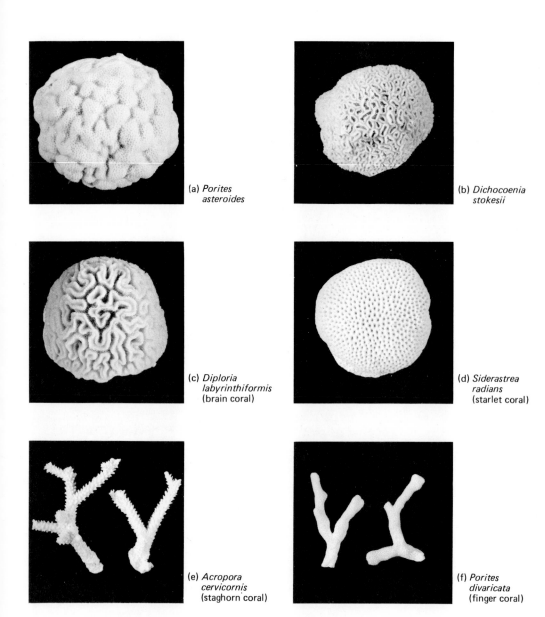

(a) *Porites asteroides*

(b) *Dichocoenia stokesii*

(c) *Diploria labyrinthiformis* (brain coral)

(d) *Siderastrea radians* (starlet coral)

(e) *Acropora cervicornis* (staghorn coral)

(f) *Porites divaricata* (finger coral)

FIG. 18.10 *Common varieties of colonial reef-building corals from the Florida Keys area. All are about half scale; however, some forms may become much larger. (a) Porites asteroides, (b) Dichocoenia stokesii, (c) Diploria labyrinthiformis (brain coral), (d) Siderastrea radians (starlet coral), (e) Acropora cervicornis (staghorn coral), and (f) Porites divaricata (finger coral). (Photos by R. D. Havira.)*

tant constituent of most reefs. The calcium carbonate skeleton secreted by corals may, in the massive colonial varieties, reach a few meters in diameter (Fig. 18.10). There are also more delicate branching types and small solitary corals. Contrary to popular belief, not all corals live on shallow tropical reefs; there are several species that live in Arctic waters and some that live at considerable depths. The vast majority do, however, live in warm waters within the zone of light penetration.

Worms.   The sea floor is occupied by a variety of worms, including the flat worms and segmented worms. These phyla contain thousands of species, most of them quite small; however, some annelids are large. The polychaete worms are very common marine annelids which may be either sessile or free-moving. Some of the free-moving species are pelagic, but most live on the bottom where they crawl on or burrow in the bottom sediment. Most sedentary polychaetes live in permanent burrows or in tubes which they construct themselves. Some of the polychaetes, such as the family Serpulidae, secrete a calcareous tube (Fig. 18.11) which comprises the major framework of some small reefs. Polychaetes feed in a variety of ways: there are filter feeders, detritus feeders, and others that use their specially adapted proboscis to capture small animals.

FIG. 18.11 *Serpulid worm in its calcareous tube. (After Barnes, 1968, p. 219.)*

Bryozoa and Brachiopoda. During the geologic past, the phyla Bryozoa (moss animals) and Brachiopoda (lamp shells) were quite important benthic animals; however, both are somewhat limited in modern seas. Both are sessile groups and have external skeletons, but this is where the similarities end. Bryozoans are tiny colonial organisms which secrete a calcareous skeleton into which the animal can completely withdraw. They commonly encrust any fairly hard surface or form delicate, branching colonies. Brachiopods are bivalves that attach themselves to the substrate by means of a short muscular stalk (Fig. 18.12). They are filter feeders that circulate water through the opened valves and extract particulate organic debris. One genus *(Lingula)* is a burrower that attaches its stalk to the bottom of the burrow. This genus has existed almost without change for half a billion years.

FIG. **18.12** *A modern brachiopod (lamp shell) with its pedicle attached to the substrate. About 1.5X enlarged.*

Mollusca. A large number and variety of terrestrial, fresh-water, and marine species comprise the phylum Mollusca. Most of the 80,000 extant marine species are benthic. Included in the phylum are the benthic groups, chitons, tusk shells, snails, clams, oysters, and scallops. The chitons are considered to be the most primitive of the mollusks. They have a chitonous shell which protects the soft anatomy of the animal, which is a muscular foot and a primitive digestive tract (Fig. 18.13). The animal crawls slowly over rocks in or near the intertidal zone, rasping algae with its radula, which is a scraping structure in the mouth. Individuals range from one to several centimeters in length depending on the species.

The tusk shells have a slightly tapered, elongate, calcareous shell which resembles a tusk and is open at both ends. The shell, which may reach 3 or 4 centimeters in length, is occupied by a primitive animal with a long, muscu-

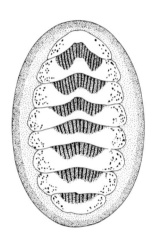

FIG. **18.13** *Diagram of one of the creeping amphineurans (chitons) which are commonly found on rocky intertidal surfaces. Approximately to scale.*

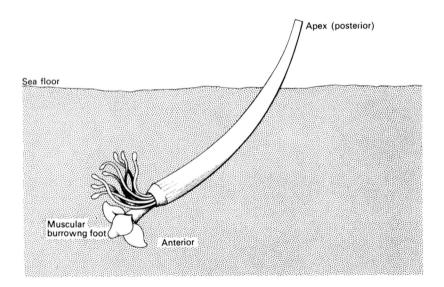

FIG. **18.14** *Typical tusk shell (scaphopod) in living position.*

lar foot which protrudes from one end of the shell and is used for burrowing (Fig. 18.14). Water and food is taken in and expelled at the small (posterior) end of the shell.

The above two groups of mollusks are not very diverse or abundant, whereas the gastropods and pelecypods contain thousands of species The snails are common inhabitants of the sea floor, where they crawl about in

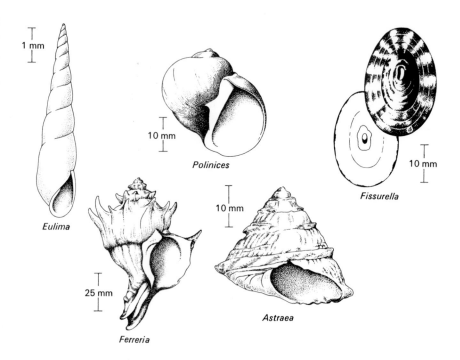

FIG. 18.15 *Various genera of gastropods (snails). Note the many shapes taken by the shells.*

search of food or burrow in the sediment. The shape and appearance of the shell (Fig. 18.15) can tell us something about the ecology of the animal. Those with massive shells having various protuberances or with bulbous shells are epifaunal, while the high, spired forms are the burrowing types. The gastropod shell is a single chamber, with the animal extending out of the aperture. Locomotion is provided by the expansion and contraction of its muscular body. Both herbivorous and carnivorous species exist, and some of the latter have a radula with which they can penetrate the shells of other animals and feed on the soft tissues.

From the standpoint of economics, pelecypods are among the most important marine invertebrates. Included in this class of mollusks are the clams, oysters, and scallops: all bivalves. Most of the pelecypods are mobile except the oysters, which exist commonly in large, reeflike associations. Scallops are the most mobile of the pelecypods, since they have a feeble jet-propulsion system which enables them to swim short distances to escape their predators.

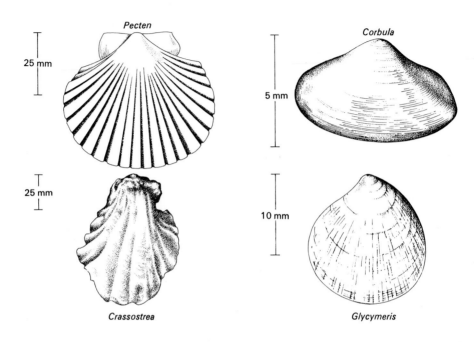

FIG. 18.16 *Selected pelecypod (clam) genera. Drawings show some of the forms assumed by this group of mollusks.*

The shape and construction of pelecypod shells (Fig. 18.16) can also give clues to their place of existence. Delicate and streamlined shells, such as that of the razor clam (Fig. 18.17), typically are burrowers, whereas a spiny shell is obviously not shaped for burrowing. Pelecypods that burrow have long inhalent and exhalent siphons (Fig. 18.18) that are used for water circulation and filter feeding. Some of the clams are specially adapted for boring into wood or rock. They accomplish this by rotating the shell or opening and closing it in such a fashion that it causes abrasion. Some secrete an acid mucus which assists in the boring process.

Arthropoda. The crustaceans contain two important groups of benthic organisms: the barnacles and the lobsters and crabs. Barnacles, which are exclusively marine, are the only sessile crustaceans. Their soft anatomy resembles that of an ostracod which is standing on its head encased in a calcareous shell with a muscular stalk (Fig. 18.19). Several pairs of **cirri** create water circulation across the mouth, thus facilitating the extraction of

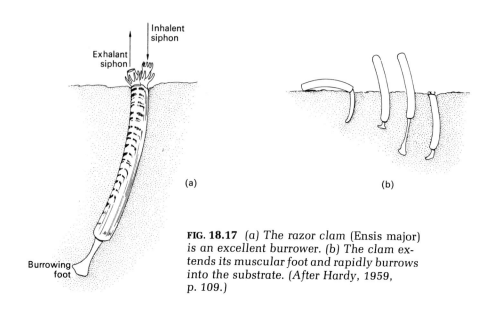

FIG. 18.17 *(a) The razor clam (Ensis major) is an excellent burrower. (b) The clam extends its muscular foot and rapidly burrows into the substrate. (After Hardy, 1959, p. 109.)*

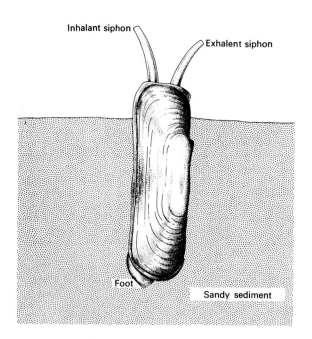

FIG. 18.18 *Infaunal pelecypod with siphons extended. (After Barnes, 1963, p. 283.)*

FIG. 18.19 *Goose-necked barna-cles, a sedentary arthropod.*

FIG. 18.20 *Encrusting barnacles typically are found on hard, intertidal sur-faces such as rocks, seawalls, and pier pilings.*

food from the water. Barnacles live throughout the world on hard substrates such as pilings, ships, and rocks, or attached to whales, sharks, crabs, and other animals. Some occupy the intertidal environment (Fig. 18.20) where they are adapted to periodic exposure to the drying effects of the atmosphere. In general, barnacles are common marine-fouling organisms that may cause considerable damage to ships or marine structures.

More than two-thirds of all species of crustaceans are in the group which includes marine crabs and lobsters (Fig. 18.21). The animal consists of a

(a)

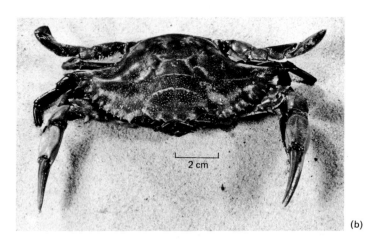

(b)

FIG. 18.21 *Two common marine crustaceans: (a)* Astacus *(a marine crayfish) and (b)* Callinectes *(a marine crab). (Photos courtesy of the Carolina Biological Supply Company.)*

jointed body with fourteen segments and a telson (terminal tail segment). The chitonous exoskeleton provides good protection, and some crustaceans also have large claws that aid in their defense as well as food-gathering. The legs serve as a means of locomotion on the substrate or in some cases for swimming. Some crabs have a pair of legs modified to serve as paddles. Crabs and lobsters are omnivorous scavengers. They frequently burrow in the mud or sand, or in the case of the unique hermit crab, they live in a vacated snail shell.

Echinodermata. These spiny-skinned animals are a marine benthos phylum composed of both vagrant and sessile forms. Within the phylum, there are four vagrant classes: common starfishes, brittle starfishes, sea urchins and sand dollars, and sea cucumbers; and one sessile group: sea lilies. All echinoderms are fairly large, the smallest being about a centimeter in diameter. The phylum is characterized by a unique pentameral (five-part) symmetry which masks its bilateral symmetry. The dominantly calcareous skeleton is covered with a variety of protuberances ranging from small tubercles to long, needle-sharp spines—thus the name spiny-skinned animals. The spines are adapted for protection, locomotion, and as sensory devices in various types of echinoderms.

The asteroids, or common stars, are probably the most familiar in the entire phylum. Although most types have five arms (Fig. 18.22), there are species that have different numbers of arms, ranging up to 40 in the genus *Heliaster* which is found on the Pacific Coast of the United States. Stars are free-moving and use their arms to crawl over the sea floor. The ventral side of

FIG. 18.22 *Starfish showing various numbers of arms. These animals have a basic bilateral symmetry which is masked by pentameral (five-sided) symmetry. (Photos courtesy of (a) L. H. Somers and (b) Charles Birkland.)*

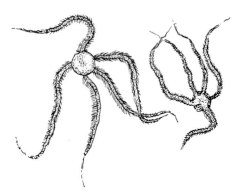

FIG. 18.23 Ophiunereis annulata, *a common member of the ophiuroids or brittle stars.*

each arm is covered with rows of small sucker feet that aid in the food-gathering process. Starfish wrap their arms around a bivalve and pull it open with their sucker feet. The stomach of the starfish is then extended into the open shell, where it digests the soft tissues of the bivalve. For this reason, starfishes are a hazard to oyster beds.

Brittle stars and basket stars resemble the common stars in that they, too, have arms radiating from a central disc. However, the arms are very long and distinctly set off from the central disc. Brittle stars are small with respect to common stars and their arms are more delicately structured (Fig. 18.23). Like the common stars, most brittle stars have five arms, but basket stars have branching arms which are flexible and resemble tentacles (Fig. 18.24). Brittle stars can move rapidly, their radiating arms acting like legs, and they are the most mobile of the echinoderms. Feeding habits of the brittle stars are significantly different from asteroids in that they are not voracious carnivores. Some feed on bottom detritus or small animals and others are filter feeders or browsers.

Sand dollars and sea urchins (Fig. 18.25) have the most extreme modification of the spiny skin, particularly the sea urchins. The body is oval or disk-shaped and is without appendages. Various forms of the spines occur even within a single species of sea urchins. The most common arrangement is short, sturdy spines on the oral (ventral) side to aid in locomotion and long, pointed spines on the aboral (dorsal) side for protection. Sea urchins may have flattened spines, stubby round spines, very long, delicate spines (Fig. 18.25a), or various color patterns, all of which aid in their protection. Sand dollars have bristlelike spines on their oral side to aid in locomotion but no spines for protection (Fig. 18.25b). They commonly burrow into the sand to conceal their presence, a habit also practiced by sea urchins that do not have any long spines. Movement is slow in these animals and is accomplished by the combined efforts of their spines and tubed feet. They slowly traverse rocky areas or sand bottoms, where they are omnivorous scavengers.

FIG. **18.24** *Complex arm development in the basket star* (Gorgonocephalus). *(Photo courtesy of L. H. Somers.)*

(a)

(b)

FIG. **18.25** *Some common echinoids. Sea urchins have a wide variety of spines, some long and needle-shaped (a), and some club-shaped, whereas sand dollars (b) have short, rather flexible spines. (Photo (a) by Charles Birkland.)*

FIG. **18.26** *A sea cucumber showing the rows of short tube feet. (Photo courtesy of the Carolina Biological Supply Company.)*

The name "sea cucumber" pretty well describes the overall appearance of the holothurians. The axis of five-sided symmetry parallels the substrate (Fig. 18.26), with the animal's mouth, surrounded by tentacles, at one end and the anus at the other. Sea cucumbers have only tiny skeletal plates incorporated into their leathery skin. They range up to a meter in length and occur in a wide variety of colors. Locomotion, provided by their small podia (feet), is sluggish. Their diet consists primarily of organic detritus supplemented by plankton.

Sea lilies (crinoids) are the only extant sessile echinoderms, although from the geological record others are known to have existed. Even in modern sea lilies, the number of species is much smaller than in ancient times. The animal resembles a plant because of its stalk with holdfasts and its flexible arms (Fig. 18.27). Like the basket stars, they may have five arms or multiples

FIG. **18.27** *A crinoid (sea lily), one of the sedentary echinoderms. (Photo courtesy of J. Engemann.)*

of five with complex branching. The mouth is located dorsally and is supplied with food by the arms, which may be several meters long. Crinoids feed on plankton and suspended organic detritus. Some species, called feather stars, are free swimming; they are concentrated in the Indian Ocean.

BENTHIC COMMUNITIES

Organisms which occupy the sea floor are rather easily discussed in terms of their associations within a given benthic environment. This is possible largely because of their slow locomotion or total immobility as compared to planktonic or nektonic organisms. Additionally, the benthos have essentially a two-dimensional distribution as compared to the other groups which may be found throughout much of the water column. The following discussion will cover only a few of the more common benthic communities.

Rocky intertidal community

On rock surfaces between the high- and low-tide marks is a fascinating community of hardy and specially adapted organisms. Essentially all are **epifaunal** in that they live on the surface as opposed to burrowing under it (**infaunal**). The rocky intertidal zone is a rigorous one because it is generally one of high physical energy (waves and currents) and also because the organisms are alternately exposed to the atmosphere and inundated by the sea.

The uppermost part of this environment is exposed most of the time and is occupied by a variety of algae, mostly blue-greens, which appear as a black crust on the rocks. A large percentage of the animals in this community are **sessile**; that is, they are attached and have no mobility. Such animals feed on plankton and suspended detritus which are carried past them by currents. Included in this category are barnacles, sponges, anemones, some bivalve mollusks, and sabellid (polychaete) worms. In terms of zonation, the barnacles are found just below the encrusting algae. Barnacles can close up tightly and thereby protect themselves from desiccation. Actually two types, acorn barnacles and goose-necked barnacles, are commonly present, the former being higher up in the intertidal zone. Below the barnacle zone are filter-feeding, bivalve mollusks (*Mytilus*), and in some areas oysters also occupy this zone. There also may be some bivalves (pholads) which can actually bore into the rock itself and thereby obtain protection from predators and pounding waves.

In addition to the above-mentioned sessile forms, there is also a variety of slow-moving, grazing animals. These organisms feed by scraping algae and detritus from the rock surface. Included in this group are chitons, limpets, some sea urchins, and some snails. The carnivores and scavengers of this community include anemones, starfish, crustaceans, and some varieties of worms.

Shallow, sandy-substrate community

Below the low-tide position and extending a few tens of meters in depth is an environment which is generally comprised of a sandy bottom and which contains an abundant and diverse benthic community. The nature of this substrate is such that it is suitable for infaunal organisms. Plants are sometimes present if the bottom is relatively stable; these would include eel grass (*Zostrea*) and turtle grass *(Thalassia)*. They are most common in coastal bays and lagoons but may be found in open water on the inner continental shelf. In low-latitude areas, varieties of calcareous algae such as *Penicillus, Halimeda,* and *Udotea* (Chapter 11; Fig. 11.5) are abundant in this environment.

The bulk of the community in this environment is comprised of infaunal animals. These include filter feeders which burrow into the substrate and extend their siphons or feeding mechanisms above the sediment surface. In addition, there are numerous burrowers which plow through the sediment and feed by grazing.

Common filter feeders include the clams, which burrow up to several times the length of the shell and extend their siphon above the sediment to catch particulate organic matter in the water (Fig. 18.17). In addition, roundworms and ghost shrimp establish burrows and feed in a similar manner. Most of the grazing animals feed by essentially plowing through the sediment and ingesting rather large quantities of it. The nutrient material is extracted and the inorganic portion is excreted as fecal pellets. Heart urchins are one of the most voracious types, although some snails and sand dollars also feed in this fashion.

Carniverous, epifaunal organisms also may be found in this community. Of particular interest are the snails. Several varieties move slowly over the sandy substrate in search of prey; once a suitable individual is located, the snail bores a hole in the shell of its victim and extracts the flesh and juices.

Abyssal community

Early studies of benthic fauna from the ocean assumed that there was no life beyond a few hundred meters depth (see Introduction). However, scientists in the late nineteenth century did collect specimens from over 2000 meters depth, and the *Challenger* Expedition provided much data on deep-water organisms. It was only relatively recently, however, that direct observation of the ocean floor was made possible, first by underwater cameras and finally by the various deep-water submersibles. It is only in the last decade or so that a comprehensive understanding of the deep-ocean community has been realized.

Among the sessile organisms which live on the relatively soft substrate are glass sponges, some soft corals, crinoids, worms, and barnacles. This portion of the community is less diverse and less abundant than that of shallow environments due to the overall scarcity of food and the disadvantages of the sessile mode of life. The mobile benthos in the community are able to move about in their search for food. Some of the common groups are the crustaceans, bottom-dwelling squid and octopods, brittle stars, and sea cucumbers.

COLLECTING THE BENTHOS

Quantitative sampling of bottom-dwelling organisms can be controlled considerably better than can the sampling of plankton or nekton. It is made possible largely by the presence of a reference surface, the bottom, and the fact that many organisms are sessile or infaunal. Vagrant animals are usually sluggish in their movements. All in all, the efforts of marine biologists to collect and study benthic organisms and their relationship to their environment have been fruitful.

Some of the most informative work on the presence and distribution of benthos has not involved collecting the organisms but observing them, either directly or indirectly. The invention and subsequent perfection of SCUBA (self-contained underwater breathing apparatus) have made it possible for scientists to study marine organisms directly and observe the details of their habits and habitat. Such observations are restricted to depths on the continental shelves, but this is where the vast majority of benthic creatures live. Areas where it is not possible or practical to use SCUBA are commonly examined by undersea photography or closed-circuit television. Photographs taken in all deep-sea environments show that there is far more benthic life there than scientists once believed. In many respects, these studies are of more value than sampling, because they show the organisms in their natural environment. Problems exist in detailed identification, however.

Dredges of various types are used to sample primarily epifaunal benthos. All these have a rigid frame with a basket of coarse wire mesh (Fig. 18.28).

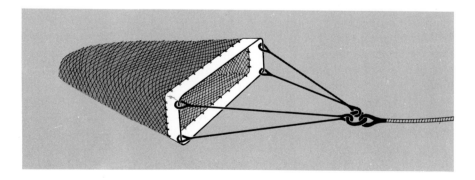

FIG. **18.28** *Basket-type dredge for collecting benthic organisms.*

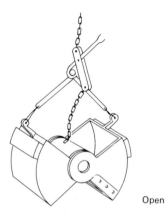

Open

FIG. 18.29 *Petersen-type quantitative bottom sampler. (From Duxbury, 1971, p. 361.)*

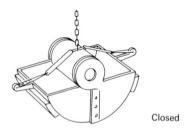

Closed

They may be conical or have a rectangular frame, and the size may range widely. Such an apparatus, dragged behind a vessel, collects bottom organisms and nonliving material such as shells and sediment. A great variety of sizes and shapes can be utilized, depending on the sampling program. Dredges have some disadvantages, such as the inability to retrieve burrowing organisms or quick-moving animals. For these reasons quantitative sampling of any significance cannot be done with dredges. Beam trawls, described in the previous chapter, also may be used for collecting benthos.

Semiquantitative samples may also be obtained by using a grab-type sampler, such as the Petersen clamshell. The apparatus consists of hinged jaws which are dropped in the open position; when retrieved they close, thereby taking a bottom sample (Fig. 18.29). The grab samplers are constructed to sample a known area, commonly 1/10 of a square meter. Shallow infaunal animals may be taken by the grab-type sampler, but it suffers from some limitations. Quick-moving animals can avoid it, as with dredges, and more important, large shells or rock fragments may be caught between the jaws and prevent them from completely closing. If this happens, the smaller organisms can be washed out of the sampler during its ascent to the vessel.

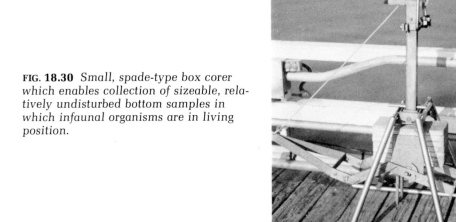

FIG. **18.30** *Small, spade-type box corer which enables collection of sizeable, relatively undisturbed bottom samples in which infaunal organisms are in living position.*

For many studies, quantitative sampling of organisms provides only part of the desired information. One may have a need for collecting organisms, especially infauna, in living position. In this fashion the relationship of organisms to one another and to their substrate can be determined. A box-coring device (Fig. 18.30) provides an excellent sampling mechanism for such data. The box-shaped core penetrates the substrate and obtains a block of sediment with its organisms intact. These box corers range in size depending on the size and capabilities of the vessel from which they are operated.

SELECTED REFERENCES

Barnes, H., 1959, *Oceanography and Marine Biology—A Book of Techniques,* New York: Macmillan.

Barnes, R. D., 1968, *Invertebrate Zoology* (2nd edition), Philadelphia: W. B. Saunders.

Buchsbaum, Ralph, 1948, *Animals Without Backbones* (revised edition), Chicago: University of Chicago Press.

Dawson, E. Y., 1966, *Marine Botany—An Introduction,* New York: Holt, Rinehart and Winston, Chapters 6–10. Excellent general book on marine botany by a world authority. Well-written, well-illustrated, and current. Appendixes contain particularly useful data on herbaria, marine laboratories, etc.

Hedgpeth, J. W. (ed.), 1957, *Treatise on Marine Ecology and Paleoecology,* Geological Society of America Memoir 67, *Ecology,* Vol. 1. Comprehensive treatment of benthos, especially the chapter on bottom communities by Gunnar Thorson (Chapter 17). Extensive annotated bibliography is most complete to that date.

McConnaughey, B. N., 1974, *Introduction to Marine Biology* (2nd edition), St. Louis: C. V. Mosby. Excellent and well-illustrated book covering all of the various marine communities.

Menzies, R. J., R. Y. George, and G. T. Rowe, 1973, *Abyssal Environment and Ecology of the World Oceans,* New York: Wiley. The most comprehensive treatment to date on the biology of the deep ocean.

Moore, H. B., 1958, *Marine Ecology,* New York: Wiley, Chapters 9–11. Thorough but poorly illustrated treatment of benthos in the shallow areas. General discussions are lacking, since most data are for selected taxa.

Ricketts, E. F., Jack Calvin, and J. W. Hedgpeth, 1968, *Between Pacific Tides* (4th edition), Stanford, California: Stanford University Press. This classic on intertidal benthos was recently revised and expanded by J. W. Hedgpeth, one of the leading marine invertebrate specialists in the world. Illustrations are superb, as is the comprehensive systematic index.

GEOLOGICAL OCEANOGRAPHY

PART **V**

MARINE SEDIMENTS AND PROPERTIES 19

The vast majority of the material that constitutes the bottom of the oceans, seas, and their coastal areas is mineral in nature. The study of oceanography is not complete without an appreciation and understanding of marine sediments, their composition, distribution, and origin. Although sediments dominate the ocean floors, igneous rocks are also present in abundance at certain places, for example, adjacent to island arcs and on oceanic ridges.

Sediments may be classified on the basis of texture and composition. Both parameters may indicate the source of the sediment and the depositional environment. However, texture reflects mainly the interaction of the sediment with water motion during transportation and at the depositional site, whereas mineral composition reflects primarily the source of the sediment. In the case of marine sediments, we usually reserve a special, more detailed type of classification for deep-sea deposits, because most of them would fall into the same category if we used the general classification of sediments. The classification will be included in the chapter on deep-sea sediments (Chapter 23). Sediments found in continental margin areas or associated with reefs are classified in the same manner as continental sediments or sedimentary rocks.

SEDIMENT TEXTURE

Size and shape of grains
The texture of a sediment involves a variety of characteristics, the most important of which is grain size. There is a tremendous range in particle size, with diameters ranging from microns to meters, and some system must therefore be used to describe grain size. Because of the tremendous range, it is not practical to establish uniform intervals or we would have literally hundreds of such intervals. A logarithmic scale is the practical answer.

In 1922 Chester K. Wentworth devised a scale which has come to be the standard size scale, although a few modifications of his original have taken place. The scale is in millimeters and includes names for the various range in

TABLE 19.1 *Primary sediment grain-size categories.*

Grain name	Size (mm)
Boulder	> 256
Cobble	64–256
Pebble	4–64
Granule	2–4
Sand	$\frac{1}{16}$–2
Silt	$\frac{1}{256}$–$\frac{1}{16}$
Clay	< $\frac{1}{256}$

particle diameter. Everyone has used terms such as sand, clay, pebble, or boulder; each of these has a definite quantitative meaning (Table 19.1).

All grains in a given sediment sample are rarely the same size or diameter. Consequently it is often difficult to look at a sample, even under a microscope, and obtain an average size of the particles. However, it is necessary to determine this average in order to give the sediment a name on the scale; is it coarse or medium sand? To obtain this data, the sediment is either sieved or settled in a column of water. In either situation, the particles of sediment are separated into very narrow size ranges. These data give a distribution of particle sizes for the sample. They can be plotted as a histogram, or more commonly as a cumulative curve, from which one can obtain such statistical values as the **mode, median,** and **mean.**

Another feature of sediment texture that is quite important is the standard deviation, or **sorting value**. This is merely a quantitative expression of the range in grain size, or the deviation from the mean. It is sometimes helpful in determining environments of deposition, some of which are characterized by a particular sorting value.

Roundness and **sphericity** are other textural properties that may be used to describe grain shape. Like grain size, both are properties of individual grains rather than the sediment as a whole. The best general definitions of the two are that roundness is the degree of smoothing of the corners on a grain, and sphericity is a measure of the degree to which the shape of the particle resembles a sphere. To illustrate, a tennis ball and a hot dog are both rounded, but only the tennis ball is spherical (Fig. 19.1a,b). On the other hand, a cube is rather spherical but poorly rounded (Fig. 19.1c). Roundness is of some importance in analyzing marine sediments in that it may be related to the environment of deposition.

The combination of grain size, sorting and roundness in sediments is used in R. L. Folk's concept of textural maturity. This measure of the physical energy in the depositional environment is a relative type of maturity without

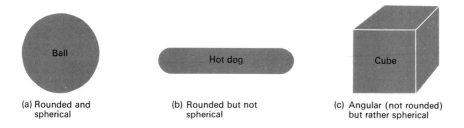

(a) Rounded and
   spherical

(b) Rounded but not
   spherical

(c) Angular (not rounded)
   but rather spherical

FIG. **19.1** *Examples of roundness and sphericity.*

any relation to absolute time. Three basic steps are necessary as a sediment proceeds toward textural maturity. These are (1) removal of fines (clay), (2) attainment of good sorting, and (3) rounding of the grains (Table 19.2). It is possible to apply quantitative measurement to sorting and rounding, for which the reader is referred to textbooks in sedimentation and sedimentary petrology.

TABLE **19.2** *Textural maturity.* (*After R. L. Folk, 1951, "Stages of Textural Maturity in Sedimentary Rocks,"* Journal of Sedimentary Petrology **21**, *128.*)

Type	Description
Immature	> 5% clay, poor sorting, angular grains
Submature	Little or no clay, poor sorting, angular grains
Mature	Good sorting, angular grains
Supermature	Good sorting, round grains

A typical "dumped" sedimentary deposit such as glacial till, or in marine geology, an underwater slide deposit, is termed "immature." In other words, the material was not winnowed and worked by currents as deposition took place, but was deposited rapidly down a steep slope or, in the case of glacial till, was transported by ice which did not allow sorting to take place. Deltaic deposits are generally in the submature category, whereas beach sediments are an excellent example of textural maturity. Beaches are constantly being changed and shifted by waves, currents, and winds. As a result, the beach is well sorted, whether it is composed of pebbles, coarse sand, or very fine, sand-sized particles. Some beaches even reach the supermature stage, when the particles are rounded by continual wave and current transport. Such a level of maturity is more easily achieved in beaches composed of relatively soft, calcium carbonate particles than those composed of quartz.

Fabric

Knowledge of the size and shape of grains composing sediment is quite important, but it is sometimes equally important to know about how these grains are arranged with respect to one another. This is the **fabric** of the sediment. Many things affect the fabric, such as size, shape, and sorting of the grains. The medium in which or by which the grains are deposited is also a factor in determining sediment fabric.

Because many variables are involved, perhaps the best way to approach the subject is to treat each one systematically. Grain shape is the most obvious factor to affect fabric. If a handful of toothpicks is dropped into a container one at a time, and a handful of marbles dropped in a second container one at a time, the result would be completely dissimilar arrangements of particles (Fig. 19.2). This is analogous to comparing the arrangement of clay mineral or mica flakes (toothpicks) with that of quartz grains (marbles). It is not uncommon for clays to have 75 percent or more of their volume occupied by air or water. This empty space, called porespace, is a measure of the **porosity** of a sediment. A quartz sand might have 30 percent or more porespace. If we carefully arranged the toothpicks in Fig. 19.2 so that they had the same orientation, the porespace would be greatly reduced. A clay might be compacted under pressure and the same result achieved; the grains would tend to align themselves.

(a) Toothpicks   (b) Marbles

FIG. 19.2 *Effect of grain shape on packing. Because of their elongate or tubular shape, some grains commonly contain much pore space (toothpicks), whereas grains that tend to be equidimensional (marbles) display much less pore space.*

By combining marbles with small shot and ping-pong balls, we would also reduce the porespace. The small particles could occupy spaces between the larger ones (Fig. 19.3), and so it is evident that sorting is a factor in porosity also. A poorly sorted sediment will have less porosity than a well-sorted sediment of the same average grain size.

(a) Coarse grained and sorted sediment

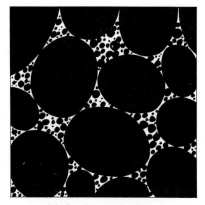

(b) Poorly sorted sediment

FIG. **19.3** *Addition of fine sediment to a coarse, well-sorted sediment (a) will cause reduction in porosity (b).*

Permeability is the ability of a material to transmit fluid. It is affected by grain size, sorting, and porosity. Most people have poured water on the beach or in a sand box and watched it disappear rapidly. The water was being transmitted through the sand because the sediment was well sorted and moderately coarse grained. The same experiment on the clayey soil of a yard or garden would yield different results. A poorly sorted or very fine-grained sediment has low permeability.

SEDIMENT TRANSPORT

The oceans are receiving tremendous quantities of sediment from erosion of the continents. Conservationists are constantly making us aware of erosion and the large amount of material carried to the sea by rivers. As evidence of the solid particles being eroded, many rivers have deltas which are rapidly growing in size. Likewise, great estuaries and coastal lagoons may be filled with sediment in a few thousands of years. Some of the sediment is being carried to the beaches and continental shelves, and some even makes its way to the ocean floor.

Waves and currents distribute sediment throughout the ocean basins and their margins. Much of the land-derived (**terrigenous**) sediment never reaches more than a few miles from shore. Sediment transportation is greatest along the beach and in the surf zone. Here the waves are breaking or at least feeling bottom and disturbing the sediment. Sediment is stirred up and, for a short period of time, perhaps only a few seconds, it is in suspension. During this time, tidal and longshore currents carry the sediment grains

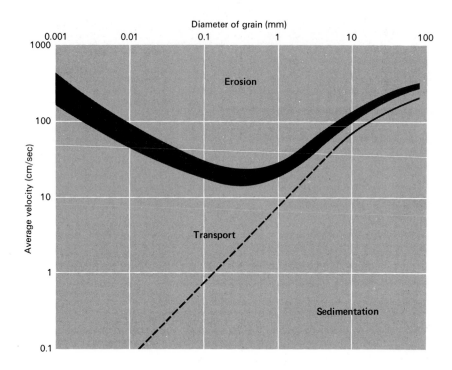

FIG. 19.4 *Hjulström's curve, which shows the relationship between current velocity and grain size with respect to erosion, transportation, and deposition of sediment. (After F. Hjulström, 1939, "Transportation of Detritus by Moving Water," Recent Marine Sediments, Amer. Assoc. Petrol. Geol., p. 10.)*

in a direction generally parallel to the shore. Most sediment transport occurs in this manner. In addition, some sediment is moved offshore by gentle currents or by wave activity. Fine particles such as silt and clay may remain in suspension for long periods of time and reach the ocean floor, or at least the outer continental shelf.

Texture and fabric of sediments are factors in, and results of, their transport. Obviously the size of grains is a limiting feature, but so is the shape of grains. A classic illustration of the relation of particle size to velocity is that of Hjulström (Fig. 19.4). Although his data were obtained from the fluvial environment (rivers), they are applicable to currents in general. The most striking fact is that it takes greater current velocity to pick up clay particles than silt. The reason is the flat shape and cohesion of clay particles as compared to silt or sand.

Sorting values of a sediment may also have an effect on the movement of the sediment by waves or currents. For instance, a poorly sorted sediment composed of clay through pebble-sized particles will be stable because the coarser particles will serve as a protective baffle for the fine material. A well-sorted sand would be transported when subjected to a current equivalent to that experienced by the poorly sorted sediment.

SELECTED REFERENCES

Blatt, H., G. V. Middleton, and R. C. Murray, 1972, *Origin of Sedimentary Rocks*, Englewood Cliffs: Prentice-Hall. Probably the best single book available that covers the total spectrum of sedimentology.

Folk, R. L., 1971, *Petrology of Sedimentary Rocks*, Austin, Texas: Hemphill's Bookstore. Excellent treatment of statistics as applied to sediment studies. Basically a laboratory manual.

Krumbein, W. C., and L. L. Sloss, 1963, *Stratigraphy and Sedimentation* (2nd edition). San Francisco: W. H. Freeman, Chapters 4–7. Good general treatment of sediments and sedimentary rocks. Emphasizes the eventual incorporation of sediments into the rock record.

Pettijohn, F. J., 1975, *Sedimentary Rocks* (3rd edition), New York: Harper and Row. One of the standard textbooks on sedimentary rocks. New edition much more comprehensive than previous ones.

Pettijohn, F. J., P. E. Potter, and R. Siever, 1972, *Geology of Sand and Sandstone*, New York: Springer-Verlag. Excellent modern treatment of sediments, with considerable emphasis on depositional environments.

Twenhofel, W. H., 1932, *Treatise on Sedimentation* (2nd edition), Baltimore: Williams and Wilkins. One of the classic and comprehensive treatments of sediments covering all aspects of their study. Obviously it is considerably out of date.

20 COASTAL SEDIMENTARY ENVIRONMENTS

A fairly diverse group of depositional environments is associated with marine coastal areas. These are intermediate between terrestrial and marine conditions because they are affected by both. The chemistry and organisms of these environments are not like that of either the terrestrial or marine conditions and are transitional between them in virtually all respects.

For a variety of reasons, these environments are of great significance to a complete study of oceanography or marine geology. Of primary concern for the geologist is understanding these modern environments in order to interpret ancient analogs found in the rock record. For practical reasons involving coastal engineering, maintenance of shipping routes to harbors, and recreational facilities, it is necessary to understand the materials and processes present in these transitional environments. The bulk of the world's coastal areas is comprised of three such transitional areas: (1) beaches, which are by far the most widespread, (2) coastal bays, and (3) deltas. Each of these environments has its own characteristics and distinct processes. All three have been extensively studied because of their importance and also because they are accessible and easy to investigate compared to true oceanic areas.

BEACHES

Without a doubt, beaches are one of the most fascinating of all modern sedimentary environments. In strict definition, the beach includes the zone of unconsolidated material between the mean low-water line and the change in material or physiography on the landward side. Commonly the upper boundary is marked by a zone of perennial vegetation, dunes, cliffs, or other coastal features. Beaches border coastal areas of all types and are present in all parts of the world. There is great variation in the geometry and composition of beaches in terms of both geography and time. Some are only wide enough for a beach towel, whereas others may be hundreds of meters wide.

Although the beach itself extends only to the level of mean low tide, it is convenient to discuss the nearshore area and surf zone along with the beach.

The processes and materials associated with the beach are largely the same as, or closely related to, those in the adjacent shallow-water environment.

Beach nomenclature

A fairly extensive series of special terms is used for the various parts of the beach and nearshore environment (Fig. 20.1). The **nearshore** area is the zone adjacent to the beach and extending to the seaward limit of the longshore bar and trough topography that is present where the bottom gradient is not steep. This zone may be only a hundred meters wide or nearly a kilometer in width, with the number of bars and troughs depending on the width of the zone, slope of the bottom, and availability of sediment.

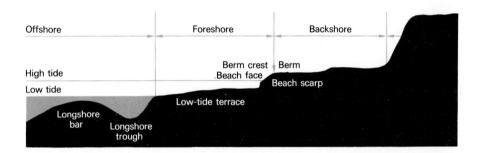

FIG. 20.1 *Major subdivisions of the beach and nearshore area. (After Shepard, 1972, p. 124.)*

The intertidal zone of the beach makes up most of what is called the **foreshore**, and it is, in general, a plane, sloping surface. The foreshore includes the seaward-sloping portion of the beach (Fig. 20.1) and is adjacent to the relatively horizontal **berm**. The berm changes size and shape as wave characteristics change. The third major beach zone, the **backshore**, is normally above the influence of water except for storms and extends from the berm crest to the coastline where there is a physiographic change.

Beach materials

If a number of people were asked to describe the composition of sediment on beaches, the vast majority would say that it is "light-colored sand." Statistically, most beaches are composed of sand, commonly quartz. Beach sands of shell fragments (calcium carbonate) are common in low latitudes (Fig. 20.2a) particularly, or the quartz and shell fragments may be mixed, as is the case along much of the Gulf of Mexico. There are many other materials which comprise beaches that are different both texturally and mineralogically.

(a)

(b)

FIG. 20.2 *(a) Beach on the Gulf of Mexico (Chandeleur Islands, off Louisiana) with a storm ridge that is composed almost totally of oyster shells. (Photo courtesy of J. C. Kraft.) (b) Small pocket beach at Cape Ann, Massachusetts that is composed almost entirely of well-rounded boulders.*

Adjacent to volcanic areas, basaltic or olivine-rich sands are common, as in Hawaii or on the western coast of Central and South America. Mud beaches also occur, but are not common. Shingle beaches are composed of flat, disk-shaped pebbles derived from bedrock of slate or shale. All possibilities of grain size exist (Fig. 20.2b), from mud to boulders, and a great variety of compositions occur.

Beach sediment is in general well sorted, and if we consider the dominant sand texture, it is usually well rounded. This sorting characteristic is perhaps one of the most conspicuous features of beach sediment. Regardless of the mean grain size, the range of particle sizes is narrow. For the past several years, many sedimentologists have been studying ways of statistically distinguishing beach sands from those of dune or fluvial environments. Although many statistical comparisons seem to indicate some significant separations, there is no standard test that the majority of geologists trust.

The source of beach sands is primarily the adjacent geologic province, or whatever area is drained by rivers that enter the ocean nearby. Most commonly the composition of the beach reflects the materials which comprise the coastal area. In addition to the runoff carried by rivers, there is a large contribution by wave activity which erodes cliffs along the coast. A minor amount of sediment may be derived from underwater erosion of Pleistocene deposits on the continental shelf, subsequently transported toward shore.

If we consider the major source, rivers, there is considerable range in the effect of this runoff on beaches, particularly with respect to the configuration of the shore line. For example, an irregular coastal plain with rivers entering many embayments would not supply much beach material. An example is the Gulf coast of Texas. The contrary situation is exemplified by the California coast, where there is at least moderate relief adjacent to the coast and few rivers enter sheltered bays but supply directly to the beaches. Here there is a large supply of potential beach sediment.

Beach profiles

Regardless of the shape of a beach profile at a particular location, it will undoubtedly change with time. One of the most impressive characteristics of beaches is their constant state of flux. This is rarely noticed unless one spends a moderate length of time at the same beach, especially through different seasons of the year.

At any given time, the profile may be a fairly uniform slope, a terrace, or a bar and trough topography. There is a somewhat predictable relationship between each of these profile types and the texture of the beach sediment. Most of the wide and gently sloping beaches are fine to medium sand (Fig. 20.3), such as those along much of the Gulf coast, Atlantic coast, and Oregon. Many of these are so firm and well packed that automobile traffic is possible. The coarser sand beaches are soft and difficult to walk over. Berms are quite

FIG. 20.3 *Wide and gently sloping beach on Mustang Island, Texas. Grain size is fine sand.*

common, thus providing a nick point on the profile. Beaches of gravel or shingle commonly have a ridge of coarse material (Fig. 20.4) that is piled up by wave activity and may reach a few meters in height.

In general, the coarser the beach sediment, the steeper the gradient (Table 20.1; Fig. 20.5). The reason for this relationship is that permeability

FIG. 20.4 *Gravel beach ridge that rises about 4 to 5 meters above the flat marsh area behind at Rye, New Hampshire.*

TABLE **20.1** *Relationship between beach sediment and slope of beach face. (From Shepard, 1972, p. 127.)*

Type of sediment	Grain size	Average slope
Very fine sand	1/16– 1/8 mm	1°
Fine sand	1/8– 1/4 mm	3°
Medium sand	1/4– 1/2 mm	5°
Coarse sand	1/2– 1 mm	7°
Very coarse sand	1– 2 mm	9°
Granules	2– 4 mm	11°
Pebbles	4– 64 mm	17°
Cobbles	64–256 mm	24°

FIG. **20.5** *Steep foreshore slope of coarse sand beach. (Photo courtesy of L. H. Somers.)*

increases with grain size. The well-sorted nature of beach sediments in general also contributes to permeability. Water in the **swash zone** rushes back and forth over the fine sand with little loss of water to the sediment because of low permeability. Consequently, there is about the same energy in the backwash as in the uprush of the water, and the slope of the beach is low. In coarse sediment like gravel, most of the water is lost to the sediment because of high permeability, and backwash is almost nonexistent. This situation gives rise to steep gradients.

There is significant difference between summer and winter beach profiles in most areas. During the summer the beach is broad and sand covered

(a)   (b)

FIG. 20.6 *Seasonal changes in the beach at LaJolla, California. (a) Summer beach showing accretion of sand due to low wave activity. (b) Winter beach with little sand due to erosive power of large winter waves. (Photos courtesy of F. P. Shepard.)*

(Fig. 20.6a), with a fairly steep beach face, but during the winter, the beach is cut back and much of the sand is gone (Fig. 20.6b). This widespread change is caused by the distinct difference in water conditions between the calmer summer months and the more stormy winter months. Small summer waves carry sediment toward the coast, where it accumulates on the beach. Large, high-energy waves generated by winter storms erode the beach and much of the sediment is carried out to the nearshore area and beyond. In actuality, the beach profile shape is related to wave and storm activity, not to seasons.

Although not actually a part of the beach, the adjacent nearshore area is closely related to it. Beyond the limit of low tide there is commonly a gently undulating topography of longshore bars and troughs. The number of bars depends on the slope of the nearshore bottom. Along the Texas coast there are typically two or three bars, but steeper areas such as the Virginia and Delaware coasts may have only one or even none. They are remarkably continuous and can be traced for several kilometers. In many areas the bars are conspicuous on aerial photographs (Fig. 20.7). Relief is about a meter or so, and the spacing and distance from shore depend on the slope. The first bar may be about 100 meters or less from shore and the second at 200 meters or so. The entire bar and trough topography is usually confined to a width of

(a)

(b)

FIG. 20.7 *Aerial photographs showing bar and trough topography, where (a) bars are visible through the water and (b) bars are delineated by breaking waves.*

about 500 meters or less. This corresponds approximately to the width of the surf zone under storm conditions.

The development of this regular and gently undulating topography seems to be the result of wave activity, and the bars in turn affect the waves. Orbital paths of water in waves pick up sediment and it accumulates parallel to shore, where the wave breaks and loses its competence. The waves form again and repeat the process, forming an additional bar closer to the strand line. Under normal conditions, waves do not break over the deepest bar, which was formed by storm waves. Tidal changes may enable waves to break over a deep bar at low tide but be unaffected during high tide. The nearshore profile is remarkably stable, at least during most of the year. It has been shown that during the winter months, the bar and trough topography may be somewhat smoothed out due to constant presence of large waves.

Sediment transport in the beach and nearshore zone
The sources of beach materials were discussed earlier. It is also important to consider their distribution and transportation along and away from the coast. By far the most significant factor in this transport is the longshore currents (Chapter 6), which are nearly ubiquitous in coastal areas. As sediment is carried to the coast from land or in small amounts from deeper water, it is picked up by longshore currents and transported essentially parallel to shore. Such currents have the ability to carry sand-size particles in the **bed load**, primarily by saltation and rolling of the individual grains. By far the greatest part of sediment transport takes place over bars of the nearshore area, but there is also longshore movement of sediment in the surf zone. Because waves nearly always strike the shore at an angle, there is a net longshore transport caused by the uprush and backwash in the swash area (Fig. 20.8).

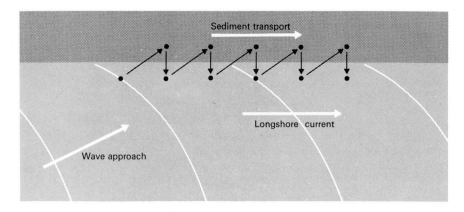

FIG. 20.8 *Path of a sediment grain caused by uprush and backwash of waves.*

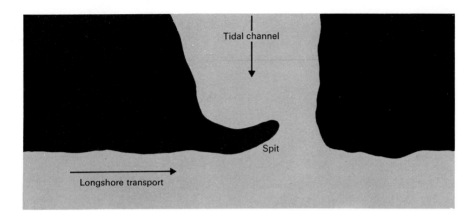

FIG. 20.9 *Spit development as the result of interruption of longshore currents.*

Mass transport of sediment by longshore currents is commonly called the "river of sand," with the channel bounded by the outer edge of the surf zone and the shoreline. There are many coastal features that may interrupt the "river of sand" and the longshore transport of sediment. The interruption can be caused by headland areas which physically block the longshore transport or by bays and estuaries in front of which longshore currents are absent. **Sand spits, tombolos,** or **baymouth bars** may be formed in these areas (Figs. 20.9 and 20.10).

Man-made coastal structures also cause problems for longshore sediment transport. Jetties and groins act like dams and cause sediment to pile up behind them (Fig. 20.11). Breakwaters shelter the beach from wave activity, and consequently longshore currents are absent behind the breakwater. As a result sediment builds up in this area (Fig. 20.12) where the "river" is broken. This demonstrates that attempts of man to alter the course of longshore transport and wave activity on the coast may create as many problems as they solve.

Even though there are many circumstances under which sediment movement is temporarily halted in its longshore path, the material stays in the beach and nearshore area. Obviously, sediment does not remain in this zone permanently. It may be removed in either the seaward or landward direction. Onshore winds blow beach sands inland and form coastal dunes. In this way sediment is recycled and may again make its way to the sea via runoff. There are two important ways for sediment to be carried from the beach and surf zone toward the ocean basin. First, rip currents carry sediment across the nearshore area, depositing it in deeper water. Second, many

(a)

(b)

FIG. 20.10 *(a) This spit was formed by the interruption of longshore currents and the consequent accumulation of sediments. (b) Tombolo connecting a granitic island to the mainland just east of Gloucester, Massachusetts.*

FIG. 20.11 *(a) Diagram showing sediment buildup behind a jetty, ▶ which acts like a dam preventing longshore transport of sediment. (b) Large amount of sediment accumulated behind a jetty as the result of longshore transport along the coast. (c) Sediment accumulating at groins in the Delaware Bay. (Photo (c) courtesy of J. C. Kraft.)*

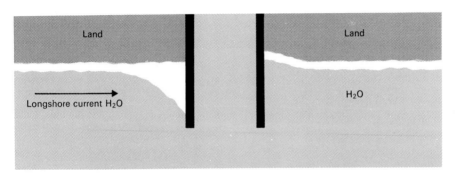

(a)

(b)

(c)

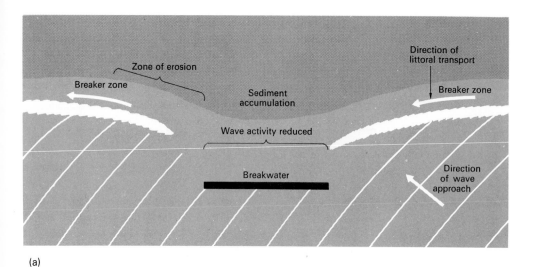

(a)

(b)

FIG. 20.12 *(a) Diagram showing sediment accumulation behind a breakwater due to the interruption of littoral sediment transport. The breakwater prevents waves from entering shallow water and thus no significant longshore currents are present behind the breakwater. (b) Breakwater effect caused by a sandbar on Padre Island, Texas. Note the accumulation of beach sand behind the sandbar and compare with (a).*

submarine canyons extend within short distances of the shore and the heads of canyons trap sediment moving alongshore. It is then carried seaward in the canyon via gravity.

Beach structures

A small trench dug through the berm or foreshore slope of a beach illustrates the presence of small-scale stratification (Fig. 20.13) in the beach sediments. Such stratification is most developed in sand beaches of at least moderately heterogeneous composition. The coarse beaches of shingle and gravel, or completely homogeneous beaches, do not show well-developed stratification. Most sand beaches are at least moderately diverse in either texture or composition, or both. Layering in the beach slopes gently seaward near the angle of the foreshore slope. The stratification is caused by concentration of heavy mineral grains or distinct differences in mean grain size. Although sorting values for the beach are nearly the same at any given time, the mean grain size may change considerably because it is a response to wave activity.

FIG. 20.13 *Trench in a beach showing the nature of the stratification incorporated in the beach sediment. The dark layers represent heavy mineral accumulations and are an indication of storm conditions and beach erosion.*

In areas of considerable variety in the size of sediment available, this is particularly evident. As a result these textural differences may be preserved in the beach stratification, at least until a storm destroys or partially destroys the berm.

There are many types of sedimentary structures found on the surface of beaches, and some of these become preserved in rocks. There they may provide information of their environment of deposition for geologists studying earth history. Preservation of these features is rare, however, because most of them are only present on the beach for short periods of time due to the everchanging nature of this environment.

Next to stratification, **ripples** are the most prevalent sedimentary structures on beach and nearshore surfaces. Both wave-formed and current-formed types are present in the nearshore area. The sand bottom of gently sloping nearshore areas is corrugated with ripples (Fig. 20.14), most of them several centimeters from crest to crest. On the beach itself, tidal currents or currents caused by turbulent backrush of waves may create ripples. The size ranges widely, with some being more than a meter in wave length. Crests of ripples generally are roughly parallel to contours of the beach slope.

FIG. 20.14 *Wave-formed ripples in the intertidal beach environment. (Photo courtesy of M. O. Hayes.)*

Swash marks are tiny ridges of sediment or nonmineral debris which mark the furthest encroachment of waves on the foreshore. As the water rushes up the beach face, it is carrying sediment, but when the swash reaches its furthest advance, much of the water is absorbed into the sediment and the rest flows down the beach face as backwash. When this happens, much of the sediment that was being carried by the water is deposited on the beach in the form of arcuate ridges which are a few millimeters to a centimeter or so in height and width (Fig. 20.15). Swash marks are concave in the seaward direction, and when preserved in the rock record are proof of the shoreline environment.

FIG. 20.15 *Swash marks on the beach face.*

It is not uncommon during times of quiet water to find triangular mounds of coarse sand or gravel along the strand line. These regularly spaced features are called **beach cusps** (Fig. 20.16), and although they have been studied for some time, their origin is unclear.

Large pebbles, cobbles, or large pieces of debris cause local changes in the currents caused by tides or wave swash. As a result, beach structures such as **current crescents** (Fig. 20.17) and **sand shadows** may be formed. The crescentic excavation around pebbles and the small sand ridges in the lee of pebbles indicate the direction of water or air movement. Actually, similar features are formed in streams and aeolian environments.

Rill marks are current-formed features that result from water flowing across the intertidal area, particularly at low tide. They are small channels which resemble a distributary-like pattern (Fig. 20.18) or branching plant stems.

FIG. 20.16 *Uniform beach cusps composed of sorted gravel.*

FIG. 20.17 *Current crescents, common beach structures that show current direction (top to bottom). Note also the footprints of a water bird.*

FIG. 20.18 *Rill marks on the beach. Current flowed from top to bottom in this photograph.*

COASTAL BAYS

A significant portion of the world's coasts are comprised of various types of embayments. They are caused largely by the configuration of the adjacent physiographic province and the Pleistocene sea level fluctuations. The Gulf and Atlantic coasts of the United States are bounded by coastal plains and characterized by nearly continuous embayments. In contrast, the Pacific coast is bordered by areas of high relief and characterized by few embayments.

Coastal bays include all enclosed or semienclosed water bodies adjacent to the open marine environment. The environment is significantly distinct from the shelf environment in terms of the water characteristics, fauna and flora, and the sediment and sedimentary processes. Two quite different special types of coastal bays are common: estuaries and lagoons. An **estuary** is a coastal water body which has open circulation with the sea and is fed by freshwater runoff, usually a single river. Well-known examples include Chesapeake Bay, Delaware Bay, and Puget Sound. A coastal **lagoon** is similar in shape but is not fed by significant runoff and is generally restricted from open circulation with the shelf, commonly by a barrier island. Laguna Madre along the southern Texas coast is a typical example. There are some coastal bays, such as Cape Cod Bay, that do not fall into either of these categories. These are also significant and distinct in terms of sedimentation.

Estuaries

The terrestrial environment has great influence on estuaries, both in terms of sediment supply and runoff. The salinity range is from essentially fresh water to that of the open sea and is perhaps the most characteristic feature of this transitional environment. Within the general definition, there are various types of estuaries, each produced in a somewhat different manner. Drowned river valleys, common along the Atlantic and Gulf coasts of North America, account for most estuaries and would represent the submergent coast of D. W. Johnson (see references at end of chapter). If this type is closed off from open circulation with the sea by a barrier bar or island, it represents a second type, exemplified by Galveston Bay in Texas and Pamlico Sound in North Carolina. In this type, there are openings in the barrier for tidal currents and interchange of waters. Faulting or other tectonic activity may also produce estuaries, such as San Francisco Bay and Drakes Estero in California. These are related to the San Andreas fault zone and are the result of crustal movement. Most of the estuaries falling in the three categories mentioned above have shallow depths, except some of those formed by tectonic activity. The last type of estuary, **fjords**, is characteristically deep. They are elongate, U-shaped valleys up to a few hundred meters deep and are formed by glacial activity.

Estuaries tend to be sediment traps and, in general, the rate of sediment accumulation is high. For the past several thousand years of increased sea level from melting glaciers, accumulation rates range up to 6 meters per 1000 years. Although it seems likely that most of the sediment in estuaries was derived from the adjacent land and carried in by rivers, there is considerable disagreement over this. Some investigators believe that much of the sediment is transported from the shelf into the estuaries. Others have suggested that sediment comes from coastal areas near the mouth of the estuary. Undoubtedly, each area may make some contribution, and the amount of river discharge into the estuary and the tidal range are also contributing factors. High runoff carries a great deal of sediment. Low discharge and low tidal range allow longshore currents to carry sediment to the mouth of an estuary, where it may be transported inland. High tidal range may produce strong tidal currents to transport sediment toward the estuary. In addition to these factors, many other combinations of conditions may cause marsh development around the margins as estuaries fill in with sediment.

Sediment and sedimentary structures. Three broad categories of sediment type are being accumulated in modern estuaries: (1) terrigenous sediment supplied directly or indirectly from the adjacent land source, (2) biogenic debris, largely shell material, and (3) authigenic mineral precipitates. Terrigenous sediment is by far the most abundant type in the majority of estuaries and may be various mixtures of sand, silt, and clay-sized particles. Coarser material is rare unless the area is bounded by cliffs or other high-relief features.

It is difficult to generalize about typical estuarine sediments because there is considerable change from one place to the next, and also within a single estuary. On the Atlantic and Gulf coasts, silt and clay predominate because the estuaries border areas of low relief and are characterized by a low level of physical energy. Pacific coast estuaries provide a striking contrast, with sand-size particles most abundant. These areas are influenced by intense wave activity and adjacent high-relief terrain. In both types of estuaries, longshore currents and onshore winds blowing over barrier islands provide a source of sand; this process is more common on the Atlantic and Gulf coasts.

Biogenic sediment is dominant locally, particularly in oyster reef areas such as those common in many Atlantic and Gulf areas. In addition, ostracods, brackish-water Foraminifera, and mollusks are also present but are minor constituents. Organic detritus, especially plant fragments, is a small but nearly ubiquitous nonmineral fraction of estuarine deposits. Authigenic deposits are usually composed of clay minerals or, in low-latitude areas of negligible runoff, calcium carbonate.

Sedimentary structures, like the sediments themselves, show wide variety. Many estuarine deposits are almost completely reworked by benthic organisms, particularly the infaunal types. In such areas, sediment cores have a mottled appearance which may be accompanied by some burrow structures. Those estuaries which have slow-moving waters and no appreciable benthic community typically show thin laminations which result from textural differences and color changes that reflect the oxidation state of organic debris.

High-energy estuaries where tidal currents are significant commonly contain small-scale cross bedding and ripple cross laminations. In-fillings of small tidal channels may also be evident in sediment cores. Under high tidal-current velocities, the bottom may develop underwater or intertidal dunes or sand waves such as are present in a rapidly flowing stream channel (Figs. 20.19 and 20.20).

Lagoons

Many environmental characteristics are common to both estuaries and lagoons; however, there is one major difference: Lagoons lack significant

FIG. 20.19 *Aerial view of sand waves in Parker River estuary in Massachusetts. The wave length is approximately 50 meters.*

FIG. 20.20 *Small intertidal sand waves in an estuary adjacent to Wasaw Island, Georgia. The wave length is approximately 4 meters. Note the ripples superimposed on the larger bed forms.*

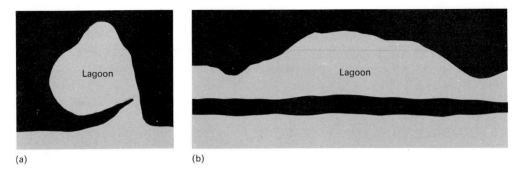

(a)   (b)

FIG. 20.21 *Diagram of lagoons formed by (a) a spit which will eventually become a baymouth bar and (b) a barrier island.*

circulation with the marine environment, whereas estuaries have such circulation and are commonly characterized by tidal current activity. Coastal lagoons have little or no source of freshwater runoff, which coupled with the lack of open circulation causes generally hypersaline conditions. These lagoons may form as the result of baymouth bar or barrier island restricting the interchange of water (Fig. 20.21).

Lagoons are low-energy environments, even more so than low-energy estuaries. Internal circulation is due solely to wind. Wind-generated currents and minor amounts of runoff provide sources of terrigenous sediment, but the overall rate of accumulation is low compared to estuaries. In areas where estuaries and lagoons are somewhat juxtaposed, as along the Texas coast, currents can supply some sediment to lagoons from the estuarine areas. Another important source of terrigenous sediment is wind-blown sand from the barrier island. Hurricanes or other intense storms transport considerable amounts of sediment over the barrier, so that one such storm might provide as much sediment to the lagoon as decades of normal conditions. Authigenic and biogenic sediment is also present in lagoons and is basically much like that of estuaries. Evaporite minerals may precipitate in lagoons of extremely high salinity. General sediment composition of lagoons is very similar to that of estuaries. Differences exist in the rate of accumulation and in the sedimentary structures associated with tidal currents.

The fauna and flora of lagoons are typical of hypersaline areas and must be able to tolerate extreme changes in conditions. Animals with skeletons that would be likely contributors to sediment include arenaceous Foraminifera and hypersaline mollusks. Ostracods and oysters, which are typical of brackish estuary conditions, are rare.

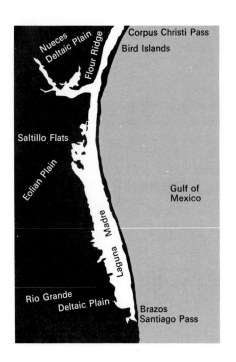

FIG. 20.22 *Map of Laguna Madre, which extends from near Corpus Christi to Brownsville, Texas. (Modified from G. A. Rusnak, 1960, "Sediments of Laguna Madre, Texas," Recent Marine Sediments, Northwest Gulf of Mexico, Amer. Assoc. Petrol. Geol., p. 155.)*

One of the classic examples of a lagoon is Laguna Madre, along the south Texas coast between Corpus Christi and Brownsville (Figs. 20.22 and 20.23). It is long, narrow, and quite shallow, less than two meters in most places. The area is semiarid, and as a result salinities are extremely high

FIG. 20.23 *View across Laguna Madre, Texas from the mainland.*

except for infrequent rain storms, which dilute the water to less than normal marine salinity. This high salinity, which may reach almost 100‰, further emphasizes the rigor which the organisms must be able to withstand.

Estuaries and lagoons are not the only kinds of coastal bays. There are some that are open to circulation with the sea but have no appreciable runoff from the adjacent land area. The water of such a bay is much like the nearshore marine area, but the bay is a relatively low-energy environment because of its protected configuration. Waves do not have much force, as they dissipate in the bay. Organisms and sediment are much like the inner shelf except for a lack of water motion to sort the sediment.

It must be emphasized that although such coastal bays may exist now, they are only temporary features if sea level remains at or near its present position. As time passes, spits will form because of longshore currents and eventually cut off the bay from open circulation, making it a lagoon.

DELTAS

The third transitional sedimentary environment to be considered is the river delta, a deposit at the mouth of a stream where it enters a standing body of water, such as the sea. A delta extends beyond the adjacent coast line into the sea, forming a protuberance. The name comes from its roughly triangular shape and was probably first applied to the Nile delta (Fig. 20.24), which exemplifies the shape quite well.

The ability of a river to build a delta at its mouth is the result of its capacity to supply sediment to the sea margin faster than waves or longshore currents can disperse it. If the amount of sediment carried by the Mississippi River is considered, it is possible to better appreciate the contribution made by rivers. In this great river, the daily sediment contribution at the delta is nearly 2×10^9 kilograms. Not all of the processes involve building of the delta, however; there is also a tearing down or destructive phase to deltaic sedimentation. Both are important to delta development.

In general, deltas are wedge-shaped masses of terrigenous sediment, largely silt and clay. All is not underwater: there is a wide area of the upper delta that is exposed to the atmosphere (Fig. 20.25) either intermittently or continuously. On this upper surface are the **distributaries** of the river, which are like tributaries in reverse, carrying water and sediment to various parts of the delta. There are also lakes, swamps and marshes, and natural levees. Sediments of a delta can be conveniently divided into three readily distinguishable types of deposits: topset beds, foreset beds, and bottomset beds (Fig. 20.25), with by far the greatest part of the delta being comprised of foreset beds. The physiography of all deltas is fairly uniform. The upper surface is nearly flat and slopes gently down below sea level to about three meters' depth, where a change in gradient occurs. At this point the slope

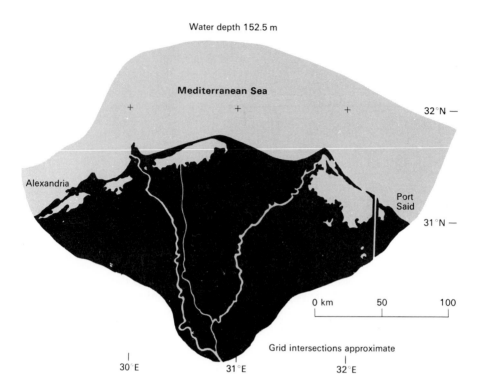

Water depth 152.5 m

Mediterranean Sea

Alexandria

Port
Said

32°N —

31°N —

0 km   50   100

Grid intersections approximate

30°E   31°E   32°E

FIG. 20.24 *Map of the Nile delta. (After M. L. Shirley (ed.), 1966,* Deltas in Their Geological Framework, *Houston Geological Society, p. 235.)*

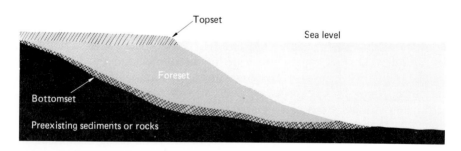

Topset

Sea level

Foreset

Bottomset

Preexisting sediments or rocks

FIG. 20.25 *Cross section of a typical delta showing relationships of the three major types of deltaic deposits. There is considerable vertical exaggeration.*

increases, but is still only a degree or two. This gradient flattens out to about
one meter per kilometer at a depth of about 75 meters.

There is a distinct textural and compositional change that closely corresponds to the physiography and the three types of deltaic deposits (Fig.
20.26). The topset beds, those nearest the land, are sandy silt with only minor
amounts of clay. Here the river loses its competence and deposits the coarser

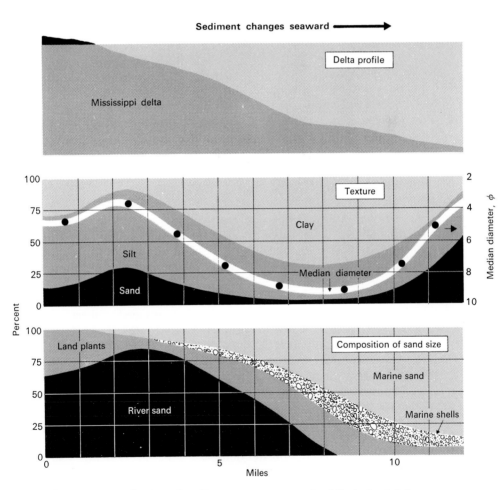

FIG. 20.26 *Distribution of sediment types across the Mississippi delta.
(After P. C. Scruton, 1960, "Delta Building and the Deltaic Sequence,"
Recent Marine Sediments, Northwest Gulf of Mexico, Amer. Assoc. Pet-
rol. Geol., p. 88.)*

part of its load. Farther out on the delta, in the foreset beds, silt and clay are deposited in subequal amounts. The beds comprising the bottomset beds, representing the farthest encroachment of the delta on the sea, are mostly silt and clay with up to 25 percent sand. This sand is not carried in directly by the river but is derived from longshore currents or other marine currents on the shelf.

Topset beds

The modern delta contains both subaerial and subaqueous topset beds. Above sea level are **natural levees, point bar deposits,** and the marshy interdistributary areas (Fig. 20.27). Sediment in the distributary channels varies with location and time, as it is a function of stream volume and velocity. The channels may be quite deep, up to 30 meters, and are continually shifting back and forth across the upper surface of the delta. As the shifting occurs, thick sand deposits form point bar deposits in the old channel. These are the coarsest sediments of the delta and comprise only about 2 percent of its total volume. Adjacent natural levees form low ridges of clayey silt along these channels. Beyond the levees are marshy areas that are much higher in clay. There is, therefore, a gradation from coarse to fine sediment from the point bar deposits to the marshes.

Below sea level there are topset sediments accumulating in two somewhat similar environments: semienclosed interdistributary bays and widespread platform deltaic deposits. Texturally the platform sediments are coarser than those in the bays. Both are high in clay, but the platform areas receive some sand as they are directly influenced by the distribution of the river.

Sedimentary structures preserved in the topset beds may help to distinguish them from other deltaic deposits. In the marshy areas, plant roots destroy stratification; however, laminations and cross laminations are preserved in channel and point bar deposits. In addition, topset beds are high in mica and wood fragments, especially in coarser sediments. There is considerable difference between the subaerial and submarine sediments from the standpoint of skeletal remains. Mollusks and Foraminifera are common in the generally marine conditions of the interdistributary bays.

Foreset beds

A break in slope occurs on the delta at a few meters depth, marking the seaward edge of the topset beds and the beginning of the more steeply sloping foreset beds. This thick accumulation of silt and clay comprises more than 75 percent of the total volume of the delta. It has the shape of a thick, clastic wedge, thickening from a feather edge to 75 meters or more at the seaward edge. These sediments are about equally divided between silt and

FIG. 20.27 *Oblique view of a portion of the Mississippi delta showing marshes, channels, and interdistributary bays. The relatively straight channels are man-made. (Photo courtesy of Russell Miget.)*

clay overall, but silt is more abundant in the upper shallow areas. They are deposited on the prograding slope as the bulk of the load in the distributaries is carried to the edge of the upper delta, where it is dumped as the streams lose their capacity.

Mica and wood fragments are more abundant here than in any other parts of the delta. Some stratification is present in the upper area, but it is commonly distorted due to slumping. Although the surface of the pro-delta slope may be inclined only a degree, it is nevertheless unstable, and much slumping as well as cutting and filling of valleys takes place. Remains of marine forams, mollusks, and other organisms are much more abundant in sediments due to the near-normal salinity compared to the submarine portion of the topset delta. The surface of the pro-delta slope has considerable relief, because gullies are cut into the sediments as new sediment cascades down the slope (Fig. 20.28).

Bottomset beds
At the base of the pro-delta slope a mixture of clays from the distributaries and marine sand combine to make up the bottomset beds of the delta. Like

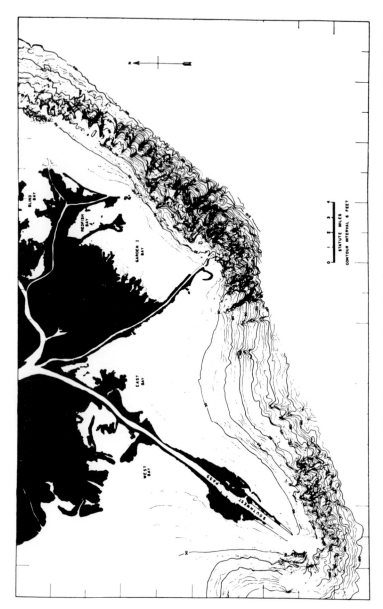

FIG. 20.28 *Bathymetric map of the Mississippi delta showing gullies on the prodelta slope. (After F. P. Shepard, 1955, "Delta-Front Valleys Bordering the Mississippi Distributaries," Bull. Geol. Soc. Amer. 66, plate 1.)*

the topset beds, they are thin and contain significant amounts of sand (Fig. 20.26). Longshore and shelf-bottom currents carry in sand which is combined with silty clays from the river to produce an odd texture. As would be expected, there is a significant increase in the amount of skeletal debris from marine organisms, with essentially a shelf fauna present. Another minor constituent is glauconite, a clay mineral that forms only in the marine environment. The abundance of bottom-dwelling organisms causes bottomset sediment to be mottled and without stratification, due to burrowing animals.

SUMMARY
Beaches, coastal embayments, and deltas have received and are receiving a considerable amount of detailed study by geologists and oceanographers. In addition to providing us with a better and more complete understanding of our modern earth, this information will help us to interpret the history and origin of similar deposits now found in the geologic record. Many such deposits are of considerable economic importance to the petroleum industry.

Of the three environments discussed, two represent areas of significant sediment influx from the adjacent land mass: coastal bays and deltas. The third, beaches, receives much of its sediment from deltas or rivers that cannot supply sufficient sediment to maintain a delta. There is a constant interplay among the three environments, particularly between the beaches and deltas. In time, estuaries receive so much sediment from the rivers entering them that they become completely filled, and sediment is then carried farther out to build a delta or to be picked up by longshore current.

SELECTED REFERENCES

Bascom, Willard, 1964, *Waves and Beaches*, Garden City, N.Y.: Doubleday, Chapters 9 and 10. Good general treatment of beach materials, configuration, and profiles based primarily on author's personal experience on the Pacific coast. Supplemented with some good photographs. A good brief introduction to beaches for the general reader.

Johnson, D. W., 1919, *Shorelines and Shoreline Development*, New York: Wiley. Old but classic book on coastal geomorphology and development. Written from the W. M. Davis concept of the geomorphic cycle. Many good illustrations of all types of coastal features.

King, C. A. M., 1972, *Beaches and Coasts* (2nd edition), London: Edward Arnold. An excellent treatment of coastal geomorphology and processes but one that suffers from lack of photographs.

Lauff, G. H. (ed.), 1967, *Estuaries*, Washington, D.C.: American Association for the Advancement of Science, Pub. No. 83. Superb and complete treatment of all aspects of estuaries with dozens of articles by current researchers. Sections III, Geomorphology, and IV, Sediments and Sedimentation, are of particular interest.

Morgan, J. P. (ed.), 1970, *Deltaic Sedimentation, Modern and Ancient*, Tulsa, Oklahoma: Society of Economic Paleontologists and Mineralogists, Spec. Pub. No. 15, Part I. Excellent papers on all aspects of deltaic processes and materials.

Shepard, F. P., 1972, *Submarine Geology* (3rd edition), New York: Harper and Row, Chapter 7. Thorough general treatment of beaches and beach processes accompanied by good illustrations. Most data come from the California coast.

Shepard, F. P., F. B. Phlegar, and T. H. Van Andel (eds.), 1960, *Recent Sediments, Northwest Gulf of Mexico*, Tulsa, Oklahoma: American Association of Petroleum Geologists. A monumental series of papers resulting from the American Petroleum Institute's Project 51, devoted to intensive study of the modern Gulf of Mexico. All articles are excellent and well illustrated. Of particular interest are those by Shepard, Rusnak, and Scruton, dealing with barrier islands, coastal bays, and the Mississippi delta, respectively.

Shirley, M. L. (ed.), 1966, *Deltas in Their Geologic Framework*, Houston, Texas: Houston Geologic Society. An excellent and up-to-date compilation of information on both modern and ancient deltas by the current experts on the subject. Statistics on the large modern deltas are included.

Zenkovich, V. P., 1967, *Processes of Coastal Development*, J. A. Steers (ed.), New York: Wiley. Very comprehensive treatment of coastal processes, with an extensive Russian bibliography.

CONTINENTAL MARGIN SEDIMENTS *21*

Much sediment that is carried to the sea by rivers eventually finds its way to the continental margin, either to pass over the area or to accumulate there. Most sand-sized sediment spends at least a moderate length of time on the beach and in nearshore environments. The continental margin provinces, coupled with the transitional environments, receive most of the terrigenous sediment that is provided from the world's land masses. Sediments of the margin provinces are quite different from those of the transitional areas and there are distinct differences among sediments in each of the margin provinces.

Within each margin province there is considerable variation in topography (Chapter 2), and also in sediment type and distribution. This is particularly illustrated on the continental shelf, whereas continental slope and rise sediments tend to be somewhat more uniform on a worldwide basis.

EFFECTS OF PLEISTOCENE EVENTS

The past million or so years have seen events and related shelf processes that have complicated the patterns of sediment distribution. Pleistocene glaciation, the underlying cause of these complications, produced the vast sea level changes that took place during that time. As the glaciers developed during the Pleistocene, they tied up tremendous quantities of water in the form of ice and snow. This lowered the sea level proportionately. The four stages of glacial advance coupled with interglacial stages of melting produced pulsations of sea level of considerable magnitude.

Just how much sea level did fluctuate as the result of the glaciers has been the subject of some disagreement among oceanographers and geologists. Most agree that the maximum range is more than 100 meters. Their calculations are based on the total volume of ice, which is computed from known geographic distribution and estimates of ice thickness, and on geologic evidence produced by now inundated shoreline deposits. Liberal sea level range estimates approach 180 meters and are based on total ice

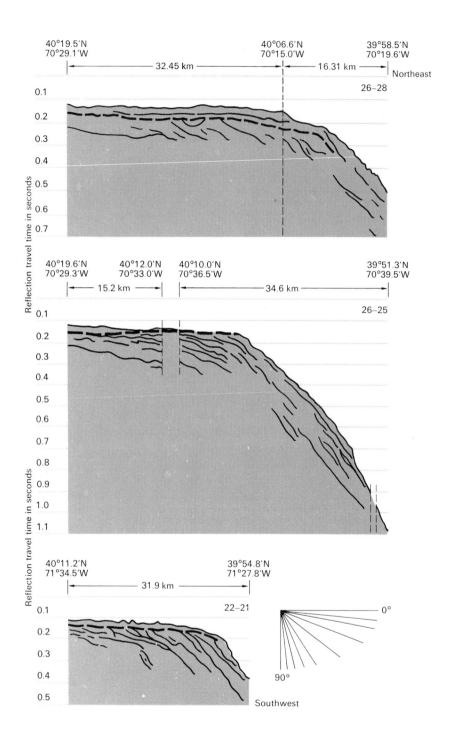

volume plus the volume of many large Pleistocene lakes that are known to have been present in many parts of the world. For instance, in the United States much of western Nevada was occupied by Lake Lahontan, and Lake Bonneville in Utah was much larger than its present remnant, the Great Salt Lake.

There are many lines of evidence on the continental shelf and the adjacent coastal areas that indicate the position of sea level at various times during the Pleistocene Epoch. One of the most prevalent and significant types of evidence is the presence of terraces which were formed during still-stands of sea level. Those above present sea level represent the time of greatest melting, and the deepest terrace marks the shoreline during the maximum glacial advance. A number of beach and shallow-water features such as reefs and oolites are found on the deep parts of the present continental shelf, also indicating former positions of sea level. Undoubtedly we are living in a rather unusual period in geologic time because of the Pleistocene glaciers and their influence on the present continental shelf and adjacent provinces.

A considerable number of continuous seismic profiles has been run across the continental margins, especially the shelf area. The profiles show a distinct **unconformity** at a depth of 50 meters or so. Below this surface are truncated dipping beds and channellike areas of significant relief (Fig. 21.1). This surface represents the erosion following the initial regression of sea level as glaciers formed. Within the overlying thickness of sediment there are distinct layers and cycles representing the regression and transgression of the sea as glaciers advanced and melted back.

Holocene transgression

Although there have been multiple transgressions and regressions of sea level over the continental shelf during the past million years, the most recent transgression is largely responsible for the present distribution of surface shelf sediments. Glaciers of the Wisconsinan Stage began melting and this caused sea level to rise beginning about 18,000 years ago. This period of transgression, termed the Holocene, has received a considerable amount of study, particularly in the Gulf and Atlantic continental shelf areas of the United States.

◀ **FIG. 21.1** *Seismic profiles of the continental shelf off the northeastern United States showing truncation of the shelf caused by Wisconsin retreat of sea level. The heavy, dashed line marks this surface. (After S. T. Knott and H. Hoskins, 1968, "Evidence of Pleistocene Events in the Structure of the Continental Shelf off the Northeastern United States," Marine Geology **6**, 21.)*

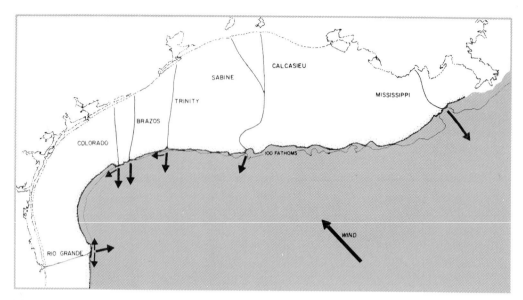

(a)

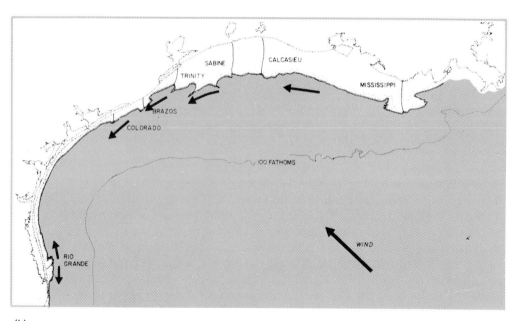

(b)

FIG. 21.2 Paleogeography of the northwest Gulf of Mexico during Holocene transgression. (a) At the low water stage, the continental shelf is almost absent and, consequently, so are prominent longshore currents. (b) As the sea level rose, a broad continental shelf was present and much sediment was transported by longshore currents, as indicated by the heavy arrows. (After J. R. Curray, 1960, "Sediments and History of Holocene Transgression, Continental Shelf, Northwest Gulf of Mexico," Recent Marine Sediments, Northwest Gulf of Mexico, Amer. Assoc. Petrol. Geol., pp. 260–261.)

At the peak of Wisconsinan glaciation, the shoreline was about 125 meters below its present level. In some parts of the world, such as the northern Gulf of Mexico, this placed the shore near or beyond the seaward edge of the continental shelf. Under these conditions, the present continental shelf was a coastal plain across which rivers transported terrigenous sediment. Unlike today, these rivers deposited their load on the relatively steep and unstable continental slope, where most of the sediment moved downslope. During this glacial maximum and sea-level minimum, longshore currents were unimportant or lacking (Fig. 21.2a) because the nearshore bottom sloped steeply. Consequently the surf zone, which is the zone of longshore transport, was narrow or nonexistent.

As time passed and the glaciers began to melt, sea level rose accordingly. This rise in sea level was accompanied by great change in the sedimentary processes caused by the prominent longshore currents (Fig. 21.2b) and deposition on the gently inclined continental shelf.

The pattern of sea level fluctuation between the beginning of the Holocene transgression and the present is a complicated one. It can, however, be divided into two distinct periods of time, each characterized by a general rate of sea-level change. From the beginning of this transgression (18,000 B.P.) until about 7000 B.P., sea level rose at an average rate of nearly 10 meters/1000 years. There were minor regressions and subsequent transgressions associated with the overall major transgression (Fig. 21.3) so that there were significant fluctuations around this average rate. About 7000 years

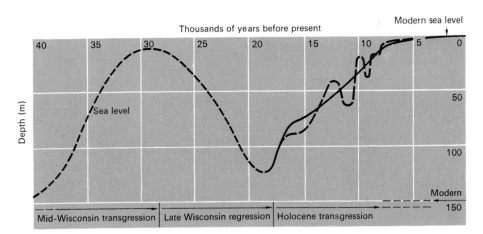

FIG. 21.3 *Sea-level positions during Late Wisconsin and Holocene transgressions. The solid line denotes the mean position, and dashed lines indicate minor fluctuations. (After J. R. Curray, 1965, "Late Quarternary History, Continental Shelves of the United States," The Quarternary of the U.S., Princeton: Princeton University Press, p. 725.)*

ago the rapid sea-level rise was abruptly slowed, with sea level at about 10 meters below its present level.

Marine geologists do not agree about the position of sea level for the past 6000 to 7000 years. There are three schools of thought: (1) sea level has been constant, (2) sea level has fluctuated a few meters about present sea-level position, and (3) although there was a considerable slowing, sea level is continuing to rise. Most recent data tends to support the third hypothesis. Actually, within this 7000-year period there seem to have been two distinct rates of sea-level increase, although both are slow. From 7000 to 3000 B.P., the rate was almost 2 meters/1000 years, but from near 3000 B.P. to the present it has slowed to about half that rate (Fig. 21.3). It should be noted that it is extremely difficult to determine the eustatic rise in sea level during this recent slow increase. Complicating factors such as postglacial rebound of the crust, coastal subsidence like that presently taking place along the Gulf of Mexico, and tectonic activity can all cause local sea-level changes and must be separated from eustatic changes.

CONTINENTAL SHELF SEDIMENTS

Surface sediments of the present continental shelf reflect the events of the Holocene transgression and in some areas reflect the more direct influences of Pleistocene glaciers. A great variety of sediment textures and composition is present on the continental shelves. In addition there is great variation from one area to the next. As a result this discussion will be centered around generalities. Details of particular continental shelf sediments around the world can be obtained from most textbooks on marine geology; a thorough description is given in *Submarine Geology,* by F. P. Shepard.

The standard introductory explanation of sediment texture distribution with respect to the shoreline describes a gradual gradation from coarse grains at the shore to fine in deeper water. Theoretically this type of distribution is also what one would expect to find as one traverses the continental shelf from the shore to its seaward edge at the continental slope. However, data from bottom sampling indicate that this theoretical distribution is in fact almost nonexistent. Observed patterns show a gradual decrease in grain size from the strand to somewhere near the midshelf area, but from there the grain size increases again. Sediments like those in the modern beach and near-shore environment are found on the outer portion of the continental shelf.

Because of the twofold nature of surface shelf sediments, it is necessary to discuss each area separately. The age, origin, and type of surface materials are distinct in the two areas.

Inner shelf

Modern sediments are currently being distributed over only a part of the continental shelf surface. They are confined, for the most part, to the shallow inner part of the shelf, although some areas of the outer shelf do receive modern sediments. There is a fairly simple explanation for this sediment distribution if one examines the major bottom currents associated with the continental shelves. Such currents are the most important means of sediment dispersal, although ice rafting, organisms, and sediment floating due to water-surface tension also make minor contributions.

Longshore and rip currents are acting almost continuously in the shallow areas of the shelf. The depth to which these are effective and the maximum distance from shore that modern sediments reach vary considerably from area to area. Throughout most of the world modern sediments accumulate to at least a few miles from shore. The type of coast and the size of the rivers carrying the sediment are significant factors. Along much of the Texas coast a series of barrier islands traps sediment in the bays behind them and prevent it from being carried out on the shelf. Along the California coast, on the other hand, most of the available sediment finds its way to the continental margin area.

Deltas exert considerable influence on modern shelf sediments and processes. For example, the Mississippi delta is the dominant feature in sedimentation of the present shelf in the northwestern Gulf of Mexico. The delta encroaches across the entire width of the shelf, thus blocking longshore sediment transport from the eastern Gulf area. As a result of this great wedge of fine sediments, longshore currents carry silt and clay westward, covering most of the continental shelf for more than 200 kilometers. Beyond this area only the **shoreface** areas receive sediments.

One of the few areas of the world where the theoretical coarse to fine sediment distribution has been found is in the northwestern Gulf of Mexico just below the shelf area off Corpus Christi, Texas. The reason for such a distribution is a unique pattern of wind and water circulation coupled with the concave shape of the coast. Winds and longshore currents are nearly parallel to the coast but in opposite directions. The currents meet south of Corpus Christi (Fig. 21.4). As a result, sediment is carried across esentially all of the shelf, the grain size diminishing due to the progressive decrease in wave and current energy. The beach opposite the current also displays a unique character. It is composed almost entirely of shells.

Sediments currently accumulating on the inner continental shelf beyond the nearshore area are commonly referred to as shelf muds. While silt and clay are quite abundant, perhaps even the predominant textures, it would be gross oversimplification to mention only these materials. Modern shelf environments receive a rather wide variety of sediment types, both from terrigenous influx and also from within the basin of accumulation.

If we consider the possible variations in sediments, the foremost factors to be considered are the composition of the source areas, the weathering in those areas, and the abundance of significant rivers to provide a means of

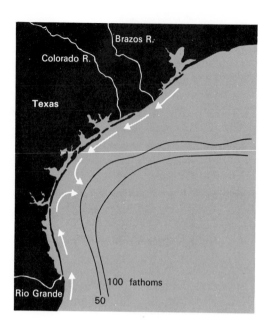

FIG. 21.4 *Currents converging off Padre Island south of Corpus Christi, Texas cause unique sediment distribution across the continental shelf.*

transportation to the sea. Even a cursory examination of the continents indicates that grossly similar source materials are available for most of the major coastal areas. Prominent local variations become masked by the large areas served by major rivers, so that there is a gross uniformity of sediment type in many areas. Climate, however, is significantly different, depending largely on latitude. Both physical and chemical weathering, which provide material for rivers to transport, are climate dependent. Climate also has a significant effect on the river's volume.

A study of inner continental shelf sediments by M. O. Hayes shows that a close correlation exists between climate and gross sediment type. Bottom sediment data from thousands of locations all over the world show that sand and mud account for nearly 85 percent of surface sediments on the inner shelf, with sand being nearly one-half of the total. Other types are rocky, gravel, coral, and shell bottoms, which occur in subequal amounts. Some of these sediment types show increases in abundance related to climatic factors. Sand is abundant at all latitudes but particularly so in temperate and arid areas except in extremely high latitudes. Mud reaches its maximum abundance adjacent to hot humid tropical areas, while rock and gravel occur most abundantly in the high latitudes. The abundance of shell is apparently not climate related (Fig. 21.5), although, as expected, coral is found only in low-latitude areas.

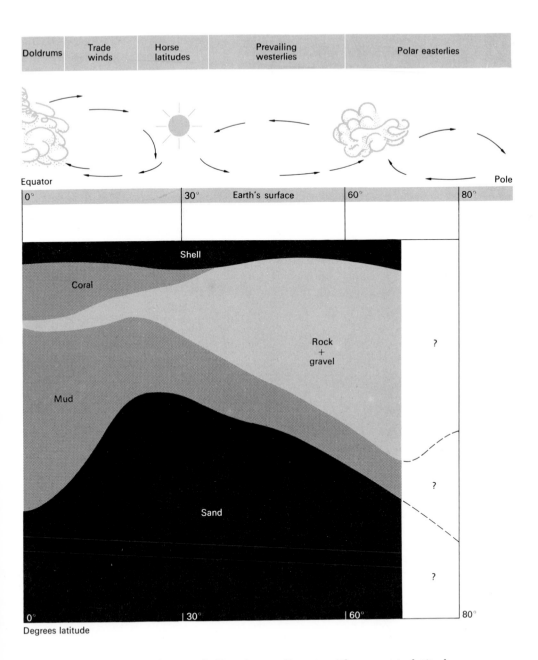

FIG. 21.5 *Composition of inner-shelf surface sediments with respect to latitude. (After M. O. Hayes, 1967, "Relationship Between Coastal Climate and Bottom Sediment Type on the Inner Continental Shelf," Marine Geology 5, 121.)*

The lack of information about areas inside the Arctic and Antarctic circles leaves a gap in the worldwide picture. However, there are enough data to speculate about the bottom type in these areas. By far the major contribution to shelf sediments is presently supplied by ice rafting of glacial sediment. These areas are covered with a poorly sorted sediment ranging in diameter from clay to boulders. It is somewhat difficult to distinguish the modern ice-rafted sediments from those deposited by Pleistocene ice sheets, which also contributed immensely to shelf sediments.

A model of sediment distribution on the continental shelf of a typical ocean was constructed by K. O. Emery (Fig. 21.6). This model is essentially restricted to present patterns of sediment accumulation. The three major sediment types display an expected general pattern which closely coincides with that of Hayes (Fig. 21.5). Glacial sediment dominates in the high latitudes and biogenic sediment, particularly coral reef contributions, dominates in the low latitudes. The middle latitudes show a predominance of terrigenous sediment derived from runoff from the land. While this is the general pattern, it should be noted that there are both latitudinal and lon-

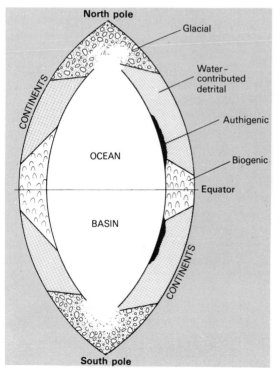

FIG. 21.6 *Idealized diagram of an ocean showing distribution of major sediment types on present continental shelves. (After K. O. Emery, 1968, "Relict Sediments on Continental Shelves of the World," Bull. Amer. Assoc. Petrol. Geol.* **52**, *446.)*

gitudinal asymmetries (Fig 21.6). Because of the circulation gyres and warm currents (Chapter 6), coral reefs and therefore biogenic sediment extends farther from the equator on the west side of the ocean basin. This circulation also serves to warm the eastern high latitudes and prevent glaciers from accumulating sediment at the same latitude as they do on the west side of the ocean. Mid-latitude upwelling (Chapter 6), such as that which occurs off the coasts of California and Peru, triggers considerable precipitation of phosphorite which appears as authigenic sediment (Fig. 21.6).

Outer shelf

The part of the present continental shelf that is mostly beyond the influence of modern sedimentation will here be termed the outer shelf. On most of the broad shelves of the word it begins a few tens of kilometers from shore. Much of the sediments presently covering the outer shelf are termed **relict sediments.** In other words, these sediments were deposited in an environment much different from the one they presently occupy. For the most part they were deposited as part of the coastal and shoreface environments that existed at these locations during periods of glacial advance when the sea level was very low; the shoreline occupied what is presently the outer shelf zone. The subsequent rise in sea level has caused these relict sediments to be left behind and they are currently not in equilibrium with their present environment. Beach sands, oolites, oyster shells, and other evidences of shallow coastal environments have been collected from this area.

A great deal of study on relict sediments, particularly by J. R. Curray of Scripps Institution of Oceanography and K. O. Emery of Woods Hole Oceanographic Institution, has shown that they are composed of much the same kinds of materials as modern sediments. This is not unexpected; however, their wide distribution, covering about 70 percent of the continental shelves, is amazing. Originally most of these sediments were terrigenous sand and mud contributed by runoff, with minor amounts of biogenic, glacial and other sediments.

Recognition of sediments as relict and out of equilibrium with their modern environment is sometimes difficult, although a wide variety of indications may be present. Most commonly the presence of coarse and sorted sediment farther from shore and in deeper water than finer, less sorted sediments is an indication. Ancient dune ridges, barrier beaches, and reefs are present on outer shelves in many areas of the world. The shells of a fair variety of animals that are normally restricted to shallow depths are sometimes found on the deep outer shelf, indicating their relict character. Oolites may also be used in much the same way, because they are formed in water only a few meters or less in depth.

Other means of obtaining information on relict sediments are possible, particularly to determine their absolute age. Calcareous shells can be dated by radiocarbon methods. Studies have also been made on pollen found in the sediments, and its chronology can yield similar results. Bones and other remains of terrestrial mammals, particularly mastodons, have been retrieved from the continental shelves.

Glaciers themselves have deposited a unique type of relict sediment off parts of North America, Europe, and Antarctica. **Glacial drift** in the form of both **till** and **outwash** is present on many shelf areas and forms a hummocky topography. In some places, currents and deep wave activity have caused some reworking of the till, in particular. As a result, some of the fines have been winnowed away, with a lag residuum in the form of a sand or gravel veneer over the till. This is common on the Grand Banks adjacent to the Maritime Provinces of Canada.

CONTINENTAL SLOPE SEDIMENTS

The continental slope is a very important zone of the ocean bottom because it is the boundary between the continental blocks (sial) and the oceanic part of the earth's crust (sima). The vast amount of data about bottom sediments that has been collected during the past few decades has been largely confined to the continental shelves and ocean floor areas. As a result, our knowledge of continental slope sediments has lagged behind and we really don't know much about them.

Some generalizations can be made from gross bottom sediment types, and comparisons with continental shelves show a predominance of mud on the slopes. Slope sediment is 60 percent mud, more than twice that of the shelf environment. The remainder is comprised of sand (25 percent), rock and gravel (10 percent), and shells (5 percent). Coarse sediments are not rare, even though the slope reaches considerable depths and is far removed from terrigenous sources.

Seismic profiles and sampling have indicated that the slope is an area of little sediment accumulation. In fact some areas are rock outcrops barren of unconsolidated sediment. This is not too surprising, considering the relatively steep gradient compared to the shelf and rise areas. The late Karl Terzaghi made many studies of slope stability and found that under certain conditions, particularly when water content of mud is high, a slope of 1° would not be stable for mud accumulation. Most of the continental slopes of the world exceed this gradient, and as a result little sediment comes to rest on the slope, or if it does, seismic shocks of even a minor nature can trigger movement of the sediment downslope in the form of a **turbidity current.**

Turbidity currents

Currents of sediment-laden water which flow downslope due to their relatively high density are called turbidity currents. These are significant transporting and depositional phenomena of the continental slope in particular, although turbidity currents are not unique to this area. The importance of these features was first proposed by F. A. Forel in 1885 and later reemphasized by R. A. Daly in 1936. Their ideas were based on a few observations in lakes and on speculation about the existence of turbidity currents. Ph. H. Kuenen and C. I. Migliorini in 1950 collaborated on a laboratory study of turbidity currents and also provided data on the sediments deposited from such a current. This study was the first real investigation of this phenomenon and paved the way for a deluge of investigations of modern and ancient sediments in light of the turbidity current as a mechanism for sedimentation.

A turbidity current is, basically, muddy water which has only a slightly greater density, 0.05 gram/cubic centimeter or so, than the surrounding water. Any mechanism such as an earth tremor or slope failure may generate a turbidity current simply by causing the sediment to go into suspension. Once this is accomplished the process of **autosuspension** takes over and gravity drives the sediment-laden water downslope. In this process, the force of gravity acting on the suspended sediment imparts energy which causes the movement downslope. Thus the sediment is really causing the water to move; a turbidity current is not moving water carrying the sediment.

Turbidity currents may be several meters thick and flow quite rapidly. Just how fast they travel has been the subject of some controversy, but it is probably at least a few kilometers/hour. It is difficult to make definite statements about their characteristics because virtually all studies have been laboratory-oriented. There have been some recent studies by diving scientists and using deep-sea submersibles to observe turbidity currents. These have been limited, and actual observations have been rare.

Though direct observation of turbidity currents in nature has not been very successful, there are detailed records of what were apparently turbidity currents on the continental slopes. In 1929, the Grand Banks earthquake shook the Newfoundland area and caused a series of breaks in the telegraph cables on the ocean bottom. The cable breaks occurred in a north-to-south sequence just beyond the earthquake epicenter. Coarse sediments have been found in the vicinity of these breaks. Many marine geologists have advocated a turbidity current, triggered by the earthquake, as the cause of these successive cable breaks. This seems logical; however, accurate records of the times these breaks took place indicate that the turbidity current was moving at a rate of 50 knots, a velocity far beyond what might be expected. By theorizing that several turbidity currents were generated by the earthquake, this high estimate of the velocity can be lowered, thus making a more plausible hypothesis. This example is a good illustration of how much conjecture must accompany a small amount of data to provide a logical solution to an oceanographic problem.

Sediments deposited by turbidity currents are called **turbidites,** and they have a set of characteristics that distinguishes them from other oceanic sediments. Because most of these deposits are found at the base of the slope on the continental rise, their discussion will be deferred to that section.

SUBMARINE VALLEYS
The ocean basins contain a wide variety of submarine valleys which occur throughout the world but are generally associated with the continental margins. Although submarine canyons are by far the most abundant and most widely studied of submarine valleys, they are only one of several types. Before going into a lengthy discussion of this type of valley, it is appropriate to briefly mention the other varieties.

Shelf channels are shallow, somewhat discontinuous valleys that occur on continental shelves. They are oriented approximately perpendicular to

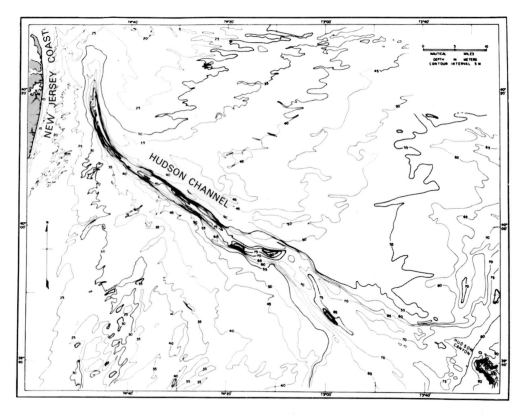

FIG. 21.7 *Hudson shelf channel on the continental shelf off New York City. (After F. P. Shepard and R. F. Dill, 1966,* Submarine Canyons and Other Sea Valleys, *Chicago: Rand-McNally, p. 8.)*

the coast and are fairly straight. Many seem to be extensions of rivers, but in few cases do they continue to connect with the head of a submarine canyon. The Hudson shelf channel (Fig. 21.7) off New York City is a well-known and typical example.

Glacial troughs are in general U-shaped, fairly deep, and wide valleys that also are common to the continental shelves. They may have a sinuous course and may have tributary troughs. Apparently these valleys were formed directly by the ice, much like fiords. One such glacial trough is present in the Gulf of St. Lawrence (Fig. 21.8).

Delta-front troughs are U-shaped and are associated with some large river deltas. They slope continuously seaward across the entire continental margin. Also associated with river deltas are **slope gullies,** the smallest variety of submarine valleys. These troughs are discontinuous and have low relief. They are common on the prograding slopes of deltas (Fig. 20.27) and are somewhat ephemeral because of the dynamic environment that they occupy.

Tectonic activity may produce grabens, or rift valleys, on the ocean bottom which are characterized by their rather straight and well defined

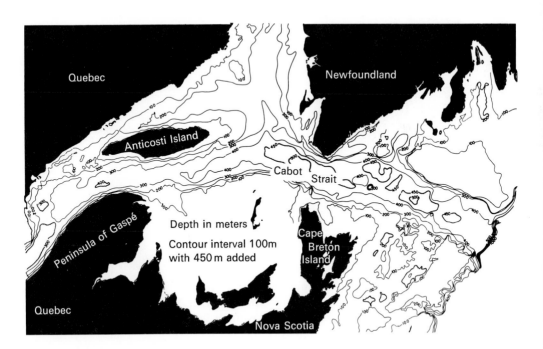

FIG. 21.8 *Laurentian trough across the Grand Banks. This is an excellent example of a glacial trough. (After Shepard and Dill, 1966, p. 9.)*

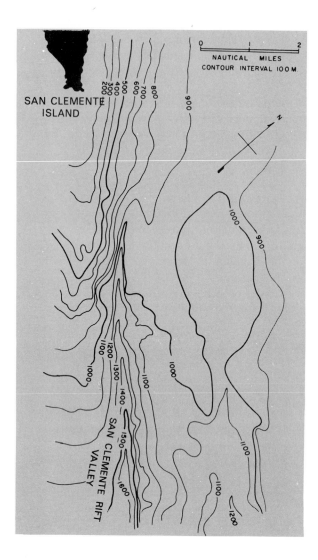

SAN CLEMENTE
ISLAND

NAUTICAL   MILES
CONTOUR INTERVAL 100 M

SAN CLEMENTE RIFT VALLEY

FIG. 21.9 *Fault valley near San Clemente Island off the southern California coast. (After Shepard and Dill, 1966, p. 12.)*

steep sides. Tributaries are lacking and they may either be V-shaped (Fig. 21.9) or have a flat bottom in the case of graben. Such features are associated with tectonically active areas such as the Aleutian Islands, California borderland, and other marginal areas of the Pacific.

Deep-sea channels are large, broad, low-relief troughs that have tributaries and are located in the abyssal areas of the ocean (Fig. 21.10).

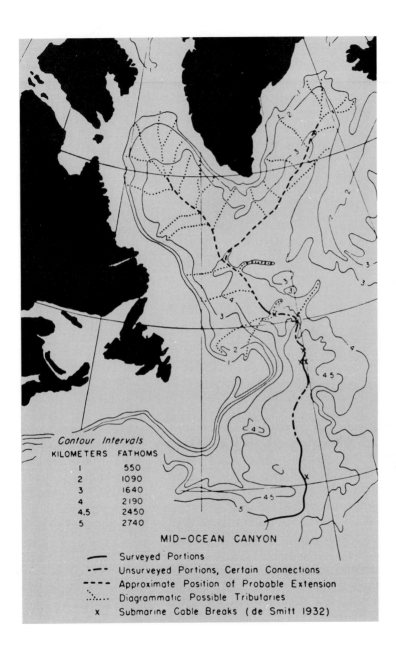

Contour Intervals

KILOMETERS	FATHOMS
1	550
2	1090
3	1640
4	2190
4.5	2450
5	2740

MID-OCEAN CANYON

—— Surveyed Portions
-·-·- Unsurveyed Portions, Certain Connections
- - - Approximate Position of Probable Extension
·.····· Diagrammatic Possible Tributaries
  x   Submarine Cable Breaks (de Smitt 1932)

FIG. 21.10 Deep-sea channel in the North Atlantic, showing locations of submarine cable breaks. (After Heezen, et al., 1959, p. 67.)

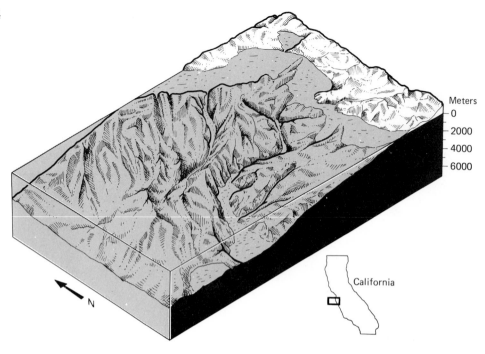

Meters
0
2000
4000
6000

California

N

FIG. 21.11 *Monterey Canyon and other nearby canyons off the central California coast. (After Shepard, 1972, p. 316.)*

The most widely distributed and thoroughly investigated type of submarine valley is the **submarine canyon.** These are steep-sided, generally V-shaped canyons that are somewhat like those formed by a mountain stream. These valleys commonly have tributaries and winding courses. Submarine canyons are basically features of the continental slope, although many extend up across part of the shelf and across the upper portion of the continental rise. The southern California borderland is particularly rich in these high-relief features (Fig. 21.11).

Origin of submarine canyons

Submarine canyons have been surveyed, sounded, and described in detail, and men have looked at them directly via SCUBA and submersibles. Tremendous collections of samples from the valley bottom and valley walls have been made with essentially two questions in mind: (1) What is the function of submarine canyons in the present marine environment? (2) How did they form? The first of these questions can be pretty well answered, but the latter is still open to speculation.

The first recognition of the features we now call submarine canyons was made less than a century ago, and for a decade or so descriptive data were gathered to better picture what they are like. At about the turn of the century, speculation began about the origin of these great underwater features. Since that time a wide variety of theories has been suggested to explain the origin of the canyons. Most of these can be and have been dismissed because of excessive inconsistencies with observable data.

Among the discredited hypotheses is the theory first advocated by D. W. Johnson, that submarine spring-sapping along the continental slope was the cause of canyon development. He based his ideas on the knowledge that some aquifers of the Atlantic coastal plain continued out under the continental shelf and were presumably truncated by the continental slope. The gently sloping aquifers would build up a hydrostatic pressure and cause springs on the slope (Fig. 21.12) which when aided by headward erosion would excavate canyons. Although there is evidence for the existence of such springs, they are not widespread throughout the world nor does it seem likely that they could erode enough material to form canyons with hundreds of meters of relief. Inconsistencies such as canyons carved in granite and lack of the coastal-plain type of aquifer negate this as a major cause for submarine canyon formation.

Tsunamis have also been suggested as a possible origin, because these phenomena impart energy to the ocean floor. This energy and the currents it generates are only of significance as the waves enter shallow water and would not explain the presence of canyons at depths of a few kilometers. Moreover, tsunamis are restricted to areas of tectonic activity, whereas submarine canyons are nearly ubiquitous. Other tectonically related phenomena, such as faulting or faulting associated with folding, have been advocated but with little substantiation. While faults are undoubtedly present on the continental margins they are not significant contributors to canyon formation. The sinuous path and tributaries of submarine canyons are inconsistent with the faulting hypothesis.

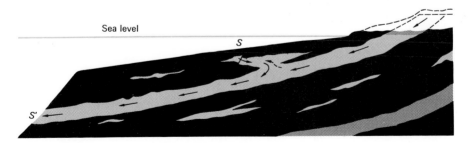

FIG. 21.12 *Cross section of an aquifer on the continental margin, showing location of spring sapping (S and S'), which was believed by D. W. Johnson to be the cause of submarine canyons. (After D. W. Johnson, 1939,* Origin of Submarine Canyons, *New York: Columbia University Press, p. 80.)*

Mass wasting similar to that which we observe on land has also been suggested as a cause for submarine canyon formation. Landslides have the ability to excavate great quantities of material, but they tend to form a valley that is more trough-shaped than the typical V-shape of submarine canyons. The sinuous course and tributaries are also incompatible with this theory.

Most marine geologists support one of two theories which have not yet been mentioned. These are erosion (1) by turbidity currents and (2) by subaerial erosion. Some scientists have recently advocated a combination of the two ideas in order to overcome some of the shortcomings of each one.

The subaerial hypothesis advocates that submarine canyons were cut during times of lowered sea level during Pleistocene glaciation. At this time sea level was considerably lower than its present level, as much as 180 meters lower. During this time nearly all the continental shelf was sub-aerially exposed, as was the upper continental slope in certain areas. Two of the most widely cited reasons for supporting this hypothesis are the similarity of submarine canyon systems to river systems, and the apparent connection between many canyons and modern river valleys.

Major objections to the subaerial erosion theory are the great depth to which submarine canyons extend and the fact that they occur throughout the world. Even the most generous estimates of sea level do not accommodate depths to 3000 meters, which is common for the seaward end of submarine canyons. Subsidence of the margin is occurring but not at a rate sufficient to account for the entire depth. Also, because submarine canyons are present all over the world, subaerial erosion could only have formed canyons during a time when arid regions were absent from the earth. It should be noted, however, that worldwide precipitation was probably higher during the Pleistocene than at present. If we assume a continuous profile between the submarine canyons and their present subaerial connections, there is a distinct nick point or abrupt change in slope in the profile. While this does not exclude subaerial erosion as a cause for original canyon development, it does indicate that significant erosion has since taken place to adjust the submarine canyon profile to present sea level. Additional evidence for recent erosion is present at places in the canyons where smoothed and polished bedrock has been observed.

The turbidity-current theory offers some explanations for inconsistencies in the subaerial-erosion theory but also presents its own problems. It is based on the idea that these special kinds of density currents have great erosive powers and occur repeatedly in the same general location. As mentioned in a previous discussion, there has been little direct observation of turbidity currents. They must not only exist but occur rather frequently in order to supply the great quantities of sediment that comprise the deep-sea fans at the base of submarine canyons. The kind of sediment found in these fans closely resembles that deposited by turbidity currents under laboratory conditions. Submarine erosion must also be taking place because of the steep-sided and smooth canyon walls.

Arguments in oppostion to the turbidity-current origin center primarily around the question of their erosive powers. To many it seems inconceivable that "muddy water" could actually erode bedrock, even granite, at a rate that would form submarine canyons in a fairly short span of geologic time. Related to this is the problem of determining just how fast turbidity currents are actually moving in the canyons. The much-discussed Grand Banks earthquake in 1929 has caused a great deal of speculation in this regard, with estimates ranging from about 15 to 50 knots. Even if we regard the lower rate as reasonable, erosion should result. In addition to the true trubidity currents, other types of downslope movement of sediment are presently occurring in submarine canyons. Submarine mudflows, slump, and sandfalls (Fig. 21.13) are known to occur and may be significant agents of erosion.

Submarine canyons, like many geological features that formed without man's observation, are probably not the result of a single process. There are ample data from present canyon processes to suggest this, as well as some internal inconsistencies in each of the two most likely theories. If instead of

FIG. 21.13 *Sandfall in a submarine canyon off southern California. (Official U. S. Navy photograph by R. F. Dill.)*

choosing one or the other we combine them, we see that their attributes are complementary and few if any serious weaknesses arise.

During the low sea level of the early Pleistocene, the continental shelf was exposed and was much like our present coastal plains. The high precipitation and runoff of the time formed many rivers across this low-relief area. The mouths of these rivers were situated near or at the shelf-slope break in the gradient. In these cases, the rivers were depositing their sediment loads on a relatively steep and unstable slope, not on a gently inclined shelf as are modern rivers. As a result, deltas were not formed; instead the sediment moved downslope in the form of a turbidity current or a similar type of mass downslope sediment transport. It should be noted that the rivers were probably abundant and carried a great deal of sediment; this is indicated by estimates of precipitation, lack of vegetative development, and weathering rates.

There is also some evidence indicating that submarine canyons have been subjected to an upbuilding as a result of sediment accumulation on the continental margin. Subbottom profiles across canyons have shown such relationships, with younger sediment layers resting over truncated layers near the tops of canyon walls.

The above combination theory provides explanations for all observable characteristics of modern submarine canyons. The canyon course, profile, and wide distribution as well as other somewhat minor factors are compatible with the idea. The nick point would be expected because the theory advocates basically two related but separate systems. Abundant sediment nearly continuously available would provide enough mass to do the excavation during a geologically short period of time. The relation of many canyons to terrestrial river valleys and shelf channels is obvious, but an explanation is necessary for the many canyons that have no apparent shelf or land counterparts. It is possible that those on the shelf have become filled or partly filled with sediment since the last sea level transgression. The terrestrial landscape changes quite rapidly, so that it is not surprising that many canyons are not located opposite a modern river.

CONTINENTAL RISE SEDIMENTS

The wedge of sediments at the base of the continental slope is a thick accumulation of mostly terrigenous sediments that have been transported down the slope by a variety of phenomena. Most of this sediment is in the form of deep sea fans at the mouths of submarine canyons. The sediment was transported via submarine canyons and, as a series of coalescing fans, forms the continental rise. Undoubtedly some of these sediments make their way out of the abyssal plains, where they are incorporated in the predominantly pelagic sediments. Pelagic sediments, particularly those of biologic origin, are there mixed with terrigenous materials.

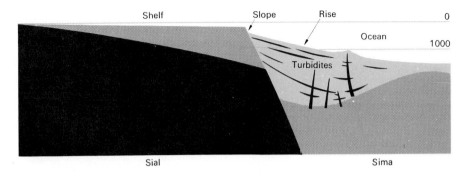

FIG. 21.14 Cross section of the continental margin showing rise sediments. (After R. S. Dietz, 1968, "Evolution of Crust from the Marine Geological Point of View," Symposium on the Primitive Earth, Miami University, Oxford, Ohio, fig. 7.)

The rise sediments are accumulating in great thickness, although their topographic expression does not indicate this thickness. Seismic studies at the base of the continental slope show subsidence in the form of a trough oriented parallel to the coast. This trough is essentially worldwide and is filled with and caused by a thick accumulation of sediment (Fig. 21.14).

The composition of these rise sediments ranges widely, as they are a conglomeration of all types and textures of shelf and nearshore sediments. Perhaps their most outstanding characteristics are the presence of unusually coarse grains and shallow-water skeletal debris. Sand, granules, and pebble-size particles are common, although most sediment is mud (silt and clay). Plant debris and shell fragments are also common.

There are various ways in which this rise sediment reaches its place of accumulation. Turbidity currents are probably the most important contributors. They are supplemented by landslides, creep, sandfalls and other types of submarine mass movement. Rafting by ice and organisms makes only a minor contribution.

Turbidites
Sediments deposited by turbidity currents are called turbidites, which have many unique properties among marine deposits. When the turbidity current moves downslope it eventually loses its capacity and competence, thereby depositing sediment. The sediment deposited by a single turbidity current forms a turbidite unit (Fig. 21.15) which may range from a few millimeters to a few meters in thickness. In general there is a direct relationship between grain size and thickness of the turbidite unit.

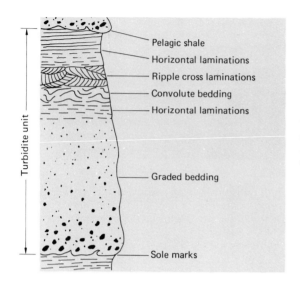

Pelagic shale
Horizontal laminations
Ripple cross laminations
Convolute bedding
Horizontal laminations

Graded bedding

Sole marks

Turbidite unit

FIG. 21.15 *Diagram of a turbidite unit showing the ideal sequence of sedimentary structures. The thickness of such a unit may range from a few centimeters to a few meters.*

One of the internal features of turbidites that is commonly used in their identification is **graded bedding.** As the turbidity current deposits its load, the coarse particles settle out first and the smaller ones last. The result is a deposit graded from coarse on the bottom to fine on the top. In many turbidites, particularly those that are totally fine grained, this grading is subtle or altogether absent. Sorting is good at any particular horizon within the turbidite, but may be poor from bottom to top. Reverse grading has been observed in some turbidites; it is attributed to one turbidity current overtaking another during deposition.

The upper part of a turbidite unit commonly contains ripple cross laminations, cross laminations, distorted or convolute laminations, and horizontal laminations. There are irregularities at the base of turbidites which are flow structures caused by the moving currents distorting the soft sediments of a previous turbidity-current deposit. Related to this process is the removal of the upper part of some turbidite units by a succeeding current or currents.

A great deal of plant debris may be incorporated in turbidite deposits, so much that some geologists advocate the continental-rise turbidites as a source of petroleum. Shell and other skeletal fragments are not abundant, but a few are found in most turbidite deposits. The presence of shallow-water Foraminifera, algae and other organisms can be used to determine the origin of the sediment.

Although the rise is not comprised entirely of turbidites, they do constitute about half of the total volume of this province. Some marine geologists feel turbidity currents account for much less of the volume and call upon normal marine bottom currents as the primary transporting agent for the coarse rise sediments. Undoubtedly all the previously mentioned transporting agents are acting on the continental slope; however, turbidity currents and associated mechanisms in submarine canyons appear to be the most important contributors to the continental rise.

SELECTED REFERENCES

Bouma, A. H., and A. Brouwer (eds.), 1964, *Turbidites, Development in Sedimentology,* New York: American Elsevier, Vol. 3. Collection of papers on both modern and ancient turbidite sediments by specialists in the field.

Burk, C. A., and C. L. Drake (eds.), 1974, *Geology of Continental Margins,* New York: Springer-Verlag. The most comprehensive volume on this subject; over 100 contributors.

Hill, M. N. (ed.), 1963, *The Sea,* New York: Wiley, Vol. 3, Chapters 20 and 27. Good general discussions of submarine canyons and turbidity currents, respectively.

Johnson, D. W., 1939, *The Origin of Submarine Canyons,* New York: Columbia University Press. Outdated but classic work on submarine canyons. Lacks good illustrations.

Pilkey, O. H. (ed.), 1968, "Marine Geology of the Atlantic Continental Margin of the Southern United States," *Southeastern Geology,* Vol. 9. Margin geology off the eastern United States, with papers by current researchers in the field.

Shepard, F. P., 1972, *Submarine Geology* (3rd edition), New York: Harper and Row, Chapters 8–11. Discussion of margin sediments and submarine valleys is extensive and greatly expanded from previous editions.

Shepard, F. P., and R. F. Dill, 1966, *Submarine Canyons and Other Sea Valleys,* Chicago: Rand-McNally. A thorough and comprehensive treatment of submarine canyons by two world experts. Largely descriptive but also discusses origin of submarine canyons.

Shepard, F. P., F. B. Phleger, and T. H. Van Andel (eds.), 1960, *Recent Sediments, Northwest Gulf of Mexico,* Tulsa, Oklahoma: American Association of Petroleum Geologists. Several papers in this volume, particularly those by Van Andel, Curry, and Shepard, are devoted to continental-shelf geology.

Stanley, D. J. (ed.) 1969, *New Concept of Continental Margin Sedimentation,* Washington, D. C.: American Geological Institute. Excellent summaries of the subject by leading researchers in the field.

22 REEFS AND ASSOCIATED SEDIMENTS

Without question, one of the most beautiful and exciting environments in the ocean is that of a coral reef. Anyone who has ever seen one will compare it with the famous natural or man-made wonders of the world. The natural history of coral reefs has been made available to all people in the past few years, at least indirectly via the many well-illustrated popular books on the subject, and the television programs and movies about reefs, most of the latter championed by Captain Jacques-Yves Cousteau.

Before beginning a complete discussion of reefs in general, it is necessary to define the term and also to emphasize that reefs may be composed of a variety of organisms other than corals. The latest U.S. Naval Oceanographic Office glossary defines a reef as "an offshore consolidated rock hazard to navigation with a least depth of 20 meters." This definition is obviously a general one and is primarily designed for navigational purposes, whereas a stricter definition requires an organic origin for reefs. Organic reefs are built by wave-resistant framework-type organisms.

NONCORAL REEF VARIETIES

A common misconception about reefs is that they are tropical features composed only of corals. Indeed, the reefs that are most commonly illustrated in books, movies, and South Pacific travel brochures are coral reefs, but there are many other kinds living now, and there have been still others in the geologic past. Any kind of organism or association of organisms that fulfllls the requirements set forth in the definition can be a reef builder.

In the early history of the earth, a variety of reef-forming organisms existed that have since become extinct. Studies of these fossilized reef structures provide sufficient data to reconstruct a reasonable picture of their environment and, in general, reef environments have remained similar throughout time. The only appreciable changes have been in the kinds of organisms which comprise them. During Paleozoic times, reefs composed of **stromatoporoids, stromatolites,** and **tabulate corals** were abundant. These groups are extinct with the exception of stromatolites. Mesozoic reefs composed of a group of pelecypods called **rudistids** were once abundant, particularly during Cretaceous times.

Modern living reefs are mostly coralline and algal; however, small reefs composed of worms, oysters, and mangroves are also known to exist or have existed in the not-too-distant past. Regardless of the composition of the reef community or the size of the reef, there are certain features common to all. The most important is the structure of fairly rigid sessile organisms, growing in massive clusters that rise above the surrounding area. The reef tends to grow or expand at the greatest rate in the direction of current and wave activity. Adjacent to the reef proper is a zone of debris, usually sloping away from the reef and derived from erosion of the reef.

Mangrove reefs

Mangrove trees (see also Chapter 18) occupy a unique niche in the marine environment and have long been known as effective sediment traps. By trapping great quantities of sediment in their root systems, which act like baffles, the mangroves can build up land. The framework in this case is the root structure of the plants (Fig. 22.1). A fossil reef of this type has been described from Key Biscayne, Florida, where the root systems have become calcified to form a resistant type of reef rock.

FIG. 22.1 *The mangrove root system serves as a sediment trap and eventually builds a wave-resistant structure.*

Worm reefs

We don't usually associate worms with reefs; however, there are certain types of polychaete annelids (Chapter 18) that are known to be reef builders. The family Serpulidae is one such group; it secretes a tube of calcium carbonate and grows in rather large masses of numerous tubes. The worm

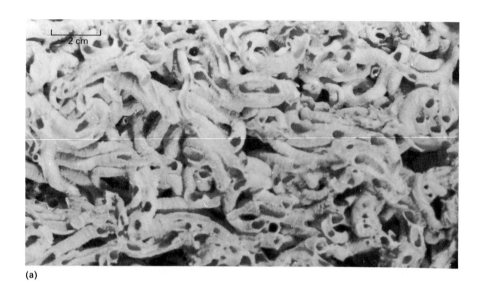

(a)

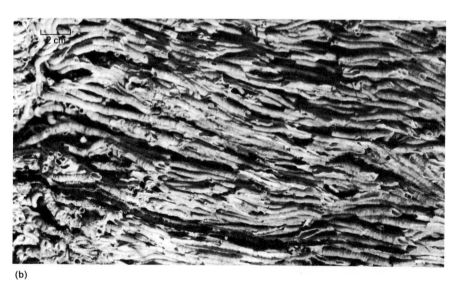

(b)

FIG. 22.2 *(a) Random and (b) oriented growth arrangements of serpulid worm tubes. (After E. W. Behrens, 1968, "Cyclic and Current Structures in a Serpulid Reef," Bull. Inst. Marine Sci.* **13**, *22.)*

itself occupies the tube, which may reach a centimeter in diameter and several centimeters in length. Like most living reef-type organisms, the framework mass is continually aggrading, so that the upper surface is covered with live organisms while the older reef below contains the framework built by previous generations.

Serpulid reefs tend to be small, usually a few tens of square meters in extent, and rising less than a meter above the surrounding substrate. Worms may grow either in an oriented fashion, with the tubes essentially parallel and vertical, or they may show a completely random growth (Fig. 22.2). The difference seems to be caused by high current and wave activity associated with the oriented serpulids.

Oyster reefs

The common edible oyster (Fig. 22.3) forms reefoid mounds, although they live in quieter waters than most reef types. Oysters commonly grow in the shallow water of brackish estuaries. They attach themselves to one another so that a framework-type structure arises, although growth is not noticeably oriented preferentially toward the direction of the currents.

Algal stromatolite reefs

There are various types of filamentous blue-green algae (Chapter 18) that have a mucilaginous coating and are adapted for life in the intertidal environment. The sticky surface of the algae causes sediment particles to adhere to the matlike algal growths, and in this manner wave-resistant structures are built. These algal colonies are most abundant in areas of calcium carbonate deposition and are restricted to the intertidal zone. As the carbonate fragments adhere to the blue-green algae, a thin layer forms, a few millimeters thick. In order to continue the building process, tiny algal filaments protrude through the carbonate layer and eventually form another layer of algae to

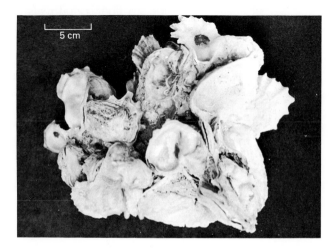

FIG. 22.3 Crassostrea virginica, *the common edible oyster, is shown here as a group of several individuals grown together such as would be found in an oyster reef.*

FIG. 22.4 *Algal stromatolites from Shark Bay, Western Australia. (Photo by B. W. Logan.)*

which particles will adhere. In this way, a mound-shaped structure, the algal stromatolite, is developed which may reach a meter or so in diameter. Such structures are commonly concentrated in small areas in modern seas (Fig. 22.4). In ancient rocks they are also common and are good environmental indicators, because of their restriction to the intertidal zone. Under conditions of low physical energy, planar stromatolite mats may form in a similar manner, but they are not reefoid.

CORAL REEFS

Corals are generally used synonymously with the term reef; however, many corals are not the reef-forming variety and the typical coral reef is likely to have nearly as much algae as it does coral. These reefs may be quite extensive, as much as thousands of square kilometers, and they may rise hundreds of meters above the sea floor. Coral reefs are restricted in their distribution because of the ecological requirements imposed by the corals and their symbiotic algae. Reef corals are called **hermatypic** organisms, signifying their association with algae, **zooxanthellae.** They are, therefore, restricted to well-lighted waters and are rarely found below 50 meters' depth. A minimum temperature for coral reefs is near 20°C and geographically they are restricted to the area bounded approximately by the thirtieth parallels (Fig. 15.1). An exception is the island of Bermuda, which is significantly beyond this latitude but is a warm-water area because of the influence of the Gulf Stream. Coral reefs require clear waters to facilitate photosynthesis. Oxygen must be plentiful in order to supply the tremendous animal population of a reef, but this is usually not a serious problem because of the abundance of photosynthetic organisms. Corals and other reef-type organisms are typically normal-salinity organisms and are stenohaline.

Reef types

The geometry and size of coral reefs and their relationship to land masses are bases for their classification. Immediately adjacent to a land mass is the **fringing reef**. These reefs colonize the shore area and may spread seaward for hundreds of meters, as in the Hawaiian Islands and other tropical land masses. They may serve as protection for the coast, which would otherwise be subjected to intense wave or tidal activity.

Barrier reefs occupy much the same position as a barrier island (Chapter 20). They are linear in shape and are separated from the land mass by a lagoon. The lagoon is sometimes too deep for hermatypic corals, although fringing reefs are sometimes found on the shoreward side of the lagoon. These barrier reefs are not a continuous solid mass but have channels allowing free or nearly free circulation between the lagoon and the open sea. Small reefs may be located on shallow areas of the lagoon. The Great Barrier Reef of Australia is the largest and perhaps the most famous of all barrier reefs. It is more than a thousand kilometers long and over 100 kilometers in width.

Atolls are circular or nearly circular in shape, surrounding a lagoon in the center. They are most abundant in the Pacific but occur elsewhere. Atolls also have channels to allow for circulation between the lagoon and open sea. The reef commonly rises hundreds to thousands of meters above the sea floor. Their origin has been the subject of some debate, although there is now general agreement.

The above three types comprise the majority of coral reefs of any consequence; however, there are varieties of small coral reefs that are commonly associated with the three main types. **Faros** are small circular reefs that resemble atolls and themselves may make up the rim of an atoll (Fig. 22.5). **Patch reefs** or **pinnacle reefs** are small masses that grow in lagoons of an atoll or behind a barrier reef. Some reefs have no appreciable rim and a generally flat upper surface. These are called **table reefs**. There are many other descriptive names that have been applied to particular kind of reefs.

One author on the Great Barrier Reef, W. G. H. Maxwell, has formulated a quite logical and workable system for reef classification. It is based on two general categories: (1) oceanic reefs, which are those that rise 100 meters or more above the sea floor, and (2) shelf reefs, which are in less deep water but that may be associated with oceanic reefs (Fig. 22.6).

Formation of atolls

Atolls are one of the most abundant of coral-reef types, and their mode of formation has been the subject of some dispute. The conditions of a circular reef with a central lagoon and the great rise from the sea floor suggest certain things about their origin. The first person to propose a comprehensive theory

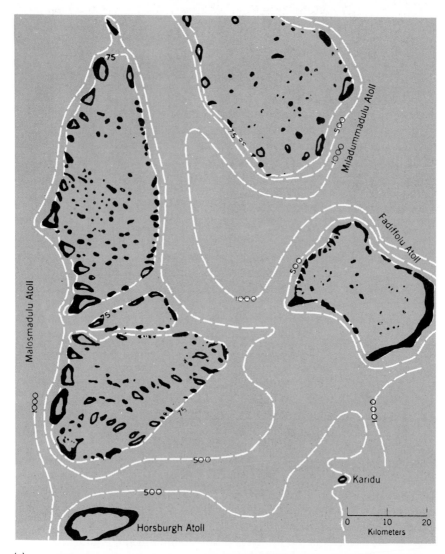

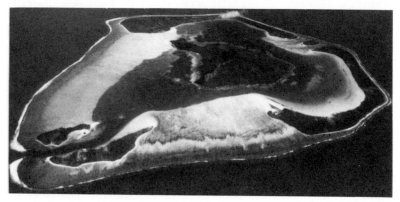

(a)

(b)

◀ **FIG. 22.5** *(a) Atolls in the Indian Ocean which have their rims composed of faros. (After Ph. H. Kuenen, 1950, Marine Geology, New York: McGraw-Hill, p. 473.) (b) Aerial photo of a Pacific atoll in the northwest part of French Polynesia.*

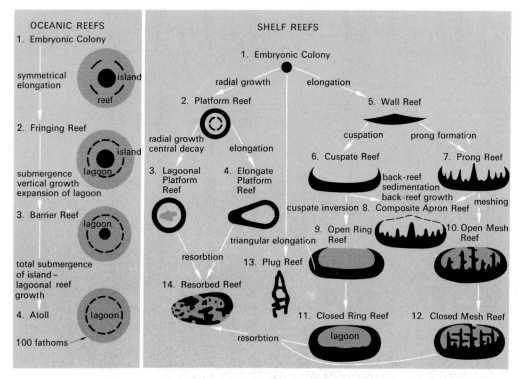

FIG. 22.6 *Classification of reefs. (After W. G. H. Maxwell, 1968, Atlas of the Great Barrier Reef, New York: American Elsevier, p. 101.)*

for atoll formation was Charles Darwin, in his book *Coral Reefs,* published in 1842. Darwin's theory was formulated as the result of his numerous observations in the Pacific Ocean during the famous voyage of *H.M.S. Beagle.* More than half a century later, alternative hypotheses were proposed that contradicted that of Darwin. Before looking at the details of these major theories on atoll formation, it is appropriate to consider the one significant difference that separates the theories into two general schools of thought. That difference is the relative change of sea level as a contributing factor. One group requires it, while the other does not.

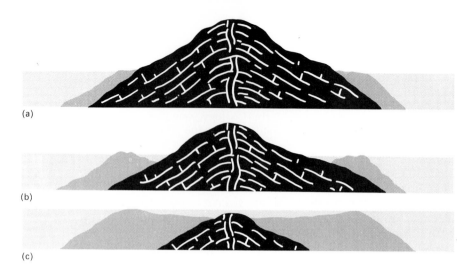

(a)

(b)

(c)

FIG. 22.7 *Evolution of an atoll as theorized by Sir Charles Darwin. The theory is based on subsidence with fringing reefs (a) evolving into barrier reefs (b) and eventually into atolls (c) as the volcanic island subsides.*

Those marine geologists that advocate that no relative movement of sea level is necessary for atoll development have advocated erosion of volcanic islands to depths of 50 meters or more. The coral reef establishes an embryonic colony on this rather deep platform and begins growing upward toward a more suitable environment. Even if one accepts this depth for coral colonization, there is little evidence to substantiate the theory. Many atoll lagoons are deeper than the depths at which corals can survive; therefore, they could not have been established on a deep platform. It is doubtful that erosion of volcanic material could take place to these depths, and then, without changing sea level, have the much less rigorous conditions required for the accumulation of carbonate debris and the establishment of coral larvae.

Darwin's theory is based on relative sea level change caused by subsidence. In this theory the initial state was the development of fringing reefs around a volcanic island (Fig. 22.7a). There was continual slow subsidence of the volcano, during which the reef grew upward and also outward because of generally favorable conditions on the outer edge. This led to a transition of the fringing reef to a barrier-type reef, with a lagoon between the reef an the volcano (Fig. 22.7b). The atoll-type reef represents continued subsidence of the island reef growth, with the island eventually disappearing completely below sea level (Fig. 22.7c). Such a theory is quite logical and fits observable data. At the time of Darwin's proposal, subsidence of the magnitude necessary for his theory could not be demonstrated; however, these data are now available in the form of geophysical studies and drilling on atolls. These show reef material to depths of more than half a kilometer. Some atoll lagoons are nearly 200 meters below sea level. These depths must be the result of considerable subsidence.

Darwin's ideas received considerable support as additional evidence of subsidence in the form of embayments in islands behind barrier reefs was noticed in many areas. R. A. Daly, however, did not agree with Darwin, although he believed sea-level changes to be important in atoll development. According to Daly's glacial-control theory, the lowering of sea level and cooling of waters during the Pleistocene caused the death of most coral reefs. In addition to temperature changes, turbidity was high because of the exposure of shelf and upper-slope muds which were subjected to wave and current attack. During the low-water stages, the crests of volcanic islands were truncated by wave erosion, producing a generally flat surface which would later become the floor of atoll lagoons.

After the return of milder, post-Pleistocene climates and more favorable conditions for coral colonization, the corals began to spread back to the truncated volcanic islands from protected areas where they existed during the Pleistocene. The outer margins of the volcanic platforms provided the necessary nutrients and clear water for corals and it was here that they began to grow, according to Daly's theory. As the sea level kept rising due to glacial melting, the coral rim grew upward, thus forming the atoll with a flat lagoon floor at a depth nearly equal to the rise in sea level. Daly's theory does not account for the large atoll with diameters of 20 to 30 kilometers which would necessitate a wave-cut platform of similar width formed during the relatively short duration of the Pleistocene. This does not seem possible. He also overestimated the flatness of lagoon floors. Altough unknown at the time, the great thickness of coral materials shown by drilling and geophysics also refutes Daly's theory of a single cause for atoll formation.

Darwin's theory of subsidence as a cause for atoll formation has been generally accepted; however, it also has some shortcomings. At the time Darwin made his observations and published his book, there was no knowledge of Pleistocene glaciation and its effects on the position of sea level. Deep drilling into atolls has provided coral samples that are Tertiary in origin.

This means that many coral reefs were thriving long before the appearance of glaciers during the Pleistocene and that sea level changes, if nothing else, must have had some effect on these coral reefs. It would seem that modern atolls are at their present position with respect to sea level largely because of glacial control, although their original formation was primarily due to subsidence.

Coral reef zonation

Most coral reefs, regardless of their size or kind, have similar environments and associated organisms. The atoll is in a sense the most mature type of reef and has most of the zones we find on other reef types. The reef itself can conveniently be subdivided into the following zones: (1) seaward slope, (2) windward reef, (3) island, (4) lagoon, (5) leeward reef, and the seaward slope back to the ocean floor (Fig. 22.8).

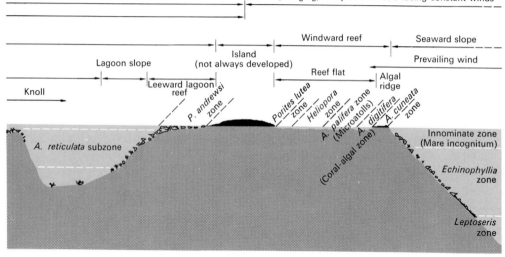

FIG. 22.8 *Atoll cross section showing typical zonation.*

Seaward slope.   There is considerable difference in the details of the seaward slope of reefs, although most are composed of reef debris at the base. This **talus** slope has a gradient that varies with the depth to sea floor and the amount of material available, but it is commonly about 30°. It is composed of skeletal material derived from erosion of the reef proper and transported down the fore-reef slope.

The extreme depth of the living reef proper is 150 meters. From this depth to 45 meters below sea level, the only hermatypic corals likely to be present belong to the genus *Leptoseris*. These are rather delicate varieties that are adapted to this nearly lightless environment. The total amount of life at this depth is only a small fraction of that in shallow waters and it is mixed with reef debris from above.

Above this zone to the approximate depth of wave base (about 15 meters) is a second slope zone characterized by a wide variety of foliate or branching corals such as *Porites, Heliopora, Millepora, Echinophyllia,* and others. The fauna is comparable to that in other environments of the lagoon and back reef at similar depths. This zone is well within the area of light penetration and there is good circulation to provide oxygen and nutrient elements for organic activity.

The uppermost zone of the seaward slope is characterized by fairly massive and resistant corals because of the intense physical energy imparted to this part of the reef by wave action. The greatest abundance of massive corals may extend to a meter or so below mean low tide and consists of various species of *Acropora*, a rigid branching coral, and various massive types such as *Monastrea, Diploria,* and others. *Acropora* is particularly adapted to areas of intense energy, and it grows in large, dish-shaped forms spreading toward the direction of wave activity (Fig. 22.9).

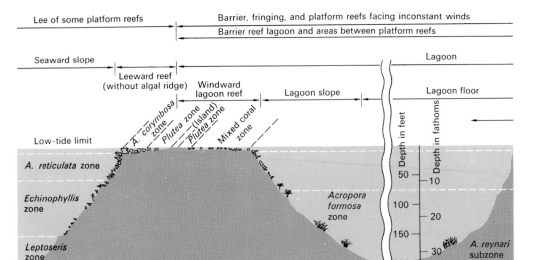

| Lee of some platform reefs | Barrier, fringing, and platform reefs facing inconstant winds |
| | Barrier reef lagoon and areas between platform reefs |

Seaward slope | Lagoon

Leeward reef
(without algal ridge) | Windward lagoon reef | Lagoon slope | Lagoon floor

A. corymbosa zone / Plutea zone / (Island) Plutea zone / Mixed coral zone

Low-tide limit

A. reticulata zone

Echinophyllis zone

Acropora formosa zone

Leptoseris zone

A. reynari subzone

Depth in feet — 50, 100, 150, 200
Depth in fathoms — 10, 20, 30

(After J. W. Wells, 1957, pp. 616–617.)

FIG. 22.9 *Oriented* Acropora palmata *in a high-energy reef area. (Photo by E. A. Shinn.)*

The upper portion of this zone is characterized by an unusual topography called a **spur and groove** system (Fig. 22.10). It is, as the name implies, a series of grooves which are oriented normal to the reef trend with flat-topped spurs or buttresses composed of the above-mentioned corals and the coralline algae *Lithothamnion*. The grooves are up to 100 meters long, a few

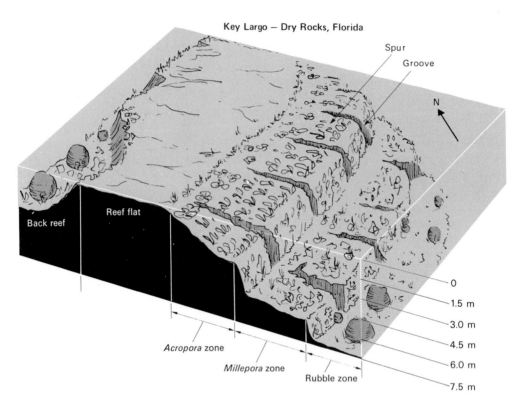

Key Largo — Dry Rocks, Florida

Spur

Groove

N

Back reef

Reef flat

0

1.5 m

3.0 m

4.5 m

6.0 m

7.5 m

Acropora zone

Millepora zone

Rubble zone

FIG. 22.10 *Typical windward spur and groove development on the upper forereef slope. (After E. A. Shinn, 1963, "Spur and Groove Formations on the Florida Reef Tract," J. Sediment. Petrol.* **33**, *295.)*

meters wide, and 2 to 7 meters deep. They serve as collecting and transporting channels for reef debris, whereas the spurs are areas of rapid reef growth. This unique growth structure is a result of the organism's growth in the direction of nutrient-laden oceanic water.

Windward reef. The rather flat and broad upper surface of the reef on the windward side is subdivided into the algal ridge and the reef flat. The algal ridge is usually narrow but may reach 100 meters in width if the entire upper algal zone is included. The ridge has a relief of about a meter and is composed largely of encrusting coralline algae (red algae) such as *Lithothamnion, Prolithon, Goniolithon*, and others. In many reefs this zone is called the *Lithothamnion* ridge. These calcareous algal crusts form a hard and resistant pavement that receives much of the wave energy imparted to the reef and lessens the impact on the adjacent reef flat behind. The coralline algae form the resistant framework of this zone but there are many other types of organisms living there. Out of necessity, all are fairly hardy types that can withstand the strong wave activity and currents. Such other algae as the green *Halimeda,* mollusks, echinoderms, Foraminifera, and some corals are also present.

The algal ridge is not a solid mass across the reef crest, as it has many channels cutting across that allow waves and tidal currents to pass through. These surge channels, as they are called, are sometimes roofed by algal growth so that caverns are formed. Blow holes, through which geyserlike spouts of water emanate, are common on the ridge and mark the location of a roofed channel.

The reef flat comprises the largest area of the reef, and as the name implies there is little vertical relief. The zone as a whole may be several hundred meters wide and can be subdivided on the basis of reef communities. The entire area is near mean low tide, but there are deeper pools and channels locally.

Immediately behind the algal ridge is a wide zone of coral and algal growth with typical framework-type corals such as *Acropora, Pacillopora, Goniastrea,* and *Seriatopora* covering about half the total area. *Lithothamnion* supplements the calcareous framework and the green calcareous alga *Halimeda* is common. The next subzone is characterized by slightly deeper water and contains microatolls. These are nearly circular, flat structures built by corals that grow radially. *Porites, Heliopora, Favites,* and *Cyphastrea* are common microatoll builders. The structures are also encrusted with coralline algae and contain a variety of life in the relatively calm central areas of the microatoll.

Throughout the reef flat, there is a fair amount of sand- and gravel-sized reef debris accumulating between living coral colonies. This material is comprised largely of coral, algal, molluscan, and foram skeletal material. This zone may also contain abundant dead but *in situ* coral heads.

Reef islands.   Most large reefs, and atolls in particular, contain a few small islands or **cays**. The size ranges widely but most are less than a few square kilometers. Cays are constructed primarily by wave action and are composed of calcareous reef debris. Their formation seems to be a response to wave refraction patterns converging at the same time and piling up this debris. The cays are usually located near the middle of the windward side of an atoll, but may also be formed on the leeward side (Fig. 22.11).

Cays are likely to be ephemeral unless they are stabilized by vegetation. Some small sand cays appear and disappear regularly due to seasonal changes in prevailing winds. Hurricanes or other tropical storms can com-

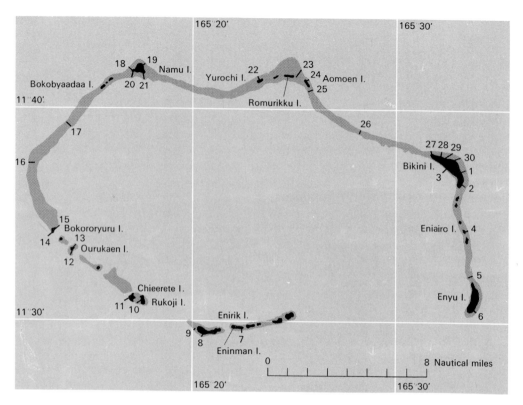

FIG. 22.11 Map of Bikini Atoll showing distribution of islands (cays). (After K. O. Emery, J. Tracey, and H. S. Ladd, 1954, "Geology of Bikini and Nearby Atolls," U. S. Geological Survey Prof. Paper 260–A, p. 16.)

FIG. 22.12 Beachrock formed in the intertidal zone of Heron Island, Australia. (Photo by M. A. H. Marsden.)

pletely destroy a cay. Scrub vegetation is established initially, and trees such as mangroves, palms, and other tropical varieties may come later. Any type of vegetation helps the cays in their fight for survival. Mangroves are particularly important because of their shelter for sediment accumulation. Another feature that aids in the resistance of cays to erosion is **beachrock** (Fig. 22.12). This rock is commonly associated with vegetated and permanent cays, where it forms in the intertidal zone. It is the result of rapid cementation of carbonate debris by calcium carbonate cement. Precipitation of the cement is facilitated by the spray of waves and by evaporation, which provides and concentrates the necessary ions. This process takes place in a short time geologically, as evidenced by the incorporation of bottles and other artifacts of modern man.

The beaches on cays are usually much like those of any other area, except that they are totally carbonate and composed of reef debris. Many beach areas are composed largely of broken sticks of coral such as *Acropora*. The sorting is typically good and there may be orientation of the coral sticks.

Lagoon. A lagoon, which has quieter water than any other part of a reef, is characterized by fantastic species diversity. Regardless of the reef type, all lagoons possess a general similarity. Within the lagoon there are three principal environments: (1) the lagoon slope, (2) the lagoon floor, and (3) the lagoon reefs, which include patch reefs, pinnacle reefs, and knolls. In atolls, the floor of the lagoon is usually 25 to 40 meters below sea level, but in other reefs, especially platform reefs, the lagoons are much shallower.

Most of the lagoon slope is covered with debris that is carried over the reef by waves and tidal currents. The substrate is not stable and consequently the sessile benthos such as corals and algae are not able to become well established in this zone. *Halimeda* and a species of *Acropora* are the only abundant living organisms in this environment.

The floor of reef lagoons is much more stable than the steep slopes and is quite flat. Bottom-dwelling organisms living in this environment are largely controlled by the depth of water, except in shallow lagoons. *Halimeda* and other calcareous green algae are abundant, along with various vagrant benthic crustaceans, echinoids, and mollusks. Waters overlying the lagoon proper tend to be quite turbid, because the fine sediments comprising the lagoon floor are easily stirred into suspension. This further restricts the fauna and flora of the environment.

Sediments comprising the lagoon floor are composed of coral, *Halimeda*, Foraminifera, and shell debris along with a high percentage (up to 40 percent) of fine silt and clay-sized particles. Much of this fine sediment is in the form of loosely consolidated **fecal pellets**. As might be expected, the lagoon-floor sediments are poorly sorted and relatively angular, due to the low level of wave activity in this environment. The fine fraction of the

5 cm

FIG. 22.13 *Parrot fish. (Photo courtesy of J. Engemann.)*

sediment may represent either direct precipitation of calcium carbonate or finely abraded skeletal debris. In reef lagoons, the latter is probably the more important contributor. The reduction of skeletal debris to silt and clay-sized particles is accomplished primarily by browsing organisms, particularly parrot fish (Fig. 22.13) and echinoids. Both of these ingest great quantities of coral and other skeletal carbonate and digest the symbiotic zooxanthellae and other organic material. The carbonate material is excreted in the form of fecal pellets after considerable size reduction in the digestive tracts of the animal.

Within the lagoon are many small reefs that rise to nearly the level of the main reef, as well as the lagoon margin reefs associated with the major reef. One of the primary differences between the lagoon reefs and the outer reef areas is the lack of coralline algae in the former; these algae require the impact of breaking waves. The lagoon provides an almost optimum environment for coral growth. Other significant contrasts between these reef environments include a lack of spur and groove development in lagoon reefs, somewhat deeper reef surface, and a discontinuous reef structure. Fish are more abundant, as are bottom-dwelling mollusks, echinoderms, and other vagrant forms.

Reef knolls, or pinnacle reefs, contain abundant corals on their upper surface, but their steep slopes are largely unstable and covered with debris, much like those of the major reef. The type of coral that dominates the reef surface depends on the water depth. Because of the calm conditions and abundant nutrients, the corals on these knolls may be tremendously large, even in the rather fragile, branching varieties. *Porites* and *Acropora* are the dominant genera. It is on these reefs and the lagoon side of the main reef that parrot fish nibble away on the corals and actually provide the most destructive process in any of the lagoon environments.

Leeward reef. Gross characteristics of the leeward reef on an atoll are much like their seaward counterparts. Generally they are less well developed, grow at a slower rate, but contain much the same fauna and flora. The differences are primarily the result of differing wave activity and differing abundance of oxygen and nutrients in the two environments.

The algal ridge is absent or poorly developed on the leeward reef because of the lack of vigorous wave activity. Cays are less common and smaller for much the same reason. Varieties of corals present are similar to those on the windward reef, but depth zonation for individual taxa may be slightly different due to the differences in water motion at those depths. Coral growth in general is somewhat more dense and luxuriant on the leeward reef due to the protection from heavy wave activity.

SELECTED REFERENCES

Bathurst, R. G. C., 1971, *Carbonate Sediments and Their Diagenesis*, New York: American Elsevier. Excellent coverage of all carbonate environments including reefs.

Cameron, A. M., *et al.* (eds.), 1974, *Proceedings of the Second International Symposium on Coral Reefs*, 2 vols., Brisbane: Great Barrier Reef Committee. Most comprehensive and up-to-date treatment of coral reefs by the world's foremost researchers in the field.

Darwin, Charles, 1842, *The Structure and Distribution of Coral Reefs*. The original classic on coral reefs containing a fantastic amount of data from Darwin's direct observations and his original theory on atoll development. Reprinted as a paperback by the University of California Press, Berkeley, in 1962.

Emery, K. O., J. I. Tracey, and H. S. Ladd, 1954, "Geology of Bikini and Nearby Atolls," U.S. Geology Survey Prof. Paper 260-A. The first and perhaps the most thorough study of Pacific atolls with special emphasis on geology. Includes many excellent photographs.

Hoskin, C. M., 1963, *Recent Carbonate Sediments on Alacran Reef*, Washington, D.C.: National Academy of Science—National Research Council. Comprehensive study of platform reef in the Gulf of Mexico, with special emphasis on the origin and distribution of sediments in the various reef environments.

Jones, O. A., and R. Endean (eds.), 1973, *Biology and Geology of Coral Reefs*, New York: Academic Press. Excellent integrated volume on coral reefs with contributions by many experts.

Kuenen, Ph. H., 1950, *Marine Geology*, New York: Wiley, Chapter 6. A thorough but somewhat out-of-date treatment of coral reefs. The theories of atoll development are well presented.

Maxwell, W. G. H., 1968, *Atlas of the Great Barrier Reef*, New York: American Elsevier. The most recent and best-illustrated book on the biggest of all reefs. Discusses growth and development and includes descriptive data.

Shepard, F. P., 1972, *Submarine Geology* (3rd edition), New York: Harper and Row, Chapter 12. Good but brief, general treatment on coral reefs and their development.

Wells, J. W., 1957, "Coral Reefs," *Treatise on Marine Ecology and Paleoecology,* J. W. Hedgpeth (ed.), Geological Society of America Memoir 67, Vol. 1, pp. 609–632. Excellent discussion of coral reef zonation and the distribution of species throughout the reef.

DEEP-SEA SEDIMENTS 23

Our knowledge of sediments which have accumulated on the sea floor is poor compared to our knowledge of the continental margins. The first samples from the ocean bottom were collected about the middle of the nineteenth century, and little progress in technology to facilitate the study of these deposits took place until about the 1940s. Since that time, great strides have been made in obtaining funds and in developing new sampling devices to better determine the character of ocean-floor sediments.

The deep sea is extremely homogeneous compared to the other marine environments, and this is reflected in its sediments. Areas of up to several thousand square kilometers may have a uniform composition. The stratigraphy is also typical of a uniform environment and is for the most part composed of thin, flat-lying, and laterally continuous layers. Most sediments reach the ocean floor in one of two general ways: They settle out of the seawater without appreciable influence from the continental margins or currents or are derived directly from land and deposited near the margins or by currents. Sediments from the former source are termed pelagic, and from the latter terrigenous. Most classifications of deep-sea sediments are built upon these two major categories.

PELAGIC SEDIMENTS
The clear and apparently sediment-free waters of the open ocean actually contain a large amount of extremely fine and dispersed sediment which may be of organic or inorganic origin. Through very slow settling, these particles eventually accumulate on the sea floor and comprise the bulk of pelagic sediments. The rate at which they accumulate is not uniform with respect to time or space, although we can be sure it is slow. A reasonable range would be from a few millimeters to a centimeter per thousand years. This is much different from any of the previously described environments, and the slow accumulation allows slow chemical reactions to take place between the sediment particles and seawater. A single grain may be exposed to the water

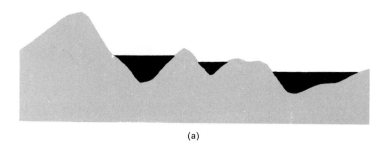

(a)

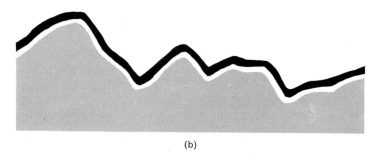

(b)

FIG. 23.1 Terrigenous and pelagic sediments in relation to ocean-bottom topography. Terrigenous sediments (a) tend to fill in low areas and lack a common depth, whereas pelagic sediments (b) are, in general, uniformly distributed over the ocean floor. Vertical exaggeration is quite large.

for many years, thereby facilitating solution, oxidation, or other chemical reactions.

Pelagic sediments accumulate somewhat differently than do the terrigenous types. Because of the way they settle down through the water, the pelagic deposits tend to follow the topography and accumulate in thin, uniform layers over the sea floor (Fig. 23.1). Terrigenous sediments have a greater tendency to be concentrated in low areas and to be thin or absent over topographic highs.

Those sediments that accumulate on the ocean floor as particles settling out of the overlying water fall into several categories. In all but one of these, the sediment settles as solid particles, whereas in the case of **authigenic** or **halmeic** sediments they crystallize directly from the seawater. The vast majority of pelagic sediments falls into the **biogenic** category, being formed of remains of organisms. Pyroclastic material from volcanos, windblown deposits, and extraterrestrial cosmic particles complete the types of pelagic materials.

Biogenic sediments

Except in the North Pacific basin, biogenic debris is the most prevalent type of pelagic sediment throughout the world (Fig. 23.2). According to a classification by F. P. Shepard, this or any other pelagic sediment is defined as having at least 30 percent of the material from which it takes its name.

Biogenic sediments are collectively referred to as pelagic **oozes** and are composed of microscopic tests and skeletal debris of planktonic organisms. These organisms include animals such as the Foraminifera, Radiolaria, and pteropods as well as the plant groups, diatoms, coccolithophores, and

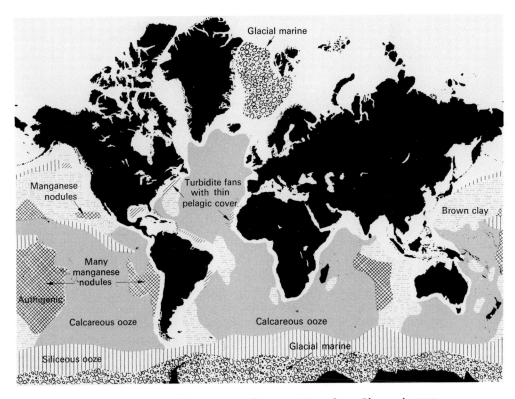

FIG. 23.2 *World distribution of deep-sea sediments. (Data from Shepard, 1972, p. 414.)*

silicoflagellates. Fish and sponges contribute small amounts to these pelagic oozes (Fig. 23.3).

The planktonic organisms that contribute to pelagic sediments are most abundant in the near-surface waters of the sea (Chapter 16). As the organisms expire they settle to the sea floor or, in the case of some, such as calcareous organisms, their tests may dissolve before they reach the sea floor.

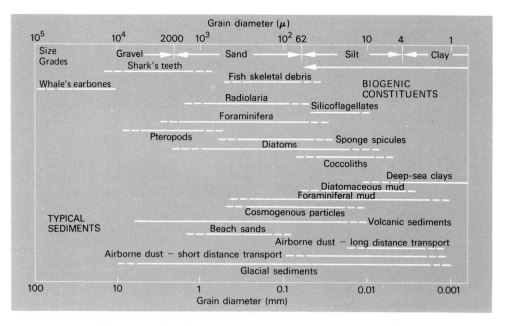

FIG. 23.3 *Grain-size distribution of common deep-marine sediments and sediment sources. (After M. G. Gross, 1972, Oceanography: A View of the Earth, Englewood Cliffs: Prentice-Hall, p. 112.)*

Calcareous oozes. Bulbous-shaped forams, pteropods, and coccoliths make up the calcareous oozes of the deep-sea environment. Calcareous oozes are predominant in the South Pacific, Indian, and Atlantic Oceans except for the high-latitude areas, particularly in the Southern Hemisphere (Fig. 23.2).

Globigerina, a bulbous planktonic foram, and related genera form the bulk of calcareous oozes. Pteropods (Fig. 23.4) and coccoliths are not abundant.

A considerable amount of work has been done on the distribution of calcareous plankton with respect to temperature. Both coccoliths and Foraminifera are apparently responsive to temperature, and certain as-

FIG. 23.4 *Pteropod debris from a modern deep-sea sediment. Each test is 1–2 millimeters long. (Photo courtesy of C. Chen.)*

semblages are found to be associated with certain water masses. The foraminifer *Globorotalia truncatulinoides* coils to the left in warm water and to the right in cold water. By studying the stratigraphy of sediment cores and relating this coiling trait to climatic changes it is possible to reconstruct environments during the past several thousand years (Fig. 23.5).

Although calcareous oozes are the most abundant of all pelagic sediments, they are much less abundant than their living counterparts floating in the near-surface waters. Calcium carbonate solubility increases greatly with an increase in pressure and with a decrease in temperature. As a result, many tests of these organisms never reach the ocean floor but go into solution while descending into deep, cold water. This is particularly so for the very tiny and delicate tests, and it explains their scarcity in sediments of high-latitude areas and extremely deep environments.

Siliceous oozes. Both animal and plant groups are included among the siliceous oozes. Diatoms (Fig. 16.1) are most abundant in a wide band between about 50° and 65° south latitude, as well as locally where upwelling is common. They, along with all other siliceous tests, are relatively unaffected by solution in the marine environment. This does not imply that there is no solution of these siliceous tests. The most small and delicate silicoflagellates and diatoms may dissolve and this silica aids in the precipitation of authigenic clay minerals and other silicates.

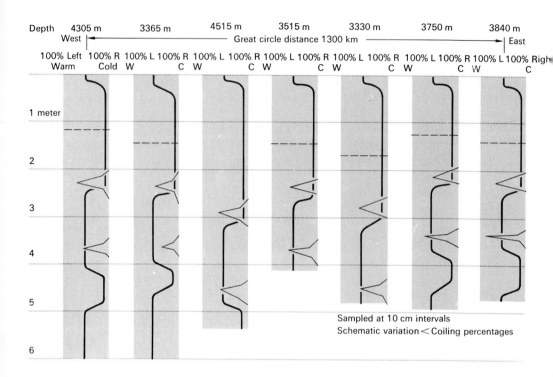

FIG. 23.5 *Changes in coiling of* Globorotalia truncatulinoides *as shown in cores from the equatorial area of the Atlantic Ocean. (After D. B. Ericson, 1963, "Cross-Correlation of Deep-Sea Sediment Cores and Determination of Relative Rates of Sedimentation by Micropaleontological Techniques," The Sea, M. N. Hill (ed.), New York: Wiley, vol. 3, p. 835.)*

Radiolaria occur throughout the oceans but are concentrated in low-latitude areas. They are present on a nearly worldwide scale, but little of the ocean floor is characterized by radiolarian ooze due largely to masking by other material.

Other biogenic debris. Other animals that live on the ocean floor or occupy the overlying waters contribute small amounts to pelagic sediments. Foremost among these are sponges, which contribute siliceous spicules, and fish, which contribute bone debris that is phosphatic in composition.

Cosmic sediments

Sediments of the sea floor contain a small number of metallic and silicate **spherules** which range from about a micron to almost 0.5 millimeter in diameter. Some occur as elemental nickel-iron, probably resulting from the breakup of metallic meteorites as they pass through the atmosphere. Cores taken from the Pacific show great increases in spherule abundance in the late Tertiary. Other metallic and magnetic spherules have been found but their origin is unknown. At least some of them appear to have a volcanic origin.

Olivine and pyrozene **chondrules** are found in association with metallic spherules in some areas. These apparently represent fragments of chondrite meteorites that broke up upon entry into the atmosphere. The silicate bodies are as plentiful as the metallic ones but are present in a wider size range.

Estimates of the total contribution of cosmic particles to the ocean floor are in the neighborhood of a few thousand tons per year; these estimates are based on their vertical distribution in sediments. This is far short of the rate of influx measured by satellites. Evidently solution and alteration render many of these particles unidentifiable.

Volcanic sediments

A wide variety but small amount of volcanic material is found in pelagic sediments. Abundance is not uniform; the Pacific Ocean receives most of it. Various types of potassium feldspar have been found in the Pacific, which presents an incongruity in the normally mafic oceanic basin. Pyroxenes are among the most widespread minerals of pyroclastic origin in the oceans and are locally altered. Volcanic glass occurs in all stages of alteration and decomposition. As yet, no positive data are available to describe the conditions under which these changes take place. **Montmorillonite** and **phillipsite** are common alteration products of volcanic ash and pumice. A large portion of the abundant zeolite minerals is also probably volcanic in origin, but some are authigenic, and it is almost impossible to distinguish between the two origins.

Eolian sediments

A fair amount of the deep-sea muds is attributed to a windblown source. There have even been a number of unquestionably fresh-water diatoms found in marine waters, with the wind as the logical transporting agent. The primary area of eolian sediment accumulation is in the North Pacific and adjacent to areas of arid climate.

Most of the pelagic sediment in the North Pacific is called brown clay, or in some books red clay, and is in part attributable to a windblown source, with altered volcanic ash as another source. The windblown particles were perhaps derived from the wide expanse of the Gobi desert. This brown clay also contains abundant quartz (Fig. 23.6) which decreases in grain size toward the center of the ocean basin. It has been suggested also that much of this eolian dust is carried in the troposphere and reaches the ocean as fallout.

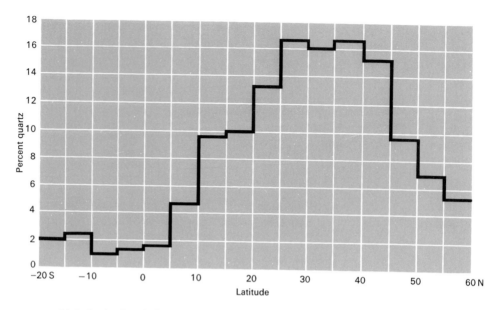

FIG. 23.6 *Latitudinal distribution of quartz in the eastern Pacific Ocean. (After Keen, 1968, p. 71.)*

Authigenic sediments

Scattered throughout pelagic sediments of all areas are small amounts of mineral matter that have crystallized from the ions in solution in seawater. With the exception of the manganese nodules (Chapter 11), all are quite minor constituents of pelagic sediments except for phillipsite. This hydrous alumino-silicate may comprise 50 percent of the pelagic sediment locally. **Zeolites** in general are also fairly common among authigenic minerals in the sea. They may form as the result of the interaction of seawater with basaltic magma, crystallize from silica-rich waters near the sediment-water interface, or be derived from alteration of volcanic debris on the sea floor.

A second common group of authigenic minerals is the phyllosilicates, which includes clay minerals. Most of these apparently have a terrigenous origin, and due to their small size (less than four microns) make their way to the open ocean before settling to the bottom. Many of the clays with a terrigenous source are changed appreciably upon reaching the marine environment. These changes include increase in crystallinity, adsorption of certain ions, and the change from one clay mineral to another.

Kaolinite, glauconite, and chlorite may form completely in the marine environment. Kaolin occurs in very delicate "books" of crystals which must have formed *in situ*. Some chlorite is so high in boron that it must have formed in the ocean, where this ion is concentrated. Glauconite is a particularly interesting clay mineral because it is known only in the marine environment and is therefore valuable in interpreting ancient environments.

TERRIGENOUS SEDIMENTS
Land-derived sediments that are transported and deposited with the aid of currents are considered terrigenous; they differ from pelagic sediments in a variety of ways. The most striking differences are their lack of biogenic materials and the relatively coarse grain size (Fig. 23.3). In general terrigenous sediments are dominant only around the periphery of the ocean floor, thus reflecting their land-derived origin. An exception is the wide band of terrigenous material of glacial origin which circumscribes the Antarctic continent (Fig. 23.2).

Terrigenous muds
A variety of silty clays and clayey silts is distributed around the margins of ocean basins, where they are commonly associated with turbidites (Chapter 21). Mineralogically these sediments are composed of quartz and various clay minerals. They reach the ocean basin by means of the normal currents that operate across the continental margin and on the abyssal floor. Over the past several years, photographs of the deep-sea floor showing such features as ripples (Fig. 23.7) indicate that substantial currents are present. Because of the association of these muds with turbidites, it is sometimes difficult to distinguish between the two, particularly in turbidites that do not exhibit the typical sedimentary features associated with their mode of deposition. Additional problems arise in distinguishing between pelagic and terrigenous muds because quartz in particular may be transported by wind over the ocean surface. The same is true to a lesser degree of volcanic fragments and clay minerals.

The clay minerals in terrigenous muds are abundant, and nearly all varieties are present. It should be noted, however, that our knowledge of the clay mineralogy of deep-sea sediments is greatly lacking. Good data are available only from spotty locations, but at the current rate of data collection this situation is improving greatly. Regardless of their general mineral content, these sediments have gross color characteristics that reflect their chemical environment, particularly the oxidation-reduction conditions. The brown and reddish sediments are the result of oxidizing conditions and the presence of ferric iron. This oxidation may occur before, during, or after deposition of the sediment. The blue, green, and black muds represent a reducing condition and contain organic matter as well as ferrous iron, particularly in the black muds. Some sulfides may be noted upon recovery of samples, but these usually oxidize rather quickly.

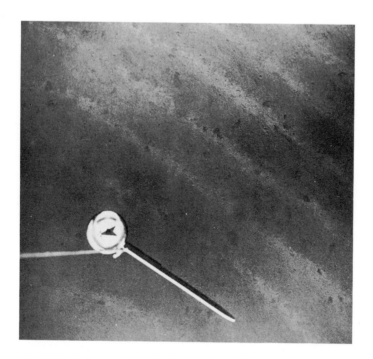

FIG. 23.7 *Soft, rounded ripples in abyssal brown clay near the floor of the Puerto Rico Trench. The water depth is 7500 meters. Ripples are about 0.25 meter in wave length; the compass indicates flow from east to west. These ripples, formed by the Antarctic Bottom Current, are the deepest known. (Photo courtesy of J. D. Hollister.)*

Illite and chlorite have been found throughout the world's ocean basins and are the dominant clays in most areas where these minerals are present. Kaolinite is also a fairly common constituent but is most abundant close to the continents. Occurrence of montmorillonite shows a general association with volcanic areas. There are chemical changes that take place as some of these clay minerals enter the marine environment. In general, however, the clays that eventually comprise terrigenous muds reflect the climate of their source area. Clay minerals are commonly formed by chemical weathering of other minerals, so that their kind and their variation within a given clay mineral are influenced by climate. Seawater can provide certain ions that have been removed during weathering and thereby also change the character of a clay mineral.

Rafted sediments

Minerals and rock fragments in general have a density that is about 2.7 times that of water, and consequently they settle readily to the bottom, where they may be moved by currents. Under some conditions, sediments can be moved about at considerable distances above the ocean floor, even at the surface. Such distribution must be accomplished with the aid of a floating or swimming medium. In other words the sediment is transported by something to a deep-sea environnent, where it is released and allowed to settle on the bottom.

Recognition of rafted sediments is based primarily on the anomalous location of rock and mineral debris on the sea floor. This may be a textural anomaly, such as pebbles or larger grains in the deep-sea area. It might be the presence of a particular rock or skeletal fragment that is far removed from its place of origin and cannot be logically explained in any other way. This does not mean to imply that rafting is a crutch for any odd sediment distribution, but it is a significant and fairly prevalent means of transporting small quantities of sediment.

Glacial marine sediments.   Ice provides a means for large-scale transport of sediments by rafting. Around the glaciated areas of the world, particularly Antarctica, there is a high concentration of rafted sediments. In some areas they comprise the dominant deep-sea deposit (Fig. 23.2). The typical glacial marine deposit is similar to terrestrial glacial till. It is unsorted and unstratified, and consists of a wide mixture of mineral and rock types. Sediment-laden ice from glaciers reaches the sea and drifts with the currents until it melts. When it does, all the sediment contained in it is released and falls to the sea floor. During the Pleistocene glaciation, ice rafting was a tremendously important means of sediment dispersal, and much of this fairly recent deposition has not yet been buried by pelagic sediments.

Other means of rafting sediments.   Although volumetrically less important than ice rafting, organisms can also play a significant role in sediment dispersal. This can be accomplished by floating plants, swimming animals, or birds. The roots or holdfasts of plants such as kelp, marine grasses, and small algae may contain a fair amount of sediment. As currents carry these plants, the sediment slowly washes out of the roots and falls to the bottom, or the plant itself sinks, carrying the sediment with it. The organic matter eventually decays, leaving the sediment behind. This type of dispersal is common among the large kelps of the world. Trees or tree stumps carried out to sea can also retain a fair amount of sediment that will settle to the ocean floor.

Animals may be agents of sediment transport in the open-ocean environment. The majority of this transportation is in the form of **gastroliths,** which are common in marine mammals such as seals, sea-lions, and walruses, and in some birds and fish. The composition of this sediment ranges widely, but the size of particle is pretty well restricted by the type of animal. Those in birds are a few millimeters in diameter, but in the large sea mammals they may be several centimeters. Birds also transport minor amounts of sediment by carrying shelled animals and dropping them away from their normal habitat.

COLLECTING DEEP-SEA SEDIMENTS

Collecting meaningful data on deep-sea sediments is one of the most difficult of all tasks in oceanographic research. The great distances between the ocean floor and the ship plus the general difficulties of operating apparatus in seawater are the major sources of these problems. There is a tremendous variety of bottom samplers, but they can be subdivided into a few groups depending on the type of sample collected. This discussion will include the general operation and function of only a few of the better-known types. The reader is referred to the list of references for a more comprehensive treatment of the subject.

Regardless of the type of apparatus or its specific function, a few general rules govern the design and construction of all bottom sediment samplers. The material used in construction should be noncorrosive whenever possible and should be durable. The apparatus should be as simple as possible with a minimum of working parts. Construction must be such that the apparatus is able to withstand rigorous impact without malfunctioning. One of the most important prerequisites is the ability of the apparatus to obtain and retrieve a bottom sample with minimum distortion and loss of the sediment.

Grab samplers and dredges

Surface samples of the bottom may be obtained by several different designs of apparatus. All fall into two broad categories: the grab sampler, which is dropped and collects a spot sample at the drop location, and the dredge, which is dragged along the surface, accumulating a sample from a traverse of the sampler. Both are of considerably utility, which one is used depends on the type of sample required.

The Petersen grab sampler was discussed previously (Fig. 18.29). This is the most widely used of the grab-type devices, but there are many similar models. The orange-peel sampler (Fig. 23.8) is also widely used and has the advantage over a normal grab sampler in that a wheel and sprocket mechanism closes the four jaws mechanically. A hemispherical sample is retained and is prevented from being washed during recovery by a canvas cover. The orange-peel device is not designed for the collection of quantitative samples.

FIG. **23.8** *Orange-peel sediment sampler in open position. (Photo courtesy of J. L. Hough.)*

Dredges may be used to sample unconsolidated sediment from the sea floor, but they are usually employed to obtain bedrock samples or large cobbles or boulders that cannot be collected with the grab-type devices. The basket dredge (Fig. 23.9) as used for rock sampling is quite similar to that

FIG. **23.9** *Heavy basket dredge for collecting large rocks and breaking off pieces of bedrock from the sea floor. (Photo courtesy of T. E. Pyle.)*

used for benthic organisms (Fig. 18.28) but is much more sturdy. A pipe dredge is also used which consists of a large cylinder closed at one end; it is used much like the basket dredge.

Dredges have an advantage over grab samplers in that they can be used while the ship is under way. In most cases, however, these devices are towed slowly or hauled up along a scarp to obtain bedrock samples. The sampler is dropped to the bottom and then hauled slowly across the sampling area until resistance is encountered. At this point, the winch retrieving the dredge is slowed until there is a release of pressure. Care must be taken to avoid breaking the cable or sampler. Any bedrock materials can usually be recognized by fresh breaks on their surfaces. The most widespread use of dredges has been on walls of submarine canyons, and the data collected have greatly aided our knowledge of those features.

Core samplers

The majority of deep-sea samples are collected by coring, which gives a vertical control on the sample as well as the surface composition. The general principle involved is to collect a relatively undisturbed bottom sample by penetrating the sea floor with a tube and to retrieve the core contained in the tube. Core samplers are of simple construction and consist of few working parts, so that they are ideal for oceanographic work. Coring must, however, be done only in fine sediments, because coarse sediment presents too much resistance to penetration. In the deep-sea area, there is no problem because of the prevalence of fine sediments.

Regardless of the specific coring device that is used, it is necessary to have a cutting edge that will cause a minimum of distortion as it penetrates the sediment. The inside angle of the cutting edge should be about 5° or less and the inside diameter of the cutting collar should be less than that of the core itself. Both of these characteristics tend to reduce friction and, as a result, limit distortion. Even though these precautions are taken, there is a significant amount of compaction in the upper several centimeters because of the soft, water-saturated character of these sediments. There is also a smearing and distortion of the layering when cores are collected and again when they are extruded from the tube (Fig. 23.10). The outer few millimeters of a core are always scraped away before careful examination of the true stratification in the core.

Most coring samplers use a liner inside the metal core barrel (Fig. 23.11); the liner may be composed of plastic, wax-coated cardboard, or some other dispensable material. The core sample is in the liner, which is easily removed from the core and then can be stored indefinitely or cut for examination. The use of a liner can eliminate the forced extrusion of the core sample, which causes compaction and distortion of the sample.

FIG. 23.10 Extruding and wrapping a sediment core for transport to the laboratory for subsequent study.

FIG. 23.11 Phleger gravity corer which will take a core about one meter long. The plastic sleeve catches the core and slides it out of the core barrel. (Photo courtesy of G. M. Manufacturing and Instrument Corporation, New York.)

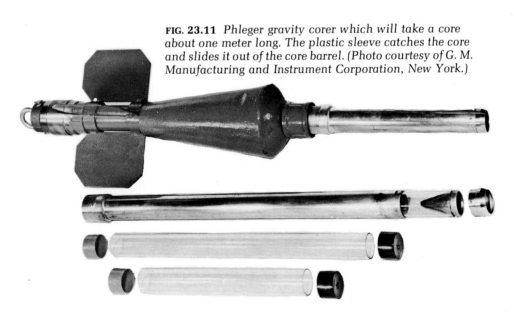

FIG. 23.12 *Heavily weighted gravity corer. (Photo courtesy of J. L. Hough.)*

Another commonly used accessory in coring devices is a core catcher that is used to prevent the core sample from slipping out of the core barrel during retrieval. There are two principal types: One consists of several prongs and the other of a hinged plate. Both of these allow the sediment core to move into the core barrel but prevent movement in the opposite direction.

Gravity corers are those which utilize weights and the free-fall method of penetration. As the coring apparatus descends it may have to be slowed somewhat in order to insure vertical descent and impact. The Phleger corer (Fig. 23.11) is widely used to obtain short cores, less than a meter in length. Good results have been obtained from very deep parts of the abyssal plain. Large gravity corers use several hundred pounds of lead weights (Fig. 23.12) on the upper end to facilitate penetration. Many also have fins above the weights to help keep the descent vertical. With such a coring device, sediment cores several meters in length can be obtained.

The invention of the piston corer by Kullenberg is one of the most significant of all contributions to marine geology. It permits the collection of longer and less deformed cores than any other coring device. The corer itself is similar to the gravity corers in that it is a long weighted tube with a cutting edge. In addition, there is a piston in the core barrel that creates a vacuum in the tube as it penetrates the sediment and in effect helps pull the sample into the tube (Fig. 23.13). The entire apparatus consists of the above-described corer plus a small pilot corer (Fig. 23.14) that leads the descent and also triggers the release of the large corer. The apparatus is let down slowly until it is quite near the bottom; the depth is determined from fathometer tracings. It is then allowed to fall freely, and as the pilot corer hits the substrate, it trips

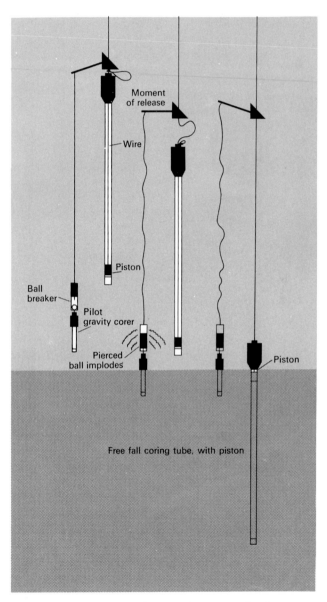

FIG. 23.13 *Piston corer showing the principle of operation. (After Shepard, 1972, p. 20.)*

FIG. 23.14 *Preparation of a piston corer and a pilot corer (right) for descent on shipboard. (Photo courtesy of R. L. Tompkins.)*

a release mechanism and the main corer is allowed to fall. The piston is connected to a wire so that as the corer enters the substrate the piston inside remains stationary and the weighted tube passes into the sediment. This has the effect of reducing friction by the vacuum created as the tube moves past the stationary piston. Cores up to 20 meters in length have been obtained with this device. The purpose of the pilot core is to provide a good sample of the upper layers, which would be badly distorted or partly washed away in an incomplete core.

There are coring devices that are particularly adapted for taking thick cores of fine sediment in shallow water. One such coring apparatus utilizes

FIG. 23.15 *A-Frame and related equipment on a coring barge designed for shallow bays.*

an A-frame mounted on a shallow-draft barge (Fig. 23.15). Cores of nearly 10 meters are possible with such a rig. One of the most efficient coring devices is the vibro-corer (Fig. 23.16). It is capable of taking a 6-meter core in seconds from nearly any type of substrate. The device is powered by an air compressor in much the same fashion as an air hammer which breaks up concrete. The vibro-corer can be used in coarse sediment and even in some lithified substrates, thereby providing significant advantage over gravity or piston corers.

Recently there has been emphasis on core samples with large diameters. Most conventional corers will take cores with diameters less than 10 cen-

FIG. 23.16 *Vibro-corer apparatus used for rapid and efficient coring in coarse or very firm sediments. The core barrel is supported by the quadripod and is forced into the substrate using compressed air to vibrate the barrel.*

timeters. This narrow diameter presents considerable problems in studying the stratification and sedimentary structures in the sediment. With the increasing study of turbidites, more information is becoming necessary in order to interpret the depositional environments of the turbidites. Large corers are collectively known as box corers (Fig. 23.17), and they exist in a variety of sizes and designs. The core area may be more than 0.1 square meter and it may be either circular or rectangular in shape. Because of the large diameter of these corers, it is necessary to have some mechanism to cover its working end and keep the core sample in place. Various box corers have

(a)

(b)

FIG. 23.17 (a) NEL-type spade or box corer being readied for lowering and (b) being hauled up from a sampling effort. (Photos courtesy of T. E. Pyle.)

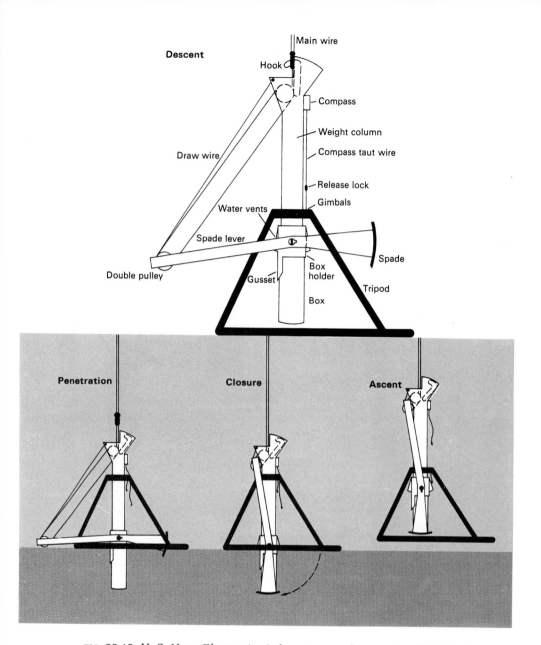

FIG. 23.18 *U. S. Navy Electronics Laboratory spade corer for obtaining large oriented cores. (After A. Richards, 1967, "Obtaining Large, Undisturbed and Oriented Samples in Deep Water," Marine Geotechnique, A. Richards (ed.), Urbana: University of Illinois Press, p. 253.)*

been designed; one example is the swinging-spade box corer (Fig. 23.18). After the corer penetrates the sediment, a spade pivots as the wire is pulled up. This spade cuts through the sediment until it is in position under the end of the corer.

SELECTED REFERENCES

Barnes, H., 1959, *Oceanography and Marine Biology*, New York: Macmillan, Chapter 1. Fairly comprehensive and well-illustrated treatment of bottom sampling devices.

Heezen, B. C., and C. D. Hollister, 1971, *The Face of the Deep*, New York: Oxford University Press. The most complete single volume on deep-ocean geology, with hundreds of excellent underwater photographs of the ocean floor.

Hill, M. N. (ed.), 1963, *The Sea*, New York: Wiley, Vol. 3, Chapters 23, 25, 26, and 33. Excellent chapters on various aspects of deep-sea sediments and processes.

Hopkins, T. L., 1964, "A Survey of Marine Bottom Samplers," *Progress in Oceanography*, M. Sears (ed.), Vol. 2, pp. 213–256. A comprehensive and up-to-date description of more than 100 bottom samplers, with an illustration of each. Also lists manufacturers.

Keen, M. J., 1968, *An Introduction to Marine Geology*, New York: Pergamon Press, Chapters 4 and 5. Brief but well-written discussion of deep-sea sediments designed primarily for the nonspecialist.

Sears, M. (ed.), 1967, "The Quaternary History of the Ocean Basins," *Progress in Oceanography*, Vol. 4. A broad treatment of ocean basins and their recent history with several papers on specific topics in deep-sea sediments.

Shepard, F. P., 1972, *Submarine Geology* (3rd edition), New York: Harper and Row.

Trask, P. D. (ed.), 1939, *Recent Marine Sediments*, Tulsa, Okla: American Association of Petroleum Geologists. Out-of-date book containing several papers that describe deep-sea sediments of various basins; also a paper by J. L. Hough on bottom-sampling apparatus.

24 MARINE GEOPHYSICS

This book deals with that part of the earth which is covered by saltwater. With the exception of the brief discussion on deep-sea sediment cores, nothing has been said about the earth's crust below the oceans. The generalized diagram of the earth (Fig. 1.3) shows three major subdivisions: crust, mantle, and core. The crust is usually divided into continental (granitic) and oceanic (basaltic) types.

One might question how information on layers deep within the earth is obtained. In all cases it must be via indirect methods. The deepest oil well is less than 8 kilometers below the earth's surface and provides the greatest depth for direct collection of data on the earth's crust. There are millions of oil wells spread over the world. However, subsurface data from the ocean are minimal compared to those from the land. The number of cores from oceans is less and the depth of penetration is considerably less than on land. Recent deep-drilling programs are, however, obtaining data to depths of several hundred meters with a specially constructed ship, the *Glomar Challenger* (Fig. 24.1).

The lack of direct data is a great handicap in our efforts to learn the character of the oceanic crust. Fortunately, there are a number of ways that we can obtain indirect information from the deep oceanic crust, and, through the efforts of trained geophysicists, learn something about this extremely remote part of the ocean basins. All the methods involve various measurements of rock properties and then interpretation of these data in order to describe the rock. The most commonly investigated aspects of geophysics are seismology, gravimetrics, magnetics, and heat flow. **Seismology** is the study of earthquakes and other crustal vibrations. By carefully studying the transmission of seismic waves through the earth's crust it is possible to learn something about its character and composition. **Gravimetrics** involves detailed measurements of the earth's gravity. Slight differences from one place to another are generally due to differences in the composition or thickness of crustal rocks. The earth's magnetic field and thermal properties also show some variation as a result of heterogeneity in the crust.

FIG. 24.1 Glomar Challenger, *drill ship used in the Deep Sea Drilling Project. (Photo courtesy of DSDP.)*

By gathering information on these gross characteristics it is possible to present a reasonable picture of the entire crust. It must also be noted that from these types of indirect data, it is possible to come up with more than one interpretation of the available information. Whenever possible, all types of geophysical measurements should be included; usually, however, seismic data are all that are available.

SEISMIC REFRACTION AND REFLECTION

One method for exploration of the oceanic crust is seismic refraction. The general techniques used in the ocean are much like those on land. The principle involved is one of measuring the velocities of sound waves as they travel through the sea and the earth's crust. Sonic energy is generated by small explosions. The frequencies of these waves must be low in order to get good penetration into the solid crust. High-frequency waves are commonly used in mapping bottom topography because they are reflected off the ocean floor.

Upon explosion, low-frequency sound waves are emitted in all directions. Those that travel toward the bottom will penetrate the crust to depths of thousands of meters. Each layer of material has different properties and will transmit seismic waves accordingly. An individual ray is refracted at the interface between two layers and then travels through the plane of the layer it

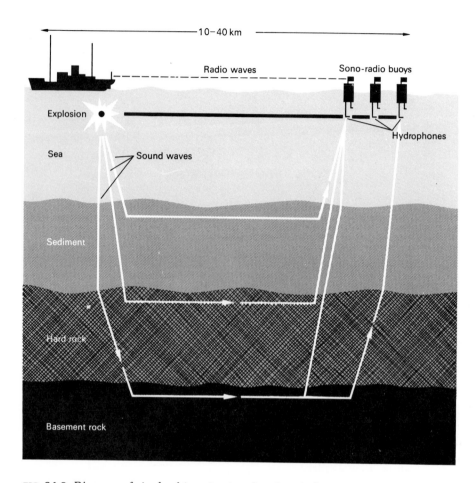

FIG. 24.2 *Diagram of single-ship seismic refraction. (After M. N. Hill, 1963, "Single Ship Seismic Refraction Shooting," The Sea, M. N. Hill (ed.), New York: Wiley, vol. 3, p. 40.)*

has just entered (Fig. 24.2). During propagation there is continual loss of energy upward until eventually it is all dissipated. The waves are picked up by sensitive hydrophones, and the data are relayed to the monitoring ship via radio buoys. These hydrophones are placed at a predetermined distance from the explosion so that the velocities of waves traveling through the layers can be accurately determined. At each of several hydrophones, the first signal will represent the sound traveling through the fastest path and so on for each successive signal.

The procedure outlined above may be accomplished in two general ways: One utilizes a single ship while the other requires two. The first oceanic seismic studies used two ships, with one dropping the explosive charges and the other towing a string of hydrophones to receive the impulses. More recently; a single ship has been used, with the hydrophones attached to floating buoys at some distance from it (Fig. 24.2). This setup eliminates reception of ship noise by the hydrophones and provides great flexibility. For instance, several sets of hydrophones can be placed over a wide area and readings obtained from a single explosion. A significant disadvantage is the time necessary to recover the buoys and hydrophones.

Seismic data on the earth's crust

Before describing the deep oceanic crust as interpreted from seismic studies, we must consider the assumptions that are inherent in this interpretation. First of all, it is assumed that each successive layer has a greater seismic velocity than the one above it. For each of the major layers, this assumption is apparently valid; however, there may be slight variations among the thin strata within these layers. In order to interpret the hydrophone data, it is assumed that all interfaces in the crust are plane surfaces. This may be true in certain areas but it is certainly invalid for others, particularly when we know there are major structural features in the oceanic crust. The third assumption is that the layers are laterally homogeneous, so that when a seismic wave is traveling within a layer, it is moving at a constant velocity. This assumption is unlikely in many parts of the ocean basins, again because of major structural features. It is up to the scientist who interprets the data to indicate the reliability of his interpretations, and if more than one valid interpretation exists, all should be pointed out.

Seismic refraction data have provided a fairly good picture of the general character of the oceanic crust. Data collected from all ocean basins indicate that there is general uniformity among the various ocean basins but also some distinct differences. There are four distinct and easily recognizable layers under the sea floor, three of them above the Mohorovicic discontinuity (Table 24.1).

TABLE 24.1 *Layers of the oceanic crust.*

Layer	Thickness (km)	Velocity (km/sec)	Composition	Density (g/cc)
Seawater	4.5	1.5	Seawater	1.0
Layer 1	0.45	2.0	Unconsolidated sediments	2.3
Layer 2	1.75	5.0+	Sedimentary and/or volcanic rock	2.7
Layer 3	4.7	6.7	Basalt	3.0
Mohorovicic discontinuity				
Layer 4	—	8.1	Ultramafic rocks	3.3

The thin first layer is composed of a variety of unconsolidated sediments (Chapter 23) and exhibits variety in composition and thickness. Although the minerals comprising this layer are themselves higher in density than the 2.3 grams/cubic centimeter that is shown for the entire layer, their lack of compaction and cementation results in a low sound velocity and low overall density.

Table 24.1 indicates that layer 2 is considerably higher in velocity than the first layer and is nearly the density of the continental crust. These two characteristics of this layer are compatible with a great variety of rock types; sedimentary, metamorphic, and volcanic igneous. On the basis of our knowledge of the other layers and the ocean basins in general, the most likely choices are lithified or consolidated sediments and volcanic igneous rocks. Distribution of this layer is not uniform; layer 2 is present in only one-third of the Atlantic stations but is present throughout the Pacific. This difference may be more apparent than real, because some seismic techniques pick up layer 2 readily whereas others miss it. There are also some thickness differences within the Pacific Ocean basin that seem to be related to volcanic areas. This and other variations in the layer indicate that it is probably composed of a variety of rock types.

Layer 3 is a thick basaltic or oceanic zone in the crust which yields rather uniform velocity data. The velocity (6.7 kilometers/second) could indicate a hard dense limestone, but this is quite unlikely due to the way in which limestone forms. The composition is probably intermediate between granite and dunite. The Mohorovicic discontinuity separates layer 3 from the mantle (layer 4). The character of this discontinuity is not known. Until recently, it was thought to be a compositional change; however, many geologists now believe that it represents a phase change between rocks that are chemically similar.

Continuous seismic-reflection profiling

One of the most significant advances in our study of the ocean floor has been the development of the continuous-recording reflection instrument. Except for deep crustal studies, this is now the most widespread seismic method used in marine geophysics. With the aid of this apparatus, marine geologists have acquired a great deal of information about the continental margins (Chapter 2). This instrument "draws" a cross section to depths of a few hundred meters below the ocean bottom that shows the geometry and spatial distribution of sediment and rock layers (Fig. 24.3).

The seismic profiler is similar to a standard depth recorder except that it emits low-frequency sound impulses that are much less attenuated as they travel through sediment or rock. As a result, the echogram shows both the bottom profile and the layers of sediment or rock below the bottom (Fig.

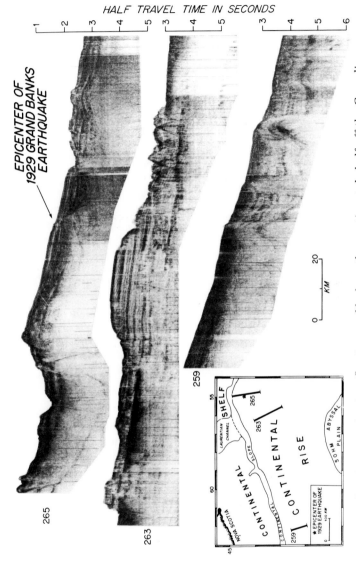

FIG. 24.3 Continuous seismic-reflection profile from the continental shelf off the Canadian Maritime Provinces. (Photo courtesy of K. O. Emery and J. D. Hollister.)

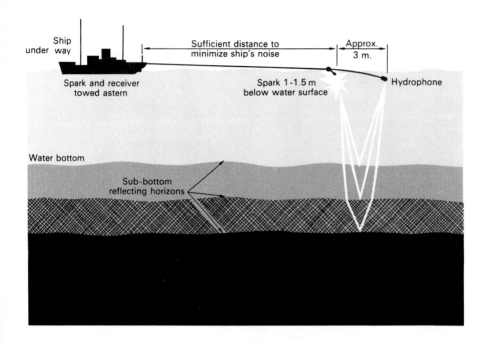

Ship
under way

Sufficient distance to
minimize ship's noise

Approx.
3 m.

Spark and receiver
towed astern

Spark 1-1.5 m
below water surface

Hydrophone

Water bottom

Sub-bottom
reflecting horizons

FIG. 24.4 *Sketch of apparatus used in continuous reflection profiling.
(After J. B. Hersey, 1963, "Continuous Reflection Profiling," The Sea,
M. N. Hill (ed.), New York: Wiley, vol. 3, p. 47.)*

24.4). Both sparkers which produce an electric spark, boomers which produce an impulse by electrically forcing two metal plates apart, and air guns
which emit a bubble impulse are used as sound sources. If great detail is
sought, a relatively high frequency is desired, whereas very low frequencies
will allow increased penetration and therefore a deeper record. Due to
advances in this technique, seismic reflection has almost totally replaced the
refraction method. Advantages to the reflection technique include the safety
of not using explosives and also not being harmful to fish and other animals
in the sea.

OCEANIC GRAVITY SURVEYS

We usually think of the acceleration due to gravity as being constant over the
earth's surface; however, there are small local differences. In order for the
pull of gravity to be equal at all points on the earth's surface, the earth would
have to be spherical and of uniform density; such is not the case. The first
recognition of gravitational differences took place when accurate pendulum
clocks were transported from Europe to low-latitude areas. The clocks were

slower in their new location. The change in the rate of pendulum movement in this case was due to both the equatorial bulge of the earth, making the distance to the earth's center farther than in Europe, and the effect of the earth's rotation. The pendulum would respond in a similar manner to changes in mass. It is actually an excellent gravimeter.

Present-day shipboard gravimeters, which can be used while the ship is underway, are of two basic types; one utilizes the pendulum principle and the other the pull of gravity on an accurately calibrated spring. Although the general principles involved in gravimeters are simple, the instruments are large and rather complicated (Fig. 24.5). Much of the instrument's function is to eliminate movements of the ship itself, both vertical movements, due to surface waves, and listing from side to side. Because of the small differences in acceleration due to gravity, such movements would yield grossly inaccu-

FIG. 24.5 *Shipboard gravimeter. (Photo courtesy of Askania.)*

rate data. In order to eliminate these movements, early studies were carried out in submarines, which by descending below the surface could eliminate ship motion. A simple pendulum apparatus was used, but the method was rather inefficient because gravity observations with a pendulum require a tremendous amount of time. Modern gravimeters are mostly of the spring type and substantially eliminate both of the above problems.

Airplanes have provided an additional means for obtaining gravity readings over the oceans. The same instrument and principles used on shipboard are utilized in aircraft, with the built-in advantage of speed. Some problems do exist, however, particularly in terms of navigation and positioning the aircraft with respect to the gravity readings. Elevation, ground speed, and course must be determined with a high degree of accuracy in order to obtain even broad gravity anomalies. Broad gravity anomalies are also detected and measured from satellites.

Gravity measurements have a distinct advantage over seismic data because there is no depth limit; gravity is based on the cumulative effect of the earth's mass below a certain point. This presents some disadvantage, however, in that we can determine nothing about the layering of the earth from gravity readings. As a result, seismic and gravity data nicely complement one another and provide much more comprehensive data than either one could alone.

Some gravity anomalies and interpretations

The most pronounced major gravity anomalies in ocean basins occur in the areas of oceanic ridges and in the areas of island arcs and associated deep trenches. There are other gravity anomalies associated with local features such as volcanic islands, guyots, seamounts, and some of the prominent fracture zones. Across the Mid-Atlantic Ridge, the Bouger anomaly is nearly the reverse of the ridge's topography, and gravity readings indicate a mass deficiency (Fig. 24.6). This could be explained by a thickening of the crust under the ridge, or as less dense material. Most marine geologists favor the concept of a thickened crust, with thickening occurring in both directions, comparable to continental mountain roots.

Striking gravity anomalies occur over volcanic islands and other similar features. Computed cross sections of these features (Fig. 24.7) indicate a gross geometry that is somewhat similar to, but on a smaller scale than, the oceanic ridges.

The pioneering work of Vening-Meinesz, using pendulums mounted in a submarine, provided us with our first knowledge of major negative gravity anomalies. The trenches, with negative anomalies, are commonly associated with island arcs, which are positive areas. Geophysicists have explained these negative anomalies as the result of deep water, in some cases thick sediment, and great depth to the mantle.

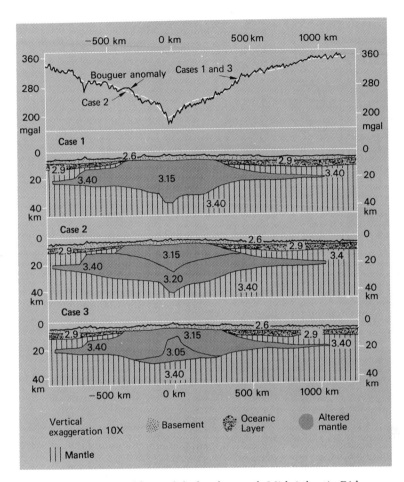

FIG. 24.6 *Three possible models for the north Mid-Atlantic Ridge which agree with both the gravity anomaly and the seismic refraction data. (After Manik Talwani, 1971, "Gravity," The Sea, A. E. Maxwell (ed.), New York: Wiley Interscience, vol. 4, p. 282.)*

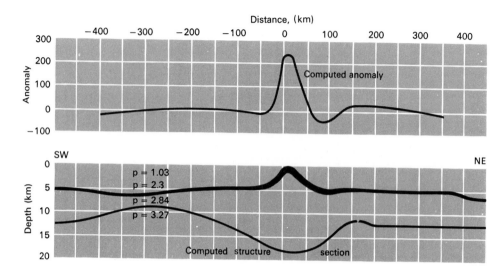

FIG. 24.7 *Gravity anomaly and structure section across the Hawaiian Islands. (After J. L. Worzel and J. C. Harrison, 1963, "Gravity at Sea," The Sea, M. N. Hill (ed.), New York: Wiley, vol. 3, p. 163.)*

MAGNETIC SURVEYS

The earth is a giant magnet, and its magnetic field has considerable influence on certain minerals that form or are deposited within the field. Magnetism in rocks is largely due to the presence of iron oxides, particularly magnetite. This mineral is fairly common in basalt and some sandstones, where it is oriented with the geomagnetic field that prevailed at the time the rock formed. Magnetite is present in basalt flows and crystallizes directly from the magma, thus giving better data on the geomagnetic field than it does in sedimentary rocks. The magnetite that accumulates in sediments settles out of water, and the grains are aligned with the geomagnetic field; however, compaction, cementation, and other diagenetic phenomena may alter the original orientation somewhat.

On land, magnetic surveys are used primarily to explore for commercially useful metallic deposits. In the oceans, similar surveys are designed to supplement seismic and gravity data in learning about the composition of the oceanic crust. This type of geophysical survey is comparatively inexpensive and involves rather uncomplicated equipment. As in gravity surveys, airplanes or ships can be used to tow the magnetometer, thus enabling wide coverage in a relatively short time. A widely used instrument for magnetic data (Fig. 24.8) consists of a wire coiled about a vessel of water or some other hydrogenous liquid. When a current is passed through the coil, it causes alignment of the hydrogen ions (protons) along the coil axis. The precession of protons over the geomagnetic field creates a weak signal that is monitored on the ship.

In ocean basins, magnetic surveys are particularly helpful in studying layer 2 and to a lesser extent layer 3. Although the percent of magnetite in basalt is low, it is possible to detect magnetic anomalies caused by small differences in composition. This method is particularly helpful in distinguishing between limestones and basalts which have similar seismic properties. Volcanic areas such as seamounts show up as highly seismic areas simply because of their great accumulation of basaltic material.

FIG. 24.8 *Proton magnetometer used in tow behind a ship. (Photo courtesy of Askania.)*

One of the most interesting patterns in the remanent magnetism of oceanic rocks is present in the eastern Pacific in the area of the great east-west fracture zones (Fig. 24.9). If the regional geomagnetic field is taken away, the result is a series of approximately north-south trending anomalies which are disrupted by the fracture zones. The cause of these striking anomalies and their offsets is now well explained by plate tectonics and sea-floor spreading (Chapter 25).

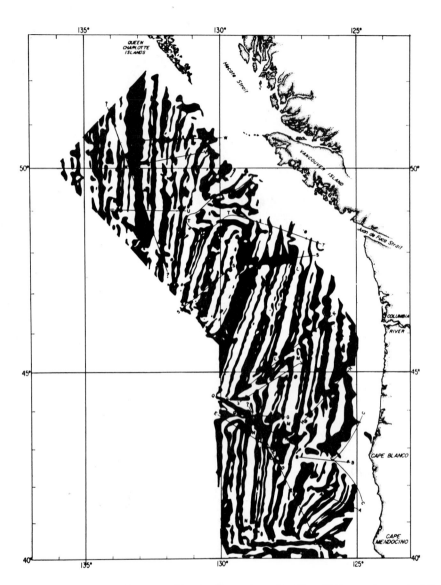

FIG. 24.9 Magnetic anomalies in the eastern Pacific off the west coast of the United States. (After A. D. Raff and R. G. Mason, 1961, "A Magnetic Survey of the West Coast of North America, 40°N to 52½°N," Bull. Geol. Soc. Amer. **72**, 1260.)

HEAT FLOW IN THE OCEAN FLOOR

The last type of geophysical data collecting that will be mentioned here is the determination of the flow of heat through the ocean floor. In the earlier discussion of continental drift (Chapter 3) it was noted that heat flow in oceanic basins is about the same as that on continents, whereas the composition of the crust suggests that heat flow ought to be higher on the continents. Not only is oceanic heat flow anomalously high, but there are significant differences throughout the ocean floor.

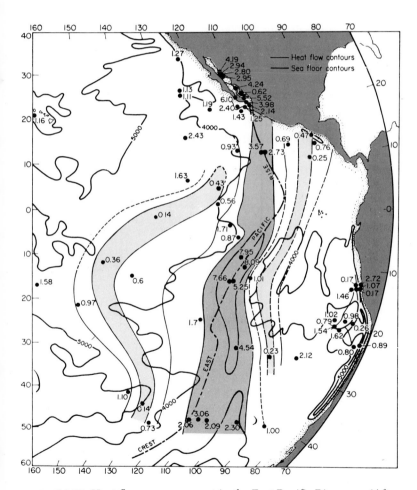

FIG. 24.10 Heat flow measurement in the East Pacific Rise area. (After E. C. Bullard, 1963, "The Flow of Heat Through the Floor of the Ocean," The Sea, M. N. Hill (ed.), New York: Wiley, vol. 3, p. 226.)

Heat-flow studies include measuring the thermal gradient and thermal conductivity in the sediments. A probe equipped with a temperature-sensitive device such as a thermocouple or thermistor is dropped, much like a coring apparatus, into the substrate. The temperature gradient that can actually be measured is only for a few meters, because of the restrictions imposed by the probe. Temperature differences over this distance are probably less than half a degree Celsius, so that the data represent only a thin skin of layer 1. There are also problems caused by the heat created when the probe penetrates the sediment. This heat may exceed the temperature difference to be measured, and so the probe is allowed to reach equilibrium temperature before data are recorded.

Much sophistication of heat-flow measurement has been accomplished by the Deep Sea Drilling Project. Heat flow has been measured in the holes drilled by this project and has yielded results comparable to the surface measurements.

To date, thousands of heat-flow measurements have been taken of the ocean basins. There is a distinct and readily explainable pattern which emerges: The highest heat-flow values are obtained from the oceanic ridges (Fig. 24.10), whereas the lowest values are found associated with the trenches. This is due to the cooling of lithosphere plates as the basaltic material moves away from the ridges.

SELECTED REFERENCES

Hill, M. N. (ed.), 1963, *The Sea*, New York: Wiley, Vol. 3, Chapters 1–11. Comprehensive treatment by experts on the various geophysical methods of exploring the ocean floor.

Hill, M. N., 1957, "Recent Geophysical Exploration of the Ocean Floor," *Physics and Chemistry of the Earth* **2**, 129–161. Good summary of marine geophysics that is somewhat out of date. Brief coverage of material similar to that in *The Sea*.

Howell, B. F., 1959, *Introduction to Geophysics*, New York: McGraw-Hill. A standard general geophysics textbook including the usual mathematics but little information on the oceans.

Jacobs, J. A., R. D. Russell, and J. T. Wilson, 1974, *Physics and Geology* (2nd edition), New York: McGraw-Hill. Well written and understandable text on general geophysics with a minimum of mathematics; somewhat out of date due to great recent advances in techniques. Considerable attention is devoted to oceans.

Maxwell, A. E. (ed.), 1971, *The Sea*, New York: Wiley Interscience, Vol. 4. Expansion and updating of volume 3 with comprehensive treatment of sea-floor spreading and plate tectonics.

PLATE TECTONICS AND SEA-FLOOR SPREADING 25

At this point, it is appropriate to discuss the modern concept of and evidence for the theory of plate tectonics and sea-floor spreading. This theory, which is well documented, is perhaps the most significant concerning the history of the earth since Darwin's theories on organic evolution. Until the late 1960s all the evidence to support the idea of shifting continents and young, mobile ocean floors was rather circumstantial. These data inferred many things that seemed to support the contention of a mobile crust, but there was no adequate proof from the deep ocean floor.

MODERN THEORY OF PLATE TECTONICS AND SEA-FLOOR SPREADING

Although Arthur Holmes's original idea in 1945 was somewhat similar, the work of Robert S. Dietz of NOAA probably represents the first of the truly modern approaches to global tectonics. In his works of 1961 and 1963, Dietz depicted the continental margins with a well-developed shelf, slope, and rise as being analogous to **geosynclines.** He showed that as the oceanic crust moved toward the continental block, there was much deformation of the continental margin, resulting in a folded coastal mountain range which was accreted to the continental block (Fig. 25.1).

The present-day theory of plate tectonics is the result of the combined efforts of several outstanding geologists and geophysicists. The current theory was actually formulated on the basis of the previously mentioned circumstantial evidence. Although the basic idea of sea-floor spreading is quite similar to Wegener's original hypothesis, the actual mechanics of the modern theory are different.

A world plot of earthquake activity shows that there is a great concentration of events along the circum-Pacific belt, in young folded mountains, and along the crest of the oceanic-ridge system (Fig. 25.2). The ridge system shows less activity as compared to the Pacific trenches. It is possible to divide the earth's surface into six major aseismic plates bounded by the narrow zones of seismic intensity (Fig. 25.3).

449

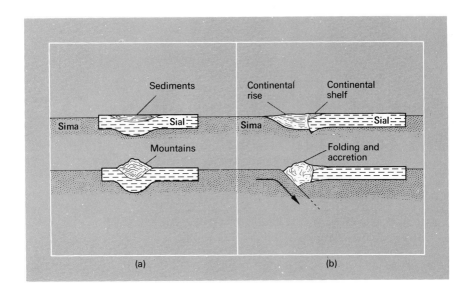

FIG. 25.1 *Schematic diagrams showing mountain building (a) by the old-fashioned view which causes continental shortening and (b) by the modern view which causes continental growth and is compatible with the concept of sea-floor spreading and plate tectonics. (After R. S. Dietz, 1970, "Continents and Ocean Basins," The Megatectonics of Continents and Ocean Basins, Helgi Johnson and B. L. Smith (eds.), New Brunswick, N.J.: Rutgers University Press, pp. 24–46.)*

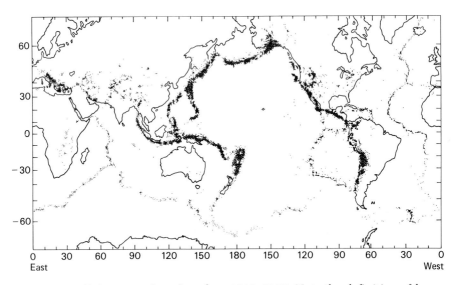

FIG. 25.2 *Epicenters of earthquakes, 1961–1967. Note the definition of large, stable blocks by the narrow and continuous seismic belts; note also that the greatest activity is at zones of convergence (trenches), whereas zones of divergence (ridges) show relatively low activity. (After L. R. Sykes, J. Oliver, and B. Isacks, 1970, "Earthquakes and Tectonics," The Sea, A. E. Maxwell (ed.), New York: Wiley Interscience, vol. 4, p. 354.)*

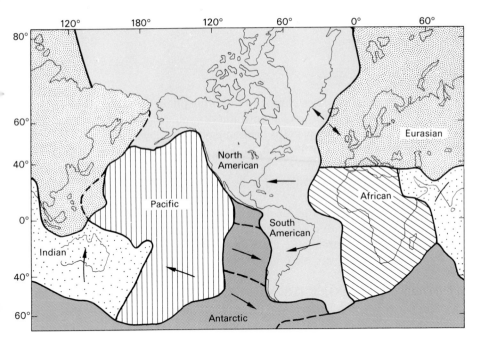

FIG. 25.3 *Plate-tectonic model showing major plates of the lithosphere. (After P. J. Wyllie, 1974, "Plate Tectonics, Sea-Floor Spreading and Continental Drift: An Introduction,"* Plate Tectonics—Assessments and Reassessments, *C. F. Kahle (ed.), Amer. Assoc. Petrol. Geol. Memoir 23, p. 8.)*

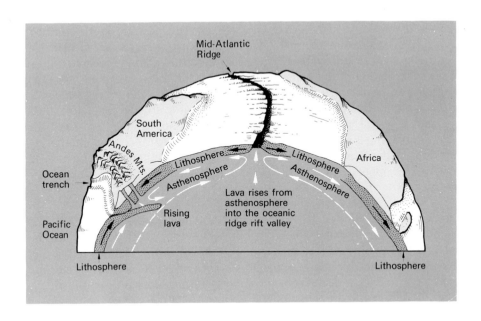

FIG. 25.4 *Schematic, three-dimensional representation of sea-floor spreading and continental drift. (After Wyllie, 1974, p. 6.)*

These rather rigid plates comprise the **lithosphere** and are constantly in motion. They are discontinuous at both the oceanic ridges and the trenches. These plates are moving over the **asthenosphere**, a layer several hundred kilometers thick that has little strength (Fig. 25.4). It is in this layer that the thermal convection that moves the plates is thought to take place. The underlying **mesosphere** is believed to be strong and static; it comprises the remainder of the mantle.

The model shown in Fig. 25.4 shows a divergence in the plates which occurs at the oceanic-ridge position. There is a convergence of plates which results in formation of trenches. **Transform faults**, which form normal to the direction of spreading, are also major tectonic elements of this model.

EVIDENCE TO SUPPORT THE MODERN THEORY

Coincident with and subsequent to the development of the modern theory, a wealth of new and diverse data in support of the concept of plate tectonics and sea-floor spreading was found. The vast majority of the data has been recovered from the sea floor, primarily by sampling and data-recovery techniques that have only recently been made available.

Magnetic anomalies.   Linear magnetic anomalies have been recognized on the crust of the ocean basins for some time. These anomalies represent periods of normal and reversed magnetism resulting from corresponding reversals in the earth's magnetic field. Early in the investigation of these features, it was recognized that they are aligned parallel to the oceanic ridges. Close examination of the patterns has shown that they are paired, with the ridge at the center of the paired system (Fig. 25.5). The most reasonable explanation for this phenomenon is that the basaltic crust is generated at the oceanic ridge and moves away from the ridge as the sea floor spreads. If the rate of spreading is the same on both sides, a symmetrical array of alternating magnetic patterns is left behind. These linear magnetic patterns have been matched on opposite sides of the oceanic ridge up to 80 million years before the present.

Earthquake epicenters.   Examination of both the global and vertical distributions of earthquake epicenters provides additional support for the modern theory of plate tectonics and sea-floor spreading. It is apparent that the greatest concentration of epicenters is located in the vicinity of oceanic trenches, whereas a lesser concentration is associated with oceanic ridges (Fig. 25.2). According to the model described earlier (Fig. 25.4), the trenches are the zones where one plate collides with another plate which descends into the asthenosphere. This collision provides the extreme concentration of seismic atcivity; however, the distribution of the epicenters is also important.

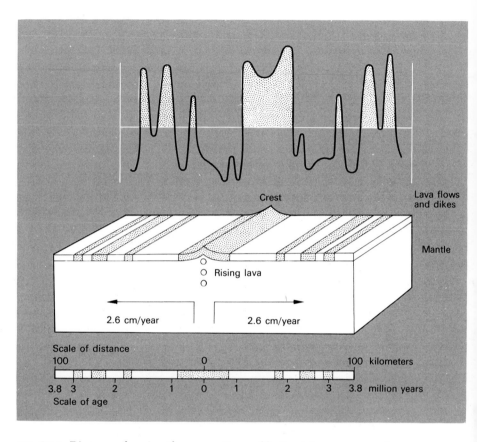

FIG. 25.5 *Diagram showing the magnetic profile (top) from a zone of the lithosphere adjacent to an oceanic ridge. Note the paired nature of the basaltic rocks and the magnetic profile. (After J. A. Jacobs, R. D. Russell, and J. T. Wilson, 1974, Physics and Geology (2nd edition), New York: McGraw-Hill, p. 388.)*

The shallow-focus earthquakes are found immediately under the trenches, whereas the deep-focus activity is under the land masses or the island arcs landward of the trenches (Fig. 25.6). These deep-focus earthquakes occur up to 600 or 700 kilometers below the earth's surface and are caused by the leading edge of a descending plate moving through the asthenosphere.

Deep Sea Drilling Project.   One of the technological marvels of the twentieth century has been the capability to drill deep holes into the ocean floor

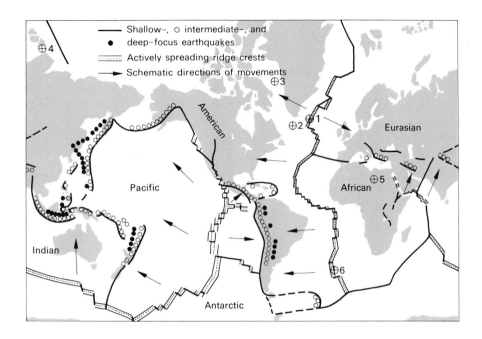

FIG. 25.6 *Global map showing the relationship of shallow, intermediate, and deep-focus earthquakes with the plate boundaries. Note that deep-focus earthquakes are landward of the collision zones or trenches. (After J. R. Heirtzler, et al., 1968, "Marine Magnetic Anomalies, Geomagnetic Field Reversals and Motions of the Ocean Floor and Continents," J. Geophys. Res.* **73**, *2119–2136.)*

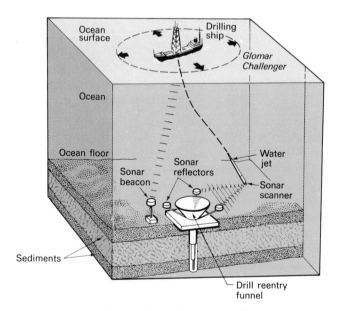

FIG. 25.7 *Sketch of the Glomar Challenger showing various aspects of its drilling technology including positioning and reentry. (After Wyllie, 1974, p. 12.)*

from a ship in about four kilometers of water. Such ships as the *Glomar Challenger* have been able to retrieve thick sections of core, many penetrating the entire sedimentary sequence and terminating in the volcanic basement. In addition, it has also become possible to pull the drill bit and pipe stem out of the holes and reenter (Fig. 25.7), although this procedure has not been an unqualified success.

In the opinion of most marine geologists, this project has confirmed the modern concept of sea-floor spreading. First of all, cores show that the oldest sediments underlying the present oceans are of Jurassic age (Fig. 25.8), which conforms well to the age expected from spreading calculations and which is compatible with the breakup of Pangea about 190–200 million years ago. This is significant not only because of the maximum age of the sediments, but also because of the areal and stratigraphic patterns depicted (Fig. 25.8). Sediments resting on the volcanic crust show a progressive increase in age away from the oceanic ridges. Also, the depth in which sediment accumulated increased away from the oceanic ridges. The project has also determined that there were long periods of time when deep-sea sediment was accumulating very slowly or not at all.

In addition to the valuable data on global tectonics, the DSDP has located large salt deposits in the Gulf of Mexico and the Mediterranean Sea. Petroleum also was found in the Gulf of Mexico, and some mineral resources were located in the Atlantic Ocean. Detailed accounts of each of the dozens of two-month cruises, as well as the complete report on the data collected, can be found in *Initial Reports of the Deep Sea Drilling Project*.

Ages of oceanic islands. The ages of all oceanic islands also provide considerable support to the modern theory of plate tectonics and sea-floor spreading. When plotted, these data show that there is a distinct progression in the ages of the islands, with the youngest on the oceanic ridge and the oldest near the outer limits of the ocean basin (Fig. 25.9). This idea was actually first proposed by J. Tuzo Wilson ca. 1961. Since then much more data and age dates have been obtained, the net result being in support of his theory. This result is, of course, in accord with the new tectonics in that the islands form in the oceanic-ridge area and move away as the lithosphere spreads toward the opposite plate boundary.

SUMMARY

The data presented above, when integrated with that in Chapter 24, provide a rather cohesive picture of the worldwide tectonic patterns that have been operating in the upper portion of the earth's volume. The presence of a quite mobile and changing lithosphere (crust) is the major aspect of this theory, which strongly contrasts with now-outdated ideas of the permanency of ocean basins.

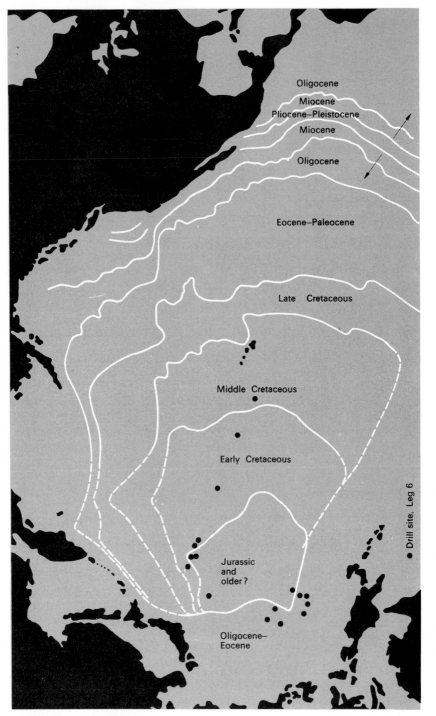

Oligocene

Miocene

Pliocene–Pleistocene

Miocene

Oligocene

Eocene–Paleocene

Late   Cretaceous

Middle Cretaceous

Early Cretaceous

Jurassic
and
older ?

Oligocene–
Eocene

• Drill site, Leg 6

FIG. 25.8 Sea-floor ages in the northwest Pacific Ocean. (From the National Science Foundation and Deep Sea Drilling Project news release, 1970.)

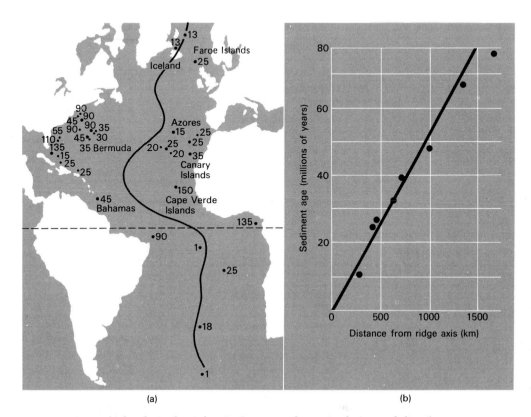

FIG. 25.9 *Ages of islands in the Atlantic Ocean as shown in their areal distribution (a) and as plotted showing age in comparison to distance from the ridge (b). These data support the concept of sea-floor spreading. ((a) After J. T. Wilson, 1965, "Convection Currents and Continental Drift," Phil. Trans. Roy. Soc. London, Sec. A, vol. 258, p. 145; (b) A. E. Maxwell, et al., 1970, Initial Reports of the Deep Sea Drilling Project, Washington, D.C.: National Science Foundation, vol. 3, p. 463.)*

Perhaps one of the best ways to summarize the modern concept is to use the Atlantic Ocean and the eastern portion of the United States as an example. Approximately 220–230 million years ago, there was a broad uplift of a portion of Pangea accompanied by some volcanic activity and faulting. This resulted in formation of a rift valley which then spread apart, with some oceanic crust appearing about 180–190 million years before the present (Fig.

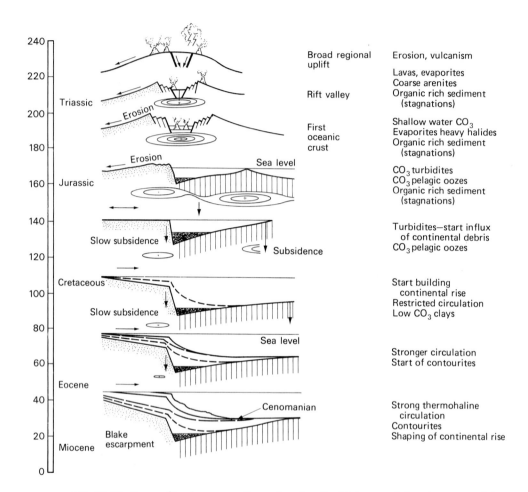

FIG. 25.10 *Generalized cross sections showing the history and development of the eastern continental margin of the United States. (After Jacobs, Russell, and Wilson, 1974, p. 75.)*

25.10). Continued spreading saw subsidence of both the oceanic and continental portions of the lithosphere. In the late Jurassic, the continental rise began to develop as turbidites accumulated at the base of the continental block. Continued subsidence and accumulation of terrigenous sediments developed the continental margin area as we see it today.

SELECTED REFERENCES

American Geological Institute, 1976, *Deep Sea Drilling Project, Legs 1–25,* Falls Church, Va: American Geological Institute. A volume containing reprints of the valuable brief reports on each DSDP leg. All the pertinent data are present without the reader having to spend great quantities of time on the complete reports.

Cox, Allan, 1973, *Plate Tectonics and Geomagnetic Reversals,* San Francisco: W. H. Freeman. Nearly 50 of the best papers published to that date on the subject. An excellent book for the geology student or professional geologist.

Johnson, Helgi, and B. L. Smith, 1970, *The Megatectonics of Continents and Oceans,* New Brunswick, N.J.: Rutgers University Press. Papers from one of the many excellent symposia on the subject of global tectonics held in the last decade.

Kahle, C. F. (ed.), 1974, *Plate Tectonics—Assessments and Reassessments,* Memoir 23, Tulsa, Okla.: American Association of Petroleum Geologists. An excellent collection of papers from world authorities, some of whom do not support the concepts of plate tectonics and continental drift.

Maxwell, A. E. (ed.), 1970, *The Sea,* New York: Wiley Interscience, Vol. 4, parts 1 and 2. Nearly 1400 pages of excellent articles covering the new global tectonics and ocean-basin evolution.

Scientific American, Inc., 1972, *Continents Adrift,* San Francisco: W. H. Freeman. A collection of 15 papers published in *Scientific American* on the subject. Good spectrum of subject matter well presented for the nonscientist.

EPILOGUE:
MAN AND THE SEA

During the past 30 years, our activities with respect to the sea have taken on new dimensions. Of course, scientific inquiry into the secrets of the sea has greatly expanded along academic lines. In addition, there has been more direct involvement with the marine environment. Much of this interest has been generated by the fact that we are beginning to depend much more on the sea for natural resources. This need is largely the result of the greatly expanding world population and increasing standards of living. In addition to increasing demands on the sea's natural resources, this growing population and its associated expanding technology is creating some other problems, the largest of which is pollution of the marine environment.

This discussion will concern itself with some of the areas in which we are developing and utilizing marine resources. One of the most important considerations is actual occupation of the ocean by people for extended periods of time. Another is greater use of the sea as a source of food. Both of these, but particularly the latter, are greatly affected by pollution of the marine environment.

MAN IN THE SEA
Ever since Alexander the Great reportedly descended into the Mediterranean Sea in a large bell jar, people have attempted various ways of directly exploring and studying the underwater world. Hardhat divers using air pumped from the surface were the only people to see much of the shallow sea floor until the development of the aqualung. In 1943, J.-Y. Cousteau and Emile Gagnan first used Self Contained Underwater Breathing Apparatus (SCUBA) in the Mediterranean. This was the beginning of our ability to move freely underwater for extended periods of time.

The aqualung, perhaps more than any other single tool, has facilitated studies of the oceans. Many experiments with various special gas mixtures have been successful in allowing descent to more than 300 meters for short times. According to most scientists, problems of decompression time and other physiological difficulties indicate that we are about at the limit of standard compressed-gas diving.

The greatest promise for future free, underwater activities by humans lies in liquid breathing. This would be accomplished by flooding an individual's lungs with a mixture of oxygen and water. Oxygen would be taken into the bloodstream through the lung walls, much as fish take in oxygen through their gills. In 1970 there were volunteers who were willing to try the technique; by 1980 this could be a fairly common method of occupying the underwater realm. Such methods have been successfully used with dogs and mice. The latter have been subjected to simulated depths of nearly 1000 meters without difficulty. The great advantage of the lung-flooding technique is that there is no nitrogen narcosis and no decompression is required.

There also has been the suggestion that a gill-type apparatus be surgically installed for purposes of diving. This sounds rather extreme and should not be necessary if the flooding technique is successful.

Habitats

In addition to short stays via diving at rather shallow depths, investigators have experimented with extended occupation at shallow depths. Foremost among many are the Sealab and Tektite projects. Both are aimed at studying the effects of prolonged underwater habitation and work by people; however, there are substantial differences between the two projects.

Sealab. Extended occupation of the shallow ocean is a must if we hope to be able to use the sea and its resources to any significant advantage. The United States Navy has carried out a multistage project to study the feasibility of submarine occupation and our ability to do physical labor underwater during extended stays. The first stage, Sealab I, took place in 1964 when a team of divers spent 11 days in a habitat nearly 70 meters below the surface. A similar operation (Sealab II) was carried out the following year, with several teams of divers spending 10 days each. Scott Carpenter, one of the original astronauts, participated in Sealab II and thus became the only person to be an aquanaut as well as an astronaut.

With all the data gathered during the first two Sealab operations, plans were formulated for a much more comprehensive program for Sealab III. The operation was originally scheduled for October 1968. Plans called for teams to spend 12 days each at about 200 meters depth. Included in the program were experiments involving salvage studies, various types of lift buoys, chemicals for preventing stirring of bottom sediments, and bacteriological studies. The human performance factor is obviously the most important aspect of all such underwater operations. The overall physiological responses of people in such an environment and their working capabilities are of utmost significance.

Sealab III had some technical difficulties which delayed the initial tests. When the trials were finally begun in February of 1969, more problems beset the project and one of the divers lost his life in an accident. This caused the halt of the project, at least temporarily.

Tektite. A project with somewhat the same aims as Sealab was carried out in the Caribbean Sea. Tektite, as the project is known, was more of a shallow-water operation than Sealab, with most of the occupation being at a depth of 16 meters. Primary emphases in the Tektite project were the physiological and psychological effects on the participants. Teams of five divers spent 14 days in the habitat and worked on various scientific research projects. During most trials, each person spent as much as several hours per day outside the habitat collecting various types of data. In addition, a small satellite habitat located at a depth of 30 meters was used in various aspects of the project.

An interesting aspect of the Tektite II operation during 1970 was the use of an all-female diving team. Each of the five women was a scientist and diver, as in previous crews. All worked on marine research during their underwater stay. According to project officials, this was one of the most successful portions of the entire program. Actually, Tektite proved to be a resounding success on nearly all counts.

The above two habitat projects are not the only ones to be undertaken in recent years, but they are perhaps the most comprehensive. J.-Y. Cousteau, a prominent investigator in all aspects of marine research, has also been active in habitat studies. Conshelf, one of the first of the modern habitat projects, was conceived and directed by Cousteau. He has also been involved with other extended underwater projects and has served as a stimulus for many others.

FUTURE RESOURCES OF THE SEA

The world's population is growing quite rapidly, nearly 2 percent per year. By the turn of the century, it will nearly double its present size. Even now there are many areas of the world where food and natural resources are inadequate. Increased population pressures will greatly compound these problems. The sea must play a role in solving some of them.

Since about 1960, there has been a number of positive steps initiated that will allow people to better utilize marine resources. These include sea farming, more modern fisheries, and utilization of new food sources. Great quantities of mineral wealth are also potentially available, either from direct mining of the sea floor or by extraction from seawater.

Nourishment from the sea

At the present time, most of the people of South America, Africa, and much of Asia are undernourished. The marine environment is presently supplying only about 10 percent of the world's food, although it covers 71 percent of the earth. We are rapidly approaching the time when the 29 percent of the earth above sea level will not be sufficient to provide 90 percent of the world's food. Not only does the increasing population require more food, but it requires additional living space that is at least partially obtained from agricultural areas.

Virtually all of the present fishing industry has out of necessity become a mechanized, scientific, large-scale operation. Nevertheless, many small and underdeveloped populations still fish by primitive methods from small boats. Such fishing can provide food for only small populations on islands and along the oceanic coasts. The great quantities of seafood needed by populous countries are obtained by large-scale operations. During recent years all major fishing countries, especially Japan and Russia, have shown marked increases in their catch, while United States production has remained constant.

In addition to the obvious methods of improving location and catching techniques, there have been other attempts to increase production and to seek new marine food sources. Sea farming, at least in the crude sense, has been going on for centuries. It began in the Indo-Pacific countries where fish is the only real source of protein. In these underdeveloped lands, sea farming is rather primitive, and techniques are passed from one generation to the next without change. Estuaries and other coastal embayments are utilized by using tidal currents to supply small fish which are then trapped and raised. Human and animal excrements are used as fertilizers to increase overall production.

Sea farms in coastal areas of Europe have been at least moderately successful at producing oysters, mussels, eels, and mullet. Japan has also raised shrimp. In the United States, oyster farming has been tried on a quite limited basis on both the east and west coasts.

One of the major problems with farming the sea is that fish, which are the desired product, are carnivores. Providing food for carnivorous marine organisms is quite difficult if it is done directly. Although in infant stages of development, the best farming possibilities seem to be to increase photosynthetic productivity. Changing the physical and/or chemical environment to facilitate plant production will ultimately provide increased nourishment for carnivorous fishes.

A project has been started in the Virgin Islands whereby deep and cold nutrient-rich water will be carried up into coastal bays in an artificial upwelling fashion. The result will be phytoplankton blooms which will increase the biomass at each trophic level.

New sources of nourishment. Human beings are without a doubt creatures of habit. This extends to their diet also. When we consider the vast area of the oceans and the fact that plant productivity and biomass per unit of area are higher in the sea than on land, there is really food for all. The lower trophic levels of the sea are in the form of algae and phytoplankton, neither of which are a part of the diet of many people. Food values of these, and also of zooplankton, are high and they can be easily obtained in large quantities.

The potential for seafood is obviously present. The most likely immediate use of this resource will be extracts and concentrates to be used as dietary supplements. It has already been suggested that certain crustaceans in the Antarctic area be used as protein additives.

Mineral resources

Nearly unlimited mineral resources lie on the ocean floor and in solution above it. At the present time, very little use is being made of these resources; however, considerable money is being spent for research and development of these marine resources.

Direct extraction from seawater. One of the most important future resources from the sea will have to be the water itself. With the increasing pressure on our fresh-water supply brought about by population growth and pollution, the sea will have to provide a source of fresh water for industry and domestic use. Research on desalination has been under way for decades with notable progress. The major obstacle that must be overcome is the cost. Currently there are about 500 installations in which fresh water is being produced for consumption, and the number is increasing annually. The first desalination plant in the United States is at Freeport, Texas, and one of the largest such plants is at Key West, Florida, where the city water supply is obtained by desalination.

Direct extraction of chemicals from seawater offers significant possibilities for the future, but vastly improved technology is needed to realize this potential. The presence in seawater of nearly all naturally occurring elements attests to this potential. Estimates indicate that some 50 quadrillion metric tons of minerals are dissolved in the oceans. Most of the valuable dissolved elements are present in concentrations of one part per million or less. Even though the total amount of any single element in the ocean is enormous, it would mean processing tremendous quantities of water in order to extract the desirable element. At the present time, salt (halite), magnesium, and bromine are produced commercially from seawater.

Mining of marine minerals. The most immediately recoverable mineral resources are the various types of nodules on the surface of the ocean floor. Phosphorite nodules are abundant on the outer continental shelves of many of the world's land masses, and manganese nodules are particularly abundant in the Pacific Ocean (Chapter 11). Recovery of these nodules is currently in the experimental stage, but indications are favorable for commerical mining in the near future. The basic recovery technique that is being tested is the use of a suction device that acts like a large vacuum cleaner. In trial runs off the Atlantic coast of the southern United States, in water nearly 1000 meters deep, an airlift suction device collected 1600 tons of nodules per day. This rate exceeded all expectations. Depth and abundance of nodules are obviously limiting factors that must be considered in such operations, but the future looks bright for nodule mining. Some estimates indicate that up to four centuries of phosphate reserves at the current rate of use are available in the oceans. Manganese reserves probably exceed these values. It is the opinion of some experts that the much deeper manganese nodules will be mined on a large-scale basis before phosphorite.

In addition to the nodule-type deposits, there is a variety of placer deposits on continental margins which are being mined commercially. Such deposits are typically in the sandy-stream deposits now on the continental shelf. These deposits include gold from the Alaskan coast, diamonds from South Africa, and tin from the coasts of England and southeast Asia (Fig. E.1).

Offshore petroleum. The present energy shortage is creating investigation of a variety of new energy sources or expansion of known sources. Nevertheless, there is little doubt that oil and natural gas will have to carry much of the energy burden for a long time. There is also little doubt that most of the future petroleum discoveries will be beneath the ocean floor of the continental margins. Initial exploration and production was off the California coast, in the Gulf of Mexico, and in Lake Maracaibo, Venezuela. Success in these areas has led to offshore exploration throughout most of the world, with much recent success in the North Sea, off the north coast of Alaska, and in the Persian Gulf.

Much increased exploratory activity has recently taken place in the eastern Gulf of Mexico and the Atlantic continental margin. Lease sales have been held and some drilling has taken place off the coast of Florida, with no success as yet. Offshore exploratory drilling and production creates many logistic, legal, and economic problems in comparison to "conventional"

FIG. E.1 *World distribution of offshore placer deposits where mining is active (•) or where there is considerable promise for commercial production (∘). (Modified from K. O. Emery and L. C. Noakes, 1968, "Economic Placer Deposits of the Continental Shelf," Tech. Bull. ECAFE **1**, 95.)*

land-based operations. As a result, costs of petroleum have escalated considerably in the 1970s.

MARINE POLLUTION

The combination of explosive population growth and rapidly expanding technology has had some significant detrimental effects on the natural environment. One of the most significant of these is aquatic pollution. Many definitions of this term have been used. In this discussion, pollution refers to the unnatural addition of materials to the environment, the vast majority of them caused by human activities. The marine environment has been greatly affected by pollution, and its impact is just beginning to be realized.

Pollution has many sources including industrial and domestic wastes, fertilizers, pesticides, accidents, and other phenomena. The greatest overall effect of these pollutants is that either directly or indirectly they harm the life in the sea.

The vast majority of these pollutants are purposely allowed to enter the ocean. With the greatly increasing concern over pollution, steps have been taken to correct these practices. Unfortunately it may be too little, too late. Most restrictions on industry and municipalities are not severe enough and are not totally enforced. For example, many industries are currently paying high fines for pollution because it is less expensive than changing their method of waste disposal.

Chemical pollution

The addition of various chemicals is perhaps the most detrimental of all types of pollution because the chemicals cover such a broad spectrum. For a long time it was thought that the marine environment could not be polluted because of its vastness. This concept is false, of course, particularly for areas along the continental slope and coastal areas. Onshore winds and coastal configurations tend to keep many pollutants from being dispersed throughout the marine environment. Some of the most polluted areas are the estuaries, particularly along the Atlantic coast of the United States.

Chemical pollutants come in a variety of forms and affect the environment in numerous ways. The most obvious are chemicals which are lethal to certain organisms. Others may be so concentrated in the tissues of organisms that they eventually kill them or are harmful to consumers. This has happened to the shellfish and fish along many areas of the Atlantic coast, causing quarantines on their consumption. For example, the common pesticide DDT is being concentrated in living tissues throughout the food chain.

One of the most widespread forms of pollution is the release of petroleum into the marine environment. During the past few years, the newspapers have frequently carried accounts of oil spills. These occur through natural seeps, which are fairly common off the southern California coast, various spills associated with offshore production, and accidents involving ships carrying oil. In addition, sludge and other oil wastes are eliminated by ships.

Oil is toxic to most organisms, but because it floats, direct contact is not made with many. It does, however, provide an impermeable layer on the surface of the water that prevents diffusion of gases, inhibits sunlight for photosynthesis, adheres to bird feathers, and ruins recreational facilities along the coast.

Various techniques are being tried to disperse and clean up oil spills. In some cases the problems are compounded, for example, by the detergents used to disperse the oil. Incendiary bombs have been used to ignite the slicks in attempts to burn them. This technique has not been very successful. There are also chemicals that degrade the oil and allow it to disperse without adding other pollutants. Some success has been achieved by using shredded polyurethane foam which absorbs the oil like a sponge. Specially equipped barges are also used to skim the oil off the ocean surface.

Sewage and other organic wastes, along with detergents and fertilizers, have similar effects on the marine environment. All these pollutants add great quantities of nutrient materials such as phosphates and nitrates. Controlled quantities of these nutrients would be beneficial by increasing productivity in the marine environment. In the present state of essentially uncontrolled introduction, immense phytoplankton and algal blooms are common. When these organisms expire, a great deal of organic material must be oxidized. This uses much oxygen and as a result deprives animals of that necessary gas.

Sewage has the same effect as the nutrients but in a more direct manner. Much of the sewage is organic matter that also must be oxidized. The combination of these two organic pollutants (nutrients and sewage) may completely exhaust the available oxygen in an area and render it a biological desert.

Pollutants may also reach the marine environment from the atmosphere. Many toxic industrial wastes are eliminated into the atmosphere and a large portion of them settle along coastal waters. Pesticides act in much the same manner. Another important problem is the addition of lead from automobile exhausts. During the past two decades, the level of concentration of all pollutants has increased greatly on an oceanwide scale.

Other types of pollution

While addition of foreign compounds is an obvious detriment to the biological community, there are also much more subtle types of pollution. One of the most widespread is thermal pollution, largely of industrial origin. Many industries use a great deal of water as a coolant or for steam power. The heated water is commonly returned to the aquatic environment. The result is considerable decrease in the dissolved oxygen. The same net effect is produced as with addition of large amounts of organic material: lower oxygen content. Animal populations are quite restricted or may be totally eliminated under an extreme temperature increase.

A rarely considered type of pollution is that of inorganic sediment. Great quantities of mineral material are introduced to the marine environment by construction, channel dredging, and tidal currents. The sediments affect the marine environment in several ways. Suspended sediment causes turbidity and thereby inhibits photosynthesis. Some sediments, particularly clay minerals, undergo modifications in their chemical composition as they pass from fresh water to salt water. Sediment influx, while it may have some effect on the biota, is not of great significance. Such influx occurs naturally as well as by human activity.

MARINE LAW

With increased use of the sea as a source of natural resources, there are also corresponding increases in the management of these resources. There must be agreement on their ownership; several attempts have been made to establish guidelines for world use of the sea.

The mobility of fish makes it difficult to establish fishing rights, and the open seas are considered to belong to all. There are differences in the territorial seas from one country to another. Most countries have limits of from 3 to 12 miles (5 to 20 kilometers), but some claim a 200-mile (335-kilometer) limit. The worldwide inconsistency points out the need for uniform regulation.

Ocean-floor resources must also be regulated, and in 1958 a United Nations convention on the oceans was held. Four topics were covered: the high seas, territorial seas and contiguous zones, fishing and conservation of living resources, and the continental shelf. Of these, the continental shelf is probably the most important because it is the area of greatest food and mineral wealth. A treaty was drawn up which allowed each country to claim the shelf to 200 meters depth. The continental margin beyond this depth would be under international trusteeship, the bordering country acting as the overseer and sharing any revenues obtained by exploiters. The areas beyond the margin would be under international control. The United States ratified the treaty in 1964.

Although this treaty is fairly comprehensive and has been adopted by many nations, there are still some legal points to be resolved. For example which are true shelf organisms? May foreign scientists investigate natural resources of the shelf? These are also difficulties in regulating the volume of fish taken. Also, how is the deep ocean to be treated with respect to mineral resources? These and other problems are being considered and proposals to eliminate them are pending.

SUMMARY

This chapter has discussed but a few of the many ways we are becoming directly involved with the sea. The first century of oceanographic activity was primarily concerned with determining the nature of the ocean and the distribution of organisms and sediments. While this effort is by no means complete, we have made tremendous progress. During the last decade or so, our efforts have turned more to the efficient utilization of the ocean's resources. Transportation and fishing were the only significant uses before the 1960s.

The future of our activities in the sea is difficult to predict; however, even the most imaginative uses may not be out of reach. Certainly, future uses of the oceans will greatly exceed those of the present.

SELECTED REFERENCES

Alexander, L. M. (ed.), 1970 *The Law of the Sea: National Policy Recommendation,* Proc. 4th Conference on the Law of the Sea Institute, University of Rhode Island. Probably the most-up-to-date and comprehensive volume of ocean law and proposed laws of the United States.

Boscom, Willard, 1969, "Technology and the Ocean," *Scientific American,* March 1969. General discussion of recent advances in shipbuilding, new materials, and other technological aspects of the sea.

English, T. S. (ed.), 1973, *Ocean Resources and Public Policy,* Seattle: University of Washington Press. Excellent collection of articles on current human involvement with marine resources.

Holt, S. J., 1969, "The Food Resources of the Ocean," *Scientific American,* March 1969. Considers present food resources, possible future uses of the ocean, and problems associated with harvesting food from the sea.

Hood, D. W. (ed), 1970, *Impingement of Man on the Oceans,* New York: Wiley. A series of articles showing how people have interacted with the oceans in all categories.

Spangler, M. B., 1970, *New Technology and Marine Resource Development,* New York: Praeger Publishers. Comprehensive treatment of how marine resources are currently being developed and expanded.

Wenk, Edward, Jr., 1969, "The Physical Resources of the Oceans," *Scientific American,* March 1969. Describes sea minerals currently extracted and suggests possible future resources. Good discussion of desalination.

Wooster, W. S., 1969, "The Ocean and Man," *Scientific American,* March 1969. Considers some of the problems of ocean management on a worldwide basis. Like some of the above articles, this one is part of *Scientific American's* special issue on the oceans.

APPENDIXES

GLOSSARY A

A

ABYSSAL. Pertaining to the ocean-floor environment below 3700 meters.

ABYSSOPELAGIC. Pertaining to that portion of the ocean below 3700 meters.

ACID. A substance that is a proton donor, because it dissociates to liberate hydrogen ions; it usually has a sour taste.

ALKALI. A strong base, usually a bitter-tasting salt such as sodium carbonate or potassium carbonate.

AMPHIDROMIC POINT. A point of no-tide on a chart of cotidal lines from which the cotidal lines radiate.

ANDESITE LINE. The postulated geographic and petrologic boundary between the sialic rocks of the Pacific Ocean margins and the simatic association of the ocean basin.

ANTIDUNES. Transient ripples formed by currents and moving against the current rather than with it.

ASEISMIC. Not subject to earthquake activity, such as a stable area of the crust.

ASTHENOSPHERE. Weak and mobile layer of the upper mantle in which thermal circulation occurs. It underlies the lithosphere and overlies the mesosphere.

ATOLL. A ring-shaped reef with a central lagoon; the reef rises from the ocean floor.

AUTHIGENIC. Pertaining to minerals that form in sediments *in situ* at or after the time of deposition.

AUTOSUSPENSION. The process whereby sediment in suspension over a slope imparts energy to the fluid, causing an increase in rate of flow.

B

BACKSHORE. The part of the beach that is usually dry and is reached only by the highest tides and storm waves; approximately equal to the supratidal zone.

BALEEN (WHALEBONE). The horny material that grows down from the upper jaw of plankton-feeding whales and acts as a strainer to filter out microscopic organisms.

BARRIER REEF. A reef that is separated from a land mass by a lagoon that is too deep for normal reef coral to grow.

BASALT.   A fine-grained, mafic, igneous rock composed primarily of calcic plagioclase and pyroxene; it comprises the simatic portion of the crust, which underlies the oceans.

BASE.   A proton acceptor that dissociates to liberate OH ions; it has a bitter taste and soapy feel.

BASIN AND RANGE.   The type of regional physiography caused by fault-block mountains and intervening basins, such as those covering much of Nevada.

BATHYAL.   Pertaining to the ocean bottom between 200 and 3700 meters depth; approximately equivalent to the continental slope.

BATHYMETRIC MAP.   Map showing bottom configuration of a water-filled basin. Points of equal depth are contoured with respect to water level.

BATHYPELAGIC.   Pertaining to that zone in the ocean lying between depths of 1000 and 3700 meters, extending to the average abyssal-plain depth.

BATHYTHERMOGRAPH.   An instrument that records a continuous trace of temperature and depth; its use is restricted to shallow water (less than 300 meters).

BAYMOUTH BAR (BAYMOUTH SPIT).   A linear sand body that forms at the mouth of a coastal embayment as a result of littoral sediment transport.

BEACH.   The zone of unconsolidated material that extends landward from low-tide level to the place where there is a marked change in material or physiographic form.

BEACH CUSPS.   Series of low mounds of beach material separated by crescent-shaped troughs spaced at more or less regular intervals along the beach face.

BEACH PITS.   Small collapse features on the berm caused by the rearrangement of sand grains.

BEACHROCK.   A friable to well-cemented rock consisting largely of carbonate fragments cemented by calcium carbonate; it forms only in the intertidal zone.

BED LOAD.   That part of the current-transported sediment that moves by traction near the bottom and is composed of relatively large particles.

BENTHIC.   Pertaining to the portion of the marine environment which is inhabited by marine organisms that live permanently in or on the bottom.

BENTHOS.   Bottom-dwelling forms of marine life.

BERM.   The nearly horizontal portion of the beach formed by deposition of material by wave action; it marks the limit of ordinary high tides.

BIOGENIC.   Pertaining to material that originates from organisms, such as deep-sea oozes or reef debris.

BIOMASS.   The total amount of living matter per unit of water surface or volume; expressed as a weight; also called *standing crop*.

BITTERN SALTS.   The group of dissolved salts including magnesium chloride and sulfate, bromides, and iodides which precipitate in the late stages of the evaporite sequence.

BOUGER ANOMALY.   The difference in gravity due to the altitude of a point and the rock between that point and sea level. This is a mass correction, whereas the free-air anomaly is a distance correction.

BRACKISH.   Pertaining to water of less-than-normal marine salinity, usually ranging between 0.5 and 17.0‰.

C

CAPILLARY WAVE.   A small wave whose velocity of propagation is controlled primarily by the surface tension of the liquid in which it is traveling.

CAY.   A low-lying island, usually composed of reef debris, that is built up by wave activity on the upper surface of the reef.

CENTRIFUGAL FORCE.   A force which impels something outward from the center of rotation.

CHLORINITY.   A measure of the mass of chloride in seawater. Defined as the weight of chlorine in grams per kilogram of seawater, after bromides and iodides have been replaced by chlorides.

CHONDRULE.   Small, rounded body of various materials, chiefly mafic; found in certain stone meteorites.

CIRRI.   Small, flexible appendages which are present on some invertebrate groups such as barnacles and annelids.

CLAY MINERAL.   A hydrous, layered aluminum silicate which may form authigenically or as a weathering product.

CONDUCTIVITY.   The facility with which a substance conducts electricity. An intrinsic property of seawater, it varies with temperature, pressure, and salinity.

CONTINENTAL BORDERLAND.   A special type of continental margin that is characterized by high-angle block faulting and high relief. A typical example is found off the southern California coast.

CONVECTION CELL.   Convective fluid movement in a mass, with the central portion moving upward and the outer regions moving downward.

CORE.   The central, dense zone of the earth, the upper boundary of which is the Gutenberg discontinuity at 2900 kilometers below the earth's surface.

CORIOLIS EFFECT.   An effect on moving particles traveling over a rotating sphere (earth); it is characterized by a deflection to the right in the Northern Hemisphere and to the left in the Southern Hemisphere.

COTIDAL LINE.   A line on a map or chart which passes through all points where high water occurs at the same time.

CRUST.   The outer shell of the solid earth, which ranges up to about 50 kilometers in thickness and the base of which is commonly defined at a density change called the Mohorovicic discontinuity.

CRUSTAL PLATES.   Those pieces of the earth's crust that are moving according to the plate tectonics concept. The earth's crust is divided into six major plates and a few smaller ones.

CUESTA.   A low ridge formed by a resistant and only slightly dipping rock unit.

CURRENT CRESCENTS.   Crescent-shaped depressions around pebbles or other large particles; caused by excavation due to currents.

D

DELTA.   An alluvial deposit, usually wedge-shaped, at the mouth of a river.

DEMERSAL.   Living on or near the bottom of the sea.

DIATOM.   Microscopic, unicellular, marine or fresh water organism which is planktonic, photosynthetic, and has a siliceous test.

DIFFRACTION.   Bending of waves around obstacles; also applies to light waves, sound waves, etc.

DIP SLOPE.   A slope of the land surface which conforms to the dip of the underlying rocks.

DISTRIBUTARY.   An outflowing branch of a river, as on a delta; the opposite of a tributary.

DOLDRUMS.   The narrow, latitudinal belt on both sides of the equator and characterized by low pressure and sluggish surface winds.

DRIFT BOTTLE.   A bottle which is released into the water and used to study surface currents; it contains a card with data about the release and space for date and location which is to be filled in by the finder and returned to an address on the card.

DROGUE.   A current-measuring device consisting of a weighted parachute and an attached surface buoy.

DYNAMIC HEIGHT.   The amount of work done when a water particle is moved vertically from one level to another.

E

EAU DE MER NORMALE.   Literally means normal seawater; it is the standard seawater which is provided for comparison in research on seawater by the Hydrographic Laboratory in Copenhagen, Denmark.

ECLOGITE.   A dense, mafic rock composed primarily of garnet and pyroxene.

EKMAN SPIRAL.   A theoretical representation of the effect that a wind blowing over a uniform ocean of unlimited size would cause, such that the surface layer would drift at an angle of 45 degrees to the right of the wind direction in the Northern Hemisphere and to the left in the Southern Hemisphere.

EPIFAUNA.   Benthic animals that live on the substrate; in contrast to infauna.

EPIPELAGIC.   Pertaining to the upper portion of the oceanic province, extending from the surface to a depth of about 200 meters.

ESTUARY.   A tidal bay formed by submergence of a river mouth or the lower part of a river valley.

EULERIAN.   Pertaining to the type of current measurement in which direct observation is made as water passes a fixed point.

EUPHOTIC ZONE.   That uppermost layer of a body of water which receives enough sunlight for photosynthesis to take place.

EURYHALINE.   Adaptable to a wide range of salinity, as certain marine organisms.

EURYTHERMIC.   The ability to tolerate wide temperature changes.

EUSTATIC. Pertaining to simultaneous and worldwide change in sea level such as that caused by melting of glaciers.

EUTROPHIC. Containing abundant nutrient matter, as certain bodies of water.

EUXINIC. Pertaining to areas of restricted circulation where sediment influx is low and organic matter is abundant.

EVAPORITE. A sediment which is deposited from aqueous solution as a result of extensive or total evaporation of the solvent.

EXPENDABLE BATHYTHERMOGRAPH (XBT). A type of bathythermograph that is thrown from a plane or ship and the temperature-depth data are transmitted back to the plane or ship. It is not retrieved.

F

FABRIC. Spatial arrangement of the elements of which a sediment or rock is composed.

FARO. Small, atoll-shaped reef only a few hundred meters in diameter; it usually forms part of a barrier reef or atoll rim.

FECAL PELLETS. Excrement of organisms, primarily invertebrates, that is composed of clay minerals and other sediment particles; may be a significant percentage of sediment or sedimentary rocʰ

FELDSPAR. A group of potassium aluminum silicates comprising the most abundant minerals in the earth's crust.

FIORD. A narrow, deep, steep-sided inlet of the sea formed by submergence of a high-relief area or excavation by a glacier.

FLUVIAL. Pertaining to rivers; produced by river action.

FORCED WAVE. A wave generated and maintained by a continuous force.

FORESHORE. The zone between the normal high- and low-tide marks.

FREE-AIR ANOMALY. The difference at any point on the earth between measured gravity and gravity calculated at sea level; a correction factor due to the position above or below sea level.

FRINGING REEF. A reef attached directly to the land mass.

FRY. Young fish, hatchlings.

G

GASTROLITHS. Polished and rounded pebbles or cobbles believed to be stomach stones that aid in digestion.

GEOMAGNETIC ELECTROKINETOGRAPH (GEK). Shipboard device for indirectly measuring currents; uses the principle that an electrolyte moving through a magnetic field will generate a current.

GEOPHYSICIST. A person engaged in the study of geophysics, or the physics of the earth.

GEOSTROPHIC MOTION. Movement defined by assuming an exact balance between the horizontal pressure gradient and the Coriolis effect.

GEOSYNCLINE.   An elongate region of considerable sediment accumulation and subsidence.

GLACIAL DRIFT.   Any sediment transported or deposited by glaciers.

GLAUCONITE.   A green mineral closely related to the micas and essentially a hydrous potassium iron silicate; it is formed only in the marine environment.

GLYCOGEN.   Animal starch used as carbohydrate storage in vertebrates; the general formula is $(C_6H_{10}O_5)_{12}$ to $(C_6H_{10}O_5)_{18}$.

GONDWANALAND.   The southern continent in some precontinental drift constructions; it was located south of the Tethys Sea and was comprised of the present-day continents of South America, Antarctica, Africa, Australia, and the subcontinent of India.

GRABEN.   An elongate block of the earth's crust, downthrown along normal faults relative to rocks on either side.

GRADED BEDDING.   Stratification in which each stratum shows a gradation in grain size from coarse at the base to fine at the top.

GRANITE.   A plutonic, igneous rock which is composed primarily of quartz and potassium feldspar.

GRAVIMETRICS.   The phase of geophysics that deals with the earth's gravity field.

GRAVITY WAVE.   A wave whose velocity of propagation is controlled primarily by gravity.

GUYOT.   A flat-topped submarine volcano.

GYRE.   Large, somewhat circular, surface-water circulation cells in the oceanic current system.

H

HADAL.   Pertaining to the greatest depths of the oceans in deep-sea trenches.

HALMEIC.   Pertaining to sediments that form out of solution in seawater; similar to authigenic sediments, but always in seawater.

HERMATYPIC.   Pertaining to the type of corals that are reef builders and have symbiotic algae.

HIGH-Mg CALCITE.   A type of calcite that contains 7–15 mole percent of magnesium (Mg) substituted for calcium in the crystal lattice.

HOLOCENE TRANSGRESSION.   The rather rapid rise in sea level that began about 18,000 years before the present as a result of the melting of the Wisconsinian glaciers.

HORSE LATITUDES.   The narrow belt about 30° either side of the equator and characterized by high pressure and light, variable winds.

HYDROSTATIC PRESSURE.   The pressure at a given depth due to the weight of the water column above that depth.

HYPERSALINE.   Pertaining to seawater that is typically above normal marine concentrations due to lack of circulation and a high rate of evaporation.

HYPOTONIC.   Having a lower osmotic pressure, due to a lower concentration of dissolved particles, than a surrounding fluid or fluid under comparison.

I

ILLITE. Group of three-layer clay minerals that are intermediate in composition between muscovite and montmorillonite; composition is that of hydrous potassium aluminum silicate.

INFAUNA. Benthic animals that burrow or bore into the substrate.

INTERNAL WAVE. A wave that occurs within a fluid, usually at a discontinuity caused by density differences.

ISLAND ARC. A curved group of volcanic islands; commonly convex toward the open ocean, they occur on the landward side of a deep-sea trench.

ISOBAR. A line on a chart or map that connects points of equal atmospheric pressure.

ISOMORPHOUS. Pertaining to minerals that have the same crystalline structure but different composition due to substitutions of certain ions within the crystal lattice.

ISOSTACY. The equilibrium assumed by the earth's crust as the result of gravity acting on rock masses.

K

KAOLINITE. Group of two-layer clay minerals of economic value with the general formula $Al_2(Si_2O_5) \cdot (OH)_4$.

KNOT. A velocity of one nautical mile per hour; it is equivalent to 51.4 centimeters/second.

KRILL. Term used by whalers and fishermen for euphausids, which are crustaceans eaten by some whales.

L

LAGOON. A shallow body of water separated from the sea by a barrier island, reef, or other similar feature.

LAGRANGIAN. Pertaining to the direct type of current measurement, tracing water movement through a water body.

LITHOSPHERE. The layer of the earth which comprises the moving plates and includes the crust and uppermost mantle to a depth of about 100 kilometers.

LITTORAL ZONE. The intertidal benthic zone.

LONGSHORE CURRENT. Current produced by waves being deflected at an angle by the shore; it runs parallel to the shore between the shoreline and the outer edge of the breaker zone.

LUMINESCENCE (BIOLUMINESCENCE; PHOSPHORESCENCE). Any emission of light at temperatures below that required for incandescence.

M

MAGNETITE. Magnetic mineral of iron oxide with the composition Fe_3O_4.

MAGNETOMETER. An instrument that measures magnetic intensity at a given location.

MANGANESE NODULE.   Nodular masses of manganese and iron oxides which precipitate on the deep ocean floor in areas of low sediment accumulation.

MANTLE (ASTHENOSPHERE).   The solid but relatively plastic layer between the crust and the core of the earth.

MEAN.   The average value in a group of data.

MEDIAN.   The middle value in a group of data; corresponds to the 50 percent mark on a cumulative curve.

MEDITERRANEAN.   Pertaining to a large body of salt water or inland sea surrounded by land with an opening to other water bodies.

MESOPELAGIC.   Pertaining to that portion of the oceanic province extending from 200 meters to 1000 meters depth.

MESOSPHERE.   The major part of the mantle that is strong and static. It lies below the asthenosphere.

MODE.   The value in a distribution curve that occurs most frequently.

MONSOON.   Seasonal wind that is present in the general area of southeastern Asia. In winter it blows from the northeast and in summer from the southwest.

MONTMORILLONITE.   The name given to a three-layered clay mineral group with the general formula $Al_2Si_4O_{10}(OH)_2$; it expands upon contact with water.

MULTINUCLEATE.   Having more than one nucleus.

N

NANNOPLANKTON (CENTRIFUGE PLANKTON).   Plankton that are too small (5 to 60 microns) to be trapped by plankton nets.

NANSEN BOTTLE.   A device used by oceanographers to collect water samples. It is lowered on a wire with both ends closed; upon reaching the desired depth, a messenger releases the bottle and it reverses, closing both ends and trapping a water sample.

NATURAL LEVEE.   The small ridge of sediment that forms at the edge of channel banks as the result of sedimentation due to flooding.

NEAP TIDE.   Tide of lowest range in the lunar cycle. Caused by the sun and moon acting at right angles to the earth, it occurs twice each lunar month.

NEARSHORE.   The area between mean low tide and the seaward extent of the bar-and-trough topography. It is approximately equal to the maximum width of the breaker zone.

NEKTON.   Animals in the pelagic environment that are active swimmers.

NERITIC.   Pertaining to the portion of the pelagic environment that extends from mean low tide to approximately the edge of the continental shelf.

NORTHEASTERLIES.   Winds caused by high pressure in the north polar region and which blow away from the pole. They are deflected by the earth's rotation to move from a northeasterly direction.

O

OCEANIC. Pertaining to that portion of the pelagic environment seaward from the edge of the continental shelf.

OFFSHORE. Pertaining to the generally flat area that extends from the seaward edge of the breaker zone to the edge of the continental shelf.

OLIGOTROPHIC. Pertaining to water bodies containing low concentrations of nutrient matter.

OLIVINE. An important mafic rock-forming mineral that is an iron magnesium silicate.

OOZE. Fine-grained, biogenic, pelagic sediment composed of at least 30 percent tests of marine organisms.

ORTHOGONAL. A line on a refraction diagram that is everywhere perpendicular to the wave crests.

OUTWASH. Sorted and stratified glacial drift.

P

PANGEA. The name of the primeval continent before continental drift took place.

PATCH REEFS. Small coral reefs without lagoons that may form part of a barrier or atoll rim.

PELAGIC. Pertaining to primary division of the sea which includes the whole mass of water in the neritic and oceanic provinces.

PELAGIC SEDIMENT. The fine-grained sediment that accumulates on the ocean floor as particles settle out of seawater without the influence of land or bottom currents.

PERIDOTITE. A plutonic, igneous rock comprised of iron-magnesium minerals and calcium plagioclase.

PERMEABILITY. The ability of sediment or rock to transmit fluids.

PETROLOGY. The branch of geology that deals with rocks, their description, and origin.

PHILLIPSITE. A hydrated aluminosilicate in the zeolite group that is common in deep-sea deposits, particularly in the Pacific Ocean.

PHOSPHORITE. Massive calcium phosphate with the composition of apatite but lacking crystal form.

PHOTOSYNTHESIS. The process in green plants whereby carbohydrates are produced from carbon dioxide and water in the presence of sunlight.

PHYSIOGRAPHIC CHARTS.

PHYSIOGRAPHY. The description of landforms; generally considered to be equivalent to geomorphology.

PHYTOPLANKTON. Photosynthetic organisms belonging to the floating or planktonic mode of existence.

PINNACLE REEF. A small coral spire which rises from the lagoon floor to near the water surface.

PLANKTON. Organisms that are drifters or feeble swimmers. Both plants and animals are included and most are small, although there is considerable range.

PLUNGING BREAKER. A breaking wave that tends to curl over and crash as it breaks.

POINT BAR. Somewhat crescent-shaped sandbars that form on the inside of stream meanders.

PORE SPACE (POROSITY). A void space in sediment or rock, usually expressed as a percentage of the total volume.

POTASH SALTS. Term used to describe the various salts of potassium, such as potassium carbonate (K_2CO_3).

PRECIPITATION. The process of separating mineral constituents from a solution.

PREVAILING WESTERLIES. Midlatitude winds that blow from the west toward the northeast in the Northern Hemisphere and toward the southeast in the Southern Hemisphere.

PRODUCTIVITY. The rate at which photosynthesis takes place.

PROFILE SECTION. Diagram showing the topography along a given traverse.

PROTECTED THERMOMETER. A reversing thermometer that is protected against water pressure by a glass shell.

R

RADIOLARIANS. Microscopic, single-celled, marine protozoans that display radial symmetry and are planktonic.

REEF. An organic wave-resistant framework. In some places, any shallow hazard to navigation is considered a reef regardless of its origin.

REFLECTION (REFLECTED WAVE). The process by which a wave is returned seaward upon contact with a steep beach, jetty, or other reflecting surface.

REFRACTION (WAVE REFRACTION). The process by which the direction of a wave is changed as it moves into shallow water at an angle to the bottom contours; the waves bend toward the orientation of bottom contours.

RELICT SEDIMENT. A sediment that was deposited in an environment other than the one it now occupies; it is out of equilibrium with its present environment.

RELIEF. The vertical differences in elevation or depth in a given area.

RESPIRATION. An oxidation-reduction process by which chemically bound energy in food is transformed to other kinds of energy upon which certain processes in all living things are dependent.

REVERSING THERMOMETER. A mercury-in-glass thermometer that records temperature when it is inverted; thereafter it retains the reading until returned to its original position.

RIFT VALLEY. An elongate valley formed by subsidence of a narrow strip bounded on each side by a fault, i.e., a graben.

RILL MARKS. A small groove, furrow, or channel in mud or sand on a beach, made by tiny streams following an outflowing tide.

RIP CURRENT. A usually strong and narrow current caused by the seaward flow of water piled up nearshore by waves.

RIPPLES (RIPPLE MARKS). Undulating surface features of various shapes produced in unconsolidated sediments by wave or current action.

ROUNDNESS. The particle shape which expresses the smoothness or roundness of its edges. Roundness may be expressed as the ratio of the average radius of curvature of the several corners or edges to the radius of curvature of the maximum inscribed sphere.

RUDISTIDS. Extinct group of pelecypods that grew with one valve attached and specially modified for a corallike reef environment.

RUNNING WAVE. The common type of wave that is propagated across the water surface.

S

SALINITY. A measure of the quantity of dissolved solids in seawater. It is defined as the total amount of dissolved solids in seawater in parts per thousand (‰) by weight when all carbonate is converted to oxide, the bromide and iodide to chloride, and organic matter is oxidized.

SALINOMETER. Any device for determining salinity, especially one based on electrical conductivity methods.

SALTATION. Sediment-particle movement in which an individual particle jumps along the bottom due to lack of sufficient current energy to keep the particle in suspension.

SAND SHADOWS. Small accumulations of sediment in the lee of pebbles or other large particles; caused by currents on the beach.

SAND SPIT. An elongate sandbar extending from the land mass and formed by littoral sediment transport.

SCUBA. Self-Contained Underwater Breathing Apparatus which enables man to move about freely at limited depths.

SEA. Waves that are under the direct influence of the wind.

SEAMOUNT. A submarine volcano.

SECCHI DISK. A usually white disk, 30 centimeters in diameter, used to measure the transparency of water.

SEICHE. A standing-wave oscillation of an enclosed or nearly enclosed water body; continues after cessation of the originating force, which may have been seismic, atmospheric, or wave-induced.

SEISMOLOGY. The study of earthquakes or other elastic waves that are transmitted through the earth.

SESSILE BENTHOS. Bottom-dwelling organisms that are attached to the substrate; they have no self-induced mobility.

SIAL.   The high-silica portion of the earth's crust that comprises most of the continental masses but is absent in ocean basins; its thickness is about 30 to 35 kilometers.

SIGMA-TEE.   An abbreviation for the density of seawater, expressed as $10^3 \times$ density -1.

SIMA.   The mafic portion of the earth's crust which is continuous and underlies the continental masses andoceans; it has a density of 3.0 to 3.3 grams/cubic centimeter.

SORTING VALUE (STANDARD DEVIATION).   A measure of the spread of distribution or uniformity of a given population sample; essentially it is the standard deviation.

SOUTHEASTERLIES.   Winds caused by high pressure in the south polar region and which blow away from the pole. They are deflected by the earth's rotation to move from a southeasterly direction.

SPECIFIC GRAVITY.   The ratio of the mass of a body to that of an equal volume of water.

SPECIFIC HEAT.   The heat capacity of a system per unit mass. It is the amount of heat required to raise the temperature of one gram of water (fresh) 1°C.

SPHERICITY.   The tendency of a particle to become equidimensional for all radii, approaching a sphere; the ratio of the surface of a sphere of the same volume as the fragment in question to the actual surface area of the object.

SPHERULES.   Tiny sphere-shaped particles that make up a small percentage of deep-sea pelagic sediments.

SPICULES.   Minute needlelike or multiradiate bodies which serve as supportive structures in soft-bodied animals such as sponges and some coelenterates.

SPILLING BREAKER.   A wave breaking in shallow water, where the wave breaks over a considerable distance.

SPRING TIDE.   Tide of the highest range in the lunar cycle, caused by sun and moon reinforcing each other. It occurs twice during each lunar month.

SPUR AND GROOVE.   Configuration of the upper part of a windward reef, composed of buttress-shaped coral ridges with deep and narrow channels between them.

STANDING CROP.   The amount of living material per unit of water surface or volume; it is expressed by weight.

STANDING WAVE.   A wave in which the water surface oscillates vertically without propagation between fixed points called nodes.

STENOHALINE.   Pertaining to certain organisms that are capable of existing only in a narrow salinity range.

STENOTHERMIC.   Pertaining to an inability to tolerate wide temperature ranges.

STRATIGRAPHY.   The branch of geology that treats the various aspects of layered rocks or strata.

STROMATOLITES.   Laminated structures of calcium carbonate that are constructed by detrital particles adhering to blue-green algae.

STROMATOPOROIDS (STROMATOPOROIDEA).   Fossils composed of laminated calcium and probably belonging to the hydrozoans.

SUBLITTORAL ZONE.   That part of the benthic environment between the mean low-tide mark and the edge of the continental shelf.

SUBMARINE CANYON.   An elongate, steep-sided valley that is found on continental margins, particularly the slope; it is usually normal to the coastline.

SURF.   Shallow-water waves that break due to over-steepening.

SURGING BREAKER.   A breaker that peaks up and then surges toward the beach, instead of spilling or plunging.

SWASH MARKS.   Thin wavy lines of fine sand or other debris left by the uprush when it recedes from its upward limit of movement on the beach face.

SWASH ZONE.   The narrow zone where the water rushes up and back on the beach.

SWELL.   Nearly sinusoidal waves that are no longer under the direct influence of the wind.

T

TABLE REEF.   Small, isolated reef possessing no lagoons and rising from a platform such as the continental shelf.

TABULATE CORALS (TABULATA).   Extinct order of colonial anthozoans characterized by horizontal tabulae in each tube.

TALUS.   The fragments of rock or debris that collect at the base of a steep slope.

TERRIGENOUS SEDIMENT.   Oceanic sediment derived from the land and transported to its site of accumulation by bottom currents

TEST.   The hard exoskeleton or supporting structure of small animals; the term is usually used only for microorganisms.

THERMOCLINE.   The portion of the temperature-depth curve in a water body that shows maximum change; usually a seasonal feature.

THERMOHALINE.   Pertaining to water circulation generated by density gradients caused by temperature and/or salinity variations.

TIDAL BORE.   High, breaking wave of water that advances up an estuary as the tide rises; it may also be generated by a tsunami.

TIDAL SPECIES.   The various astronomical frequencies that give rise to a particular tide at a given location.

TIDE.   The cyclic rising and falling of the earth's oceans due to the mutual attractions of the earth, moon, and sun.

TILL.   Unsorted and unstratified material deposited by a glacier.

TECTONICS.   The regional assembling of structural or deformational features of the earth's crust; their origin, history, and mutual relations.

TRACTIVE FORCE.   A pulling force.

TRADE WINDS.   These are prevailing winds that blow from low-latitude belts of high pressure toward the equator. Due to the earth's rotation, they are deflected so that they blow from the northeast in the Northern Hemisphere and from the southeast in the Southern Hemisphere.

TOMBOLO. A sand spit connecting two land bodies.

TROPHIC LEVEL. A successive stage of nourishment as represented by links of the food chain.

TSUNAMI. A long-period, high-velocity sea wave produced by a submarine earthquake, volcanic eruption, or landslide.

TURBIDITE. Sediment deposited by a turbidity current and characterized by graded bedding, as well as other unique sedimentary structures.

TURBIDITE CURRENT. A highly turbid density current carrying large quantities of sediment in suspension. Its flow is caused by a density gradient.

U

UNCONFORMITY. A buried surface of erosion.

UNDERTOW. The subsurface return by gravity flow of water carried toward the shore by waves or breakers.

UNPROTECTED THERMOMETER. A reversing thermometer which is not protected against hydrostatic pressure.

V

VAGRANT BENTHOS. Mobile organisms that occupy the sea floor.

VERTICAL EXAGGERATION. The deliberate increase in vertical scale of a profile diagram or cross section in order to make the illustration more clearly perceptible.

W

WATER MASS. A body of water which tends to retain its identity and is characterized by a particular set of temperature, salinity, and chemical characteristics.

WAVE BASE. The depth at which wave action ceases to stir sediment.

WAVE DRIFT. Slow current moving in the direction of wave propagation and caused by the net transport of water particles as the particles move in orbital paths.

WHITING. A phenomenon of shallow tropical waters in which water is milky; caused by tiny suspended particles of calcium carbonate.

WIND-DRIVEN. Pertaining to surface water currents generated by friction from air circulation.

Z

ZEOLITES. A group of hydrous aluminum silicates having easy and reversible loss of water. Many have high ion-exchange capacity.

ZOOPLANKTON. The animal portion of the floating or planktonic mode of life.

ZOOXANTHELLAE. Minute yellow or brown flagellates which are commonly symbionts, particularly with hermatypic corals.

CLASSIFICATION OF COMMON MARINE ORGANISMS   **B**

KINGDOM MONERA. Organisms having cells without nuclear membranes; basically unicellular.

Phylum Schizophyta. Bacteria; smallest known cells.

Phylum Cyanophyta. Blue-green algae; no motile stages.

KINGDOM PROTISTA. Possess nuclear membranes; true chromosomes and chloroplasts; primitive powers of mobility.

Phylum Chlorophyta. Green algae; food stored in form of starch.

Phylum Chrysophyta. Golden-brown algae; diatoms; color due to carotenoids; silica in cell wall; wide variety of forms.

Phylum Pyrrhophyta. Dinoflagellates; food stored as starch or oil; possess flagella.

Phylum Phaeophyta. Brown algae; large nonmotile bodies; food stored as oil or complex polysaccharides.

Phylum Rhodophyta. Red algae (coralline algae); food stored as starch.

Phylum Mastigophora. Flagellated protozoa; singles or colonies.

Phylum Sarcodina. Ameboid protozoa; have pseudopodia, mostly free-living; includes orders Foraminifera and Radiolaria.

Phylum Cilophora. Ciliate protozoa.

KINGDOM METAPHYTA. Green, multicellular plants.

Phylum Tracheophyta. Vascular plants with stems, roots, and leaves; special cells for conducting food and fluids.

   Class Angiospermae. Flowering plants; seeds enclosed by carpels; broad leaves.

KINGDOM METAZOA. Multicellular animals.

Phylum Mesozoa. Simplest multicelled animals.

Phylum Porifera. Sponges; sessile benthos adults; spicules are the only hard parts.

   Class Calcarea. Calcium carbonate spicules.

   Class Hexactinellida. Silicic spicules.

   Class Demospongia. Soft sponges, some without spicules.

489

Phylum Coelenterata.   Two-layered; radial symmetry; contain polyp and medusa stages.

    Class Hydrozoa.   Soft corals, sea whips.

    Class Scyphozoa.   Jellyfishes, polyp stage reduced or absent; planktonic.

    Class Anthozoa.   Corals, anemones; only polyp stage; sessile benthos with tentacles.

Phylum Ctenophora.   Comb jellies; planktonic; radial symmetry.

Phylum Platyhelminthes.   Flatworms; bilateral symmetry; hermaphroditic; benthic and planktonic.

    Class Turbellaria.   Free-living flatworms.

    Class Trematoda.   Flukes; parasitic.

Phylum Nemertinea.   Advanced marine worms; planktonic and benthic.

Phylum Nematoda.   Round worms; parasitic.

Phylum Entoprocta.   Stalked, sessile benthos; ciliated tentacles.

Phylum Ectoprocta.   Moss animals; branching or encrusting colonies; sessile benthos.

Phylum Brachiopoda.   Lamp shells; benthos; bivalved; important in geologic past but uncommon now.

Phylum Chaetognatha.   Arrow worms; planktonic.

Phylum Pogonophora.   Many similarities to chordates; notochord; tubular nervous system.

Phylum Annelida.   Segmented worms; bilateral symmetry; trochophore larvae; mostly benthic.

    Class Archiannelida.

    Class Polychaeta.   Tubeworms; include the family Serpulidae.

    Class Hirudinea.   Leeches; freshwater parasites.

Phylum Mollusca.   Common shell animals; bilateral symmetry.

    Class Amphineura.   Chitons; vagrant benthos.

    Class Scaphopoda.   Tusk shells; infaunal; filter feeders.

    Class Pelocypoda.   Clams, oysters, scallops; bivalves; sessile or vagrant benthos; mostly filter feeders.

    Class Gastropoda.   Snails; coiled shell on most species; infaunal or epifaunal benthos; carnivorous or scavengers.

    Class Cephalopoda.   Squids, octopuses and nautiluses; nektonic; may have internal or external shell.

Phylum Arthropoda.   Jointed appendage animals; chitinous exoskeletons; tripartite body.

    Class Xiphosurida.   Horseshoe and king crabs.

    Class Crustacea. Limy skeleton; gills; two pairs of antennae.

        Subclass Branchiopoda.

        Subclass Ostracoda.

        Subclass Copepoda.

        Subclass Cirripedia.

        Subclass Malacostraca.

Phylum Echinodermata. Spiny-skinned animals; pentameral symmetry; water vascular system.

Subphylum Pelmatozoa. Sessile benthos forms attached by stalks; classes Cystoidea and Blastoidea are extinct.

    Class Crinoidea. Sea lilies; long, flexible, branching arms.

Subphylum Eleutherozoa. Free-moving forms.

    Class Asteroidea. Common starfish; commonly five arms; tube feet; regenerative powers.

    Class Ophiuroidea. Brittle stars; arms quite flexible and distinct from central body.

    Class Echinoidea. Sea urchins and sand dollars; spines specially adapted for protection and mobility; Aristotle's-lantern structure for feeding.

    Class Holothuroidea. Sea cucumbers; pentameral symmetry obscure; soft bodies.

Phylum Hemichordata. Acorn worms; primitive nerve chord; gill slits; planktonic.

Phylum Chordata. All have notochord, dorsal nerve chord, and gills or gill slits at some stage of development.

Subphylum Tunicata. Sea squirts; chordate features in larvae only; adults benthic.

Subphylum Cephalochordata. Amphioxus; well-developed notochord; no distinct heart.

Subphylum Vertebrata. Internal skeleton; spinal column of vertebrae; brain; red blood; two pairs of appendages.

    Class Agnatha. Lampreys and hagfish; no scales or jaws; some with rasping tongue; most primitive vertebrates.

    Class Chondrichthyes. Sharks, rays and skates; cartilaginous skeletons; no gill covers.

    Class Osteichthyes. Bony fishes, scales; gill covers; swim bladder; paired fins.

    Class Amphibia. Frogs, toads, etc.; no true marine forms.

    Class Reptilia. Snakes, lizards, turtles, alligators; marine forms are in the orders Squamata (snakes) and Chelonia (turtles).

    Class Aves. Birds; no true marine forms, although many live on and near oceans and feed from the sea.

Class Mammalia. Mammals; hair; mammary glands; bear live young; homoiothermal.

Order Sirenia. Sea cows, dugongs; hind legs gone, front ones modified to flippers.

Order Cetacea. Whales, porpoises; little or no hair; appendages modified to fins.

Era	Period	Epoch	Approximate time of beginning*	Duration†
Cenozoic (recent life)	Quaternary (addition to an 18th century classification)	Recent	0.01	
		Pleistocene (most recent)	2.0–3.0	2.0–3.0
	Tertiary (third period in an 18th century classification)	Pliocene (very recent)	7	4–5
		Miocene (moderately recent)	25	18
		Oligocene (slightly recent)	40	15
		Eocene (dawn of recent era)	60	20
		Paleocene (early dawn of recent era)	68–70	8–10
Mesozoic (intermediate life)	Cretaceous (chalk)		135	65–67
	Jurassic (Jura Mts., France)		180	45
	Triassic (threefold division in Germany)		225	45
Paleozoic (ancient life)	Permian (Perm, Russian province)		270	45
	Carboniferous (from abundance of coal) or Pennsylvanian (state in U.S.) / Mississippian (from river valley)} recognized only in the U.S.		325 / 350	55 / 25
	Devonian (Devonshire, England)		400	50
	Silurian (Silures, ancient British tribe)		440	40
	Ordovician (Ordovices, ancient British tribe)		500	60
	Cambrian (Cambria, Roman name for Wales)		600	100
Precambrian‡	Several local subdivisions, but no established worldwide classification		4500+	4000+

* Millions of years before present.

† In millions of years.

‡ The Precambrian includes most of geologic time (about 87.5 percent of it), but is not well known because most evidence of it is covered, badly deformed, or lacks fossils. As a result, this great period of time is lumped under one heading.

D CONVERSION TABLES*

$$\text{Centimeter (cm)} = 0.39370 \text{ inch}$$
$$0.032808 \text{ foot}$$
$$0.01 \text{ meter}$$
$$10 \text{ millimeters}$$
$$1 \times 10^4 \text{ microns}$$

$$\text{Cubic centimeter (cm}^3\text{)} = 2.1997 \times 10^{-4} \text{ gallon}$$
$$3.531477 \times 10^{-5} \text{ cubic foot}$$
$$8.7988 \times 10^{-4} \text{ quart}$$
$$0.0021134 \text{ pint}$$
$$0.033814 \text{ ounce}$$
$$0.061023 \text{ cubic inch}$$
$$1 \times 10^{-6} \text{ cubic meter}$$
$$9.9997 \times 10^{-4} \text{ liter}$$
$$1000 \text{ cubic millimeters}$$

$$\text{Cubic yard (yd}^3\text{)} = 27 \text{ cubic feet}$$
$$202.0 \text{ gallons}$$
$$807.9 \text{ quarts}$$
$$4.6656 \times 10^4 \text{ cubic inches}$$
$$0.76455945 \text{ cubic meter}$$
$$764.54 \text{ liters}$$
$$7.6455945 \times 10^5 \text{ cubic centimeters}$$

$$\text{Degree (°)} = 1/360 = 0.0027778 \text{ circumference}$$
$$60 \text{ minutes}$$
$$3600 \text{ seconds}$$

$$\text{Fathom (fath)} = 6 \text{ feet}$$
$$1.828804 \text{ meters}$$

---

* All are U.S. values unless otherwise stated.

Foot (ft) = 1.6447×10^{-4} nautical mile
1/6 or 0.16667 fathom
0.3048006 meter
30.48006 centimeters

Inch (in.) = 2.540005 centimeters
25.40005 millimeters
0.08333 foot

Kilometer (km) = 0.53961 nautical mile
0.62137 statute mile
1093.6 yards
3280.8 feet
1000 meters

Knot = 1 nautical mile

Liter (l) = 0.21998 gallon
0.035316 cubic foot
1.056710 quarts (liquid)
33.8147 ounces (fluid)
61.025 cubic inches
0.001000027 cubic meter
1000.027 cubic centimeters

Micron (μ) = 3.937×10^{-5} inch
1×10^{-6} meter
1×10^{-4} centimeter
0.001 millimeter

Mile (nautical) = length of 1 minute of arc on the
earth's surface at the equator
1.1516 statute miles
6080.2 feet
1.85325 kilometers

Mile (mi) (statute) = 0.86836 nautical mile
1760 yards
5280 feet
1.60935 kilometers
1609.35 meters

Milliliter (ml) = 0.061025 cubic inch
0.0338147 ounce (fluid)
0.001 liter
1.000027 cubic centimeters

Millimeter (mm) = 0.039370 inch
0.001 meter
0.1 centimeter
1000 microns

Month (mo) (lunar) = 29 days, 12 hours, 44 minutes

Square kilometer (km²) = 0.3861006 square mile
247.1044 square acres

Square mile (mi²) = 2.78784 × 10⁷ square feet
2.589998 square kilometers

VELOCITY

Centimeter per second (cm/sec) = 0.02237 mile per hour
0.032808 foot per second
1.9685 feet per minute

Foot per second (ft/sec) = 0.5921 knot
0.6818 mile per hour
1.0973 kilometers per hour
30.4801 centimeters per second

Knot = 1 nautical mile per hour
1.1516 miles per hour
1.689 feet per second
51.48 centimeters per second

Meter per second (m/sec) = 2.2369 miles per hour
3.281 feet per second
3.600 kilometers per hour

INDEX

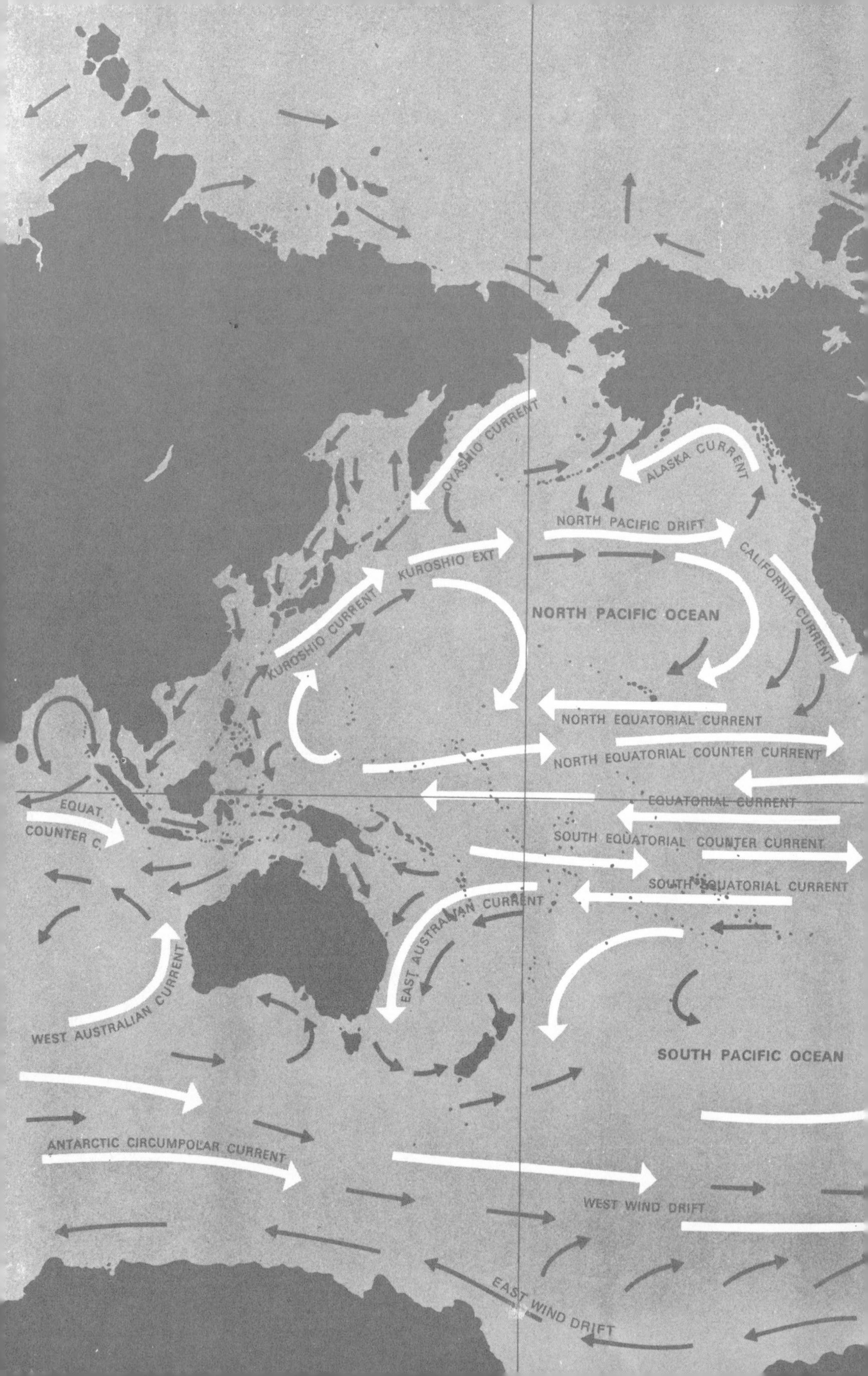

OYASHIO CURRENT
ALASKA CURRENT
NORTH PACIFIC DRIFT
KUROSHIO EXT
CALIFORNIA CURRENT
KUROSHIO CURRENT
NORTH PACIFIC OCEAN
NORTH EQUATORIAL CURRENT
NORTH EQUATORIAL COUNTER CURRENT
EQUAT.
COUNTER C.
EQUATORIAL CURRENT
SOUTH EQUATORIAL COUNTER CURRENT
SOUTH EQUATORIAL CURRENT
EAST AUSTRALIAN CURRENT
WEST AUSTRALIAN CURRENT
SOUTH PACIFIC OCEAN
ANTARCTIC CIRCUMPOLAR CURRENT
WEST WIND DRIFT
EAST WIND DRIFT